NUCLEAR DAWN

Nuclear Dawn
F. E. Simon and the Race for Atomic Weapons in World War II

KENNETH D. McRAE

Professor Emeritus of Political Science, Carleton University, Ottawa

OXFORD
UNIVERSITY PRESS

Great Clarendon Street, Oxford, OX2 6DP,
United Kingdom

Oxford University Press is a department of the University of Oxford.
It furthers the University's objective of excellence in research, scholarship,
and education by publishing worldwide. Oxford is a registered trade mark of
Oxford University Press in the UK and in certain other countries

© Kenneth D. McRae 2014

The moral rights of the author have been asserted

First Edition published in 2014

Impression: 1

All rights reserved. No part of this publication may be reproduced, stored in
a retrieval system, or transmitted, in any form or by any means, without the
prior permission in writing of Oxford University Press, or as expressly permitted
by law, by licence or under terms agreed with the appropriate reprographics
rights organization. Enquiries concerning reproduction outside the scope of the
above should be sent to the Rights Department, Oxford University Press, at the
address above

You must not circulate this work in any other form
and you must impose this same condition on any acquirer

Published in the United States of America by Oxford University Press
198 Madison Avenue, New York, NY 10016, United States of America

British Library Cataloguing in Publication Data

Data available

Library of Congress Control Number: 2013948386

ISBN 978–0–19–968718–3

Printed in Great Britain by
Clays Ltd, St Ives plc

Links to third party websites are provided by Oxford in good faith and
for information only. Oxford disclaims any responsibility for the materials
contained in any third party website referenced in this work.

OXFORDSHIRE LIBRARY SERVICE	
3302970504	
Askews & Holts	08-Jul-2014
B.SIM	£35.00

PREFACE

A house in Oxford

Franz Simon was born in 1893 in Berlin, and trained in physics. In 1933 he left Germany after the Nazi seizure of power, accepting an offer to transfer his family, his research, and part of his research team to the Clarendon Laboratory in Oxford. By this date he was recognized internationally for his experimental work on low-temperature physics. At the outbreak of war in 1939, this work had to be discontinued when scientists in countries at war were transferred to priority defensive or weapons research. In several countries, military authorities began to explore how to use nuclear energy to build a weapon of unprecedented destructiveness. Simon was well aware of these developments, and with other German scientists in exile he lived for years in fearful anxiety that Germany, with its head start on rearmament, might well become the first country to do so. In the history of physics, Simon is remembered for record-breaking advances in methods of cooling closer and closer to absolute zero. In the history of the war, he is known to a wider public for a successful method of separating the isotopes of uranium on an industrial scale.

Because of his background, Simon would become involved in one way or another with the nuclear efforts of four out of the five powers actively pursuing nuclear weaponry. As an émigré from Germany and naturalized British citizen, he became in 1940 a key figure in the successive Technical Committees of Britain's nuclear research project, Tube Alloys. In this capacity, he also became involved in the early stages of British cooperation with the American nuclear programme, making five extended visits to North America between 1942 and 1944. As an exile from Hitler's Germany, he kept a close watch on German scientific publications for teaching schedules, military casualty lists of scientists, and other signs of nuclear activity. Finally, through the USSR's systematic nuclear espionage network, Simon's contributions to the British nuclear programme became known at an early date to Soviet nuclear authorities.

The possibility of a book on Simon's life and contributions arose from a previous and entirely different chain of circumstances. In December 1999, Simon's wife Charlotte, known to family and friends as Lotte, came to the end of a long and active life at the age of 102. She had lived in the same family house, in Belbroughton Road in North Oxford, for all of the family's years in England. Active to the last, Lotte put up a brief, valiant battle against an invasive virus. Just in time, my wife Dorothea (known to family and friends as Dor) flew to Britain to her mother's side. Within a day or two, death came swiftly, mercifully, cleanly. After a few days Dor returned home in time for Christmas.

Our telephone rang one day in early March of 2000. It was Dor's sister, Kay. For years they have kept in close touch by telephone. Kay, living in London, had had the brunt of care for her mother to this point, and was also her mother's executor. She was calling to work out a timetable for clearing and selling their mother's house, and inviting us over to lend a hand. A visit from us in two stages was agreed on, a clean-up in April followed by a clear-out and sale in June and July.

Early in April, we flew to Heathrow and boarded a bus for Oxford to begin the first phase. The procedure for this visit was strangely familiar. It is a stage in the cycle of many people's lives. We had agreed with Kay to come over and help clean up the house and prepare it for sale. We would talk to city officials, solicitors, estate agents, potential buyers. We would help to sort out the accumulated contents, those from Franz's and Lotte's lifetimes and some from earlier: the clothing, furniture, books, china, silver, papers, letters, photographs, paintings, and anything else. The family agreed that the house must be sold. For Kay and Dor, the sale would clearly be emotional. It would break a connection of 67 years, close to a normal lifetime, with the same house. Because Lotte had stayed on in the house after her husband's death, the entire cultural inheritance brought from Germany in 1933 and its later development in Britain had been undisturbed by any major move.

In the clear-out, there was a minor niche for me. In the 1950s, I had helped Lotte to sort out some branches of her family tree, talking with relatives, splicing together jottings from various sources into a continuous tree, with branches reaching back three, four, or even five centuries. My concern was that documents of family genealogy or history should not be lost in the days ahead.

As the elderly approach the end of their lives, they often reveal one or the other of two diverging tendencies. Some leave their possessions in place, untouched, for their heirs to sort out. Others clear out their personal belongings, ostensibly to help their family. Meticulousness can be carried too far. Lotte was of a robustly different mind. "You will find a hell of a mess when you have to clear out *this* house," she warned her daughters on more than one occasion. We had some advance idea of the tasks that awaited.

On arrival at the house, we began an assessment. The downstairs rooms looked almost untouched since Lotte's death, in spite of Kay's periodic visits. The bed where Lotte died, bedpans, medicines, hot water bottles, an accumulation of dust.

While we were taking this in, one of the student lodgers came home, and gave us a cheery update as she headed upstairs. For the three tenants, all had been running smoothly. Soon we could hear her vacuuming her room. The continuities of life upstairs heightened the feeling that on the ground floor time had frozen since the preceding December.

We stepped out into the back garden. It too needed work. From spring to autumn this garden had been the house's crowning glory. By now it had witnessed five generations of Simon family activities. For Franz and Lotte, it had been a source of pleasure and relaxation, a haven, a reflection of their lives.

After three months of inactivity, much needed to be done here, both inside and out. For the time being, we turned to more mundane matters, and all thoughts of papers and records were sidelined.

Three weeks later, we could look back on some progress. We had almost completed a considerable physical clean-up of the house downstairs and the garden. We talked with estate agents, discussed proposals for showing and selling the house, and set up a timetable.

When time permitted, attention turned to things that lure historians: the written word, the printed word, photographs, drawings, paintings, postcards, press clippings, artefacts, souvenirs, and ornaments of various kinds. There was a vast accumulation of all kinds of print in the house, a good deal of it set aside in storage and forgotten for half a century or more.

But where to begin? The house had three main storage areas to be cleared: an air raid shelter, a garage, and a large attic loft, crammed with discarded furniture and old files, above the entire second floor. These areas could be worked on first, keeping the main living areas tidy and in good order for viewing by potential buyers. For practical reasons, the air raid shelter was the most accessible, a low-ceilinged, above-ground structure adjoined to the back corner of the house, with stout brick walls and a thick concrete roof. In the postwar years, the baffle walls were removed and a window added on the garden side, to make a sizable storage room. By a spatial equivalent to Parkinson's law, it had been filled to the ceiling over the years with overflow material from rooms in daily use. Only a slim space for walking allowed passage between entry doors at either end. The rest was piled high with garden furniture, broken chairs, boxes, books, files, artwork, suitcases, cushions, games and toys, mattresses, and bolsters from long-discarded German sofas.

The materials piled in this shelter were reminders of Britain's postwar years of austerity, when everything was scarce and nothing was discarded. Much of the accumulation dated from this period. One whole wall was stacked with large envelopes, book packs, wrapping paper, and file covers of all kinds. In the lean years, Lotte had learned frugality, and after the war she continued to practise it. Some of this material bore testimony to friendships and exchanges of long ago, a book pack from Professor Bridgman from Harvard, or a Christmas envelope from Frieda Urey.

Several envelopes addressed to Professor Simon were marked prominently OHMS. His postwar work on Britain's energy policy was well known, and we also kept a special watch for envelopes marked Tube Alloys. Of four remaining OHMS envelopes, three are stamped "SECRET" and sealed with blobs of red sealing wax. The fourth, unwaxed, carries a directive printed in capital letters above the address "TO BE HANDED TO PROFESSOR SIMON." All four were hand-delivered, and there are no postmarks. Three of the envelopes are spelled out "On His Majesty's Service" while the fourth has a gender change in recognition of the new Queen.

By this stage it was clear that even an empty envelope or folder might have something to say and should not be discarded without inspection. This message would be reinforced powerfully by an item that turned up a little later. It was a black, folio-size file folder with a spring clip, of German manufacture, old, well worn, empty like the others. On the cover, an octagonal label, marked *Steuern*, taxes, and in the same hand on the label, five names in order: Sigmund Münchhausen, Pauline, Rosa, Carl Münchhausen, Katie. The persons concerned are all on Lotte's father's side of the family. The markings on the label tell a story, and a sombre one.

Sigmund, whose name appears first, was Dor's and Kay's revered maternal grandfather, Opapa, who died in Oxford in 1950. The next two, Pauline and Rosa, were Sigmund's older sisters. They were left in Berlin when he and Omama escaped to Britain in 1938, just as the exits were closing on Jews still remaining in Germany. Both sisters had died during the war. It was understood that they had died in the Holocaust, though no details were ever volunteered by Lotte or Franz. Dor too had no details.

Onkel Carl, Sigmund's younger brother, was a former partner with Opapa in the family business, which involved the design and manufacture of fashionable ladies' hats. For years they ran an elegant shop in Friedrichstrasse, and the business had prospered. Carl worked as a designer. At some point before the storm broke, he left Berlin and retired to Switzerland, where he lived his last years in Locarno. Sigmund too retired early. In retirement he became a *Handelsgerichtsrat*, an honorary magistrate, dealing with cases in the commercial courts.

Katie was the widow of a third Münchhausen brother, Hermann, a painter, architect, and designer of avant-garde furniture. She was partly English, the daughter of a British general, and on her mother's side a Baltic German. Hermann and Katie had owned a farm near Lübeck, until an order for his arrest as a Jew and an anti-Nazi was issued in 1935. After a quick, forced sale of the farm, they escaped to England, where they farmed again in Essex. Though Hermann died during the war, Aunt Katie continued to farm in England, with some financial help from Franz Simon, until the property in Germany was recovered, after a lengthy restitution case, in 1952.

From the evidence on the label, one can reconstruct the most probable connections. The first three names—Sigmund, Pauline, and Rosa—are in black ink in Sigmund's hand. This folder is clearly Sigmund's tax file. He, as a retired businessman, must have managed his older sisters' tax matters in their retirement years. The file undoubtedly dates from Berlin days, and must have been brought or shipped to Oxford in the escape from Germany. At some later point, the two women's names were crossed out in pencil, and Carl's name is pencilled in just below, apparently in the same hand. Carl's tax or estate matters were now also being looked after by Sigmund. Later still, but this time in blue ink, Katie's name was added. This fifth name is perched at an odd angle on the cover, but the handwriting could well be Sigmund's. The label on the tax file and its troubled subtractions and additions hint at the silent anguish of Sigmund as he faced the progressive losses of his siblings.

From this slim initial evidence was born an urge to know something more of these family relatives of an earlier generation, and to understand more deeply the ordeals that they had faced. As the clear-out progressed, more evidence would build up until the initial impulse became a compulsion to know as much as family records could reveal about the horrors endured by the generations that had faced the Nazi terror.

As the work continued, more information emerged on Pauline. One day a small envelope of newspaper clippings on Hermann Münchhausen's designs yielded a brief printed obituary of Pauline, a totally puzzling surprise. How could a victim of the terror have a printed obituary in 1939, when others simply disappeared? Kay explained: "Didn't you know? Pauline committed suicide. Just before the day that she was summoned to report to the Gestapo." In 50 years of knowing the Simon family, this act of courage and independence had never been mentioned by anyone.

The 21-line obituary briefly describes Pauline's major project. As a teacher in a Jewish private school, she became interested in the welfare of her retired colleagues, many of whom had to retire into poverty without state pensions. With other social activists, she founded a retirement home, through an association which first rented and eventually owned its own house. Opened in November 1899, this foundation had marked its Fortieth anniversary just four days before Pauline's suicide. It is not difficult to see a context for her act through the guarded language of the notice. At 84 years old, she had few options.

For Rosa, family tradition held that she had been deported from Berlin to the camps but did not survive the journey, this was clarified and corrected when the Central Database of Holocaust Victims at Yad Vashem in Israel was released on the internet in November 2004. German deportation lists show that Rosa was removed from Berlin to Terezin on Transport I/65 on 15 September 1942. Her "prisoner number" in the transport was 7407. She reappears on a list of inmates at Theresienstadt Camp, which also records a place and date of death: Terezin in Bohemia, 22 November 1943. The record also notes that her body was cremated. Although Theresienstadt was not an official extermination camp, it had its own crematorium that could handle up to 200 bodies a day. These records, which list her correct date of birth, show that Rosa, the supposed chronic invalid, endured more than 14 months of camp hardship before she died—or was killed—half way through her eighty-seventh year.

In the air raid shelter, the clearing continued. One of the last discoveries, from the remotest part of the floor of the shelter, was an entire suitcase of mixed correspondence. This proved a significant find. On examination, these files turned out to be papers classified as private and returned to the family by the Royal Society archivist when Simon's professional and scientific papers were being selected for deposit at the Royal Society Library in London. After decades of neglect on the unheated concrete floor, these files were in deplorable condition from dampness and mould, but with sunshine and improvised drying racks they were made usable again, a few pages at a time.

By early June, after a break, we returned to Oxford for the second phase of the clear-out. By this date, conditions at the house were ready for the next and final stages. The lodgers upstairs were moving or preparing to move. An estate agent had been chosen. In the next fortnight the house was listed and shown to potential buyers. By early July, a buyer was found and a contract concluded, with a closing date of 1 August. The timetable was comfortable: it allowed time for a last visit by Lotte's and Franz's grandchildren and great-grandchildren to visit the house and enjoy the garden before the removal of furniture would make the house less habitable.

The next stage was to arrange for distribution of furniture and other contents of the house. At that time, five out of the six Simon grandchildren were in North America and still furnishing houses or flats of their own. The bulk of the furniture was offered to them. A shipping container was ordered for the transatlantic crossing. On arrival in North America, its contents were forwarded to several destinations. The cultural and intellectual property was shared mainly between Kay and Dor, with Kay having priority for the artistic and photographic inheritance, and Dor for papers, correspondence, and documents on family history and genealogy. Once the showings and sale arrangements were completed, the extensive photographic and documentary files from the living room, including Franz's diaries, calendars, and photographs, and Lotte's family trees, could be retrieved, sorted, and studied. Far from the chaos that Lotte had predicted, these files and photographs were packaged, labelled, and in exemplary order. Lotte—and Franz before her—had left an orderly, clearly labelled inheritance for later generations. For their tidiness and care, we owed them a similar care in preserving what should be kept.

On 18 July, the overseas movers arrived at Belbroughton Road. They were to pack china and glassware, crate the larger paintings, and wrap furniture for the container. Three days later, the container was loaded, sealed, and collected for an odyssey that would last several weeks. In the shipment with the furniture were included 15 mid-size boxes of family papers, hastily packed by ourselves, on the principle that cases of doubt could be sorted out later. They included a lifetime—and beyond—of family letters, diaries, official documents, pamphlets, offprints, clippings, genealogical records, and mixed private correspondence. In these boxes—some of them still unsorted—were included family trees on Lotte's side of the family, and also materials vital for sections of this volume. These latter included Franz's main diary and calendar, Lotte's partial transcription of the diary, and Franz's correspondence with Lotte over a quarter of a century. Ten days later, we left an empty house—stripped of furniture, carpets, and barely recognizable—and headed for Heathrow and a flight home.

Our original task of clearing had been completed, but a new one was emerging from our discoveries of the previous four months: an urge to preserve the story of Simon's life and career including the additional sources found in the clearout.

What benefits could be expected from a full-scale book on Simon's connections and roles in these programmes? I had been deeply impressed while still working on the clear-out in Oxford by his almost daily reports to Lotte during periods of their separation, and most of all—within the limits of the censorship—by the fascinating portrayal of nuclear development between 1940 and 1944. Simon is his own ideal biographer in these letters. His reports are those of a well-placed insider, and the style is one of graphic immediacy, often written within 24 hours of the events he is describing. When more detail is needed, the diary adds a level of precision to earlier literature, and occasionally corrects it. The private papers now available for the first time would justify a volume on Simon taken by themselves; when combined with his public papers already at the Royal Society in London, they produce an account that is in many ways unique.

A separate volume focused on Simon also allows additional space for themes in which he has a special interest. Thus the ups and downs of British–US collaboration can be documented at length, and assessed as a factor in Manhattan's ultimate success. While the British–American cooperation had serious shortcomings, German–Japanese collaboration was close to zero. Another example is Simon's long, brave resistance to the revival of ex-Nazi bureaucracy in post-occupation Germany, which won a few victories in individual cases but had little or no impact on British policy.

As the manuscript on Simon progressed, it became obvious that systematic comparisons of the wartime nuclear programmes were lacking, though enhanced resources for comparison had become available in the nuclear literature. It appeared feasible and desirable to assess these programmes one against another. Such an approach would assess the origins, progress, and achievements of each programme, but also trace mistakes, delays, blind alleys, or missed opportunities in each case. The importance of the nuclear race for the history of the second half of the twentieth century is self-evident.

This book accordingly addresses the more fundamental question of why the nuclear race ended with the results that it did. This question is addressed specifically in Chapter 7, "Why Manhattan?". And this is why some parts of this study have ventured beyond the biography of an individual to assess the five active nuclear programmes in their broader social, political, and military contexts. As that analysis will suggest, the outcome of the nuclear race was determined by a complex mix of factors that included: the numbers and quality of scientists involved; the organization, management, and risk-taking capacity of the project leadership; the industrial priorities accorded the projects; adequate raw materials; and the effects in wartime of enemy action.

ACKNOWLEDGEMENTS

One of the great pleasures of authorship is memories of those who have helped the project in a variety of ways and thanking them. This volume has received more help from both documentary and oral sources than can be fully acknowledged here, but it seems in order to mention a few names that stand out.

My most obvious debt is to the late Marion Armstrong, who took charge of the typescript at the inception of the work, and helped to shape a text that stood out for style and clarity. Her unexpected death in September 2010 was an irreplaceable loss to the project and all who worked with her.

On assistance with content, a few highlights stand out. Stephen and Pamela Frank revisited Stephen's boyhood memories of Berlin between 1933 and 1936 to recover Stephen's memories of growing difficulties under the Nazi regime. Hannah Howard reconstructed the details of her husband's mission to Berlin in 1945 to look for his mother, who survived the war in Berlin by taking an assumed name. I also have benefitted from the patient work of Kay Simmon Blumberg on Simon family genealogy, without which there would be little available on the Simon tree. In a much wider sense these connections link to my entire experience with the Simon family. Franz's two daughters, Kathrin Simon Baxandall and Dorothea McRae, have been constantly available for needed information, and Lotte in her prime years was always ready to reminisce about decades long past.

Among documentary sources, the largest was the Simon Papers deposited in the Royal Society Library in London. I am grateful for many kindnesses from Joanna Corden, the archivist, and from library staff. These public sources were supplemented by private family documents found in clearing the house in Oxford, particularly the almost daily letters exchanged between Franz and Lotte Simon whenever one or the other was absent from home.

When the book began to take shape I became obligated to three anonymous referees for Oxford University Press for readers' reports that were both fair but also constructive, providing enough detail to give some useful guidance to a novice biographer. In revising a longer manuscript, I was sustained by the encouragement from Robert Fox for a Simon biography. One of my former students, Mary Trueman, helped with wise advise when the task ahead appeared beyond reach. Finally, I am grateful to Sönke Adlung Physical Sciences Editor of Oxford University Press, for patience and encouragement, and also to Jessica White, Erin Pearson, and other OUP staff who contributed to stages of the book's production.

K.D.M.

CONTENTS

Illustrations — xv

1 Growing up into a world at war — 1
   Franz's war — 2
   Lotte's war — 7

2 Berlin 1919–1930 — 11

3 Breslau 1931–1933 — 18
   Discovering America — 20
   Portents of the storm — 26

4 Oxford 1933–1939 — 34
   Getting settled — 34
   Growth at the Clarendon — 37
   The exodus continues — 40
   An eye on Germany — 50
   The darkening political scene — 52
   A farewell to peace — 56

5 Any capable physicist 1939–1941 — 57
   A German "secret weapon"? — 57
   The prospect of German invasion — 61
   Transatlantic lifeline — 65
   The Maud Committee at work — 68
   The Maud Report and its reception — 75
   Beginnings: United States — 81
   Beginnings: Germany — 83
   Beginnings: USSR and Japan — 87

6 Industrial Plants ... Heretofore Deemed Impossible 1942–1945 — 89
   Mobilization and cooperation — 89
   A General takes charge — 91
   The long stalemate — 98
   The Manhattan Project in 1943 — 103
   Restarting cooperation — 105
   The barrier question — 109
   The family reunited — 112
   Manhattan: continuing uncertainties — 117

	Germany: the Army's withdrawal and its consequences	120
	Germany: the war closes in	127
	The round-up of nuclear scientists	130
	Awaiting the end	133
7	**Why Manhattan?**	**137**
	Prewar developments	137
	The missile alternative	140
	Inside Farm Hall: a version for the world	143
	Organizational factors	149
	Raw materials and priorities	152
	Effects of enemy action	154
	Nuclear projects in the Soviet Union and Japan	158
	Vicissitudes of an alliance	166
	Looking ahead	170
8	**Something reasonable again**	**175**
	Approaching absolute zero: experimentation and theory	176
	Nuclear legacy	178
	Committees, committees	181
	Public voice for science	183
9	**Security lapses**	**191**
	The Fuchs affair	191
	The view from the Soviet side	198
10	**Germany in the balance**	**202**
	Signals of survival	202
	The Pakenham letters	209
	Who should be invited to Britain?	219
	Heisenberg's visit to Oxford	224
	Would Simon visit Germany?	229
11	**A rounded life**	**235**
	Summer holidays	235
	Professional travels	237
	Climacteric year: 1956	245

Appendix: Simon's Report to Perrin about Heisenberg, 11 March 1948	255
Notes	**257**
Bibliography	**269**
Index	**273**

ILLUSTRATIONS

1 Paternal grandparents. Reproductions by Kay Simmon. (a) Henriette Simon, née Timendorfer (Gleiwitz 1832–Pless 1899). (b) Heinrich Simon (near Pless 1828–Berlin 1906).

2 Franz's father Ernst and his siblings, photographed in 1907 at Ernst's villa in Preussenpark, Berlin, with dates and places of death, and married names of women. All 12 were born in Pless, Reproduction by Kay Simmon. Front row from left: Bernhard, 1861–1924 d. Berlin; Rosalie (Lewin), 1856–1939 or 1940 d. in transport to camp; Cassilde (Preuss), 1860–1935 d. Breslau; Ernst, 1857–1925 d. Berlin; Selma (Lindenberg), 1865–1939 d. place unknown. Standing from left: Ludwig, 1866–1934, d. Berlin; Clara (Kochmann), 1871–1942 d. "in Poland," place unknown; Isidore, 1863–1919, d. Breslau; Julius, 1876–1929 d. Hanover; Joseph, 1869–1954 d. Sao Paulo; Therese (Guttfreund), 1878–1942 d. Tel Aviv; Gustav 1870–1942 d. Theresienstadt camp.

3 Ernst Simon's family, *c*.1906, From left: Ebeth, Franz, Ernst, Anna, Mimi. Seated: Anna's father Philibert Mendelssohn, a former public servant with the Prussian State Survey.

4 Franz at the door of a film shop, *c*.1913, at the threshold of a lifelong interest in photography.

ILLUSTRATIONS xvii

(a)   (b)

5 Military service in the First World War. (a) Franz, convalescent after shrapnel wound to right elbow, sustained in Vosges Mountains, 20 April 1915. (b) Franz, decorated and home on leave, Berlin, probably 1917.

6 Ernst Simon with his family, Berlin, probably 1917.

7 Lotte and Franz, early 1920s.

8 Franz with his daughters in Locarno, Switzerland, prior to his departure for a term at Berkeley, December 1931.

ILLUSTRATIONS

9 En route to Berkeley, Franz visited Einstein in Pasadena, where he took a few pictures, January 1932.

(a)   (b)

10 From Breslau to Bournemouth. (a) Breslau, 1932. From left: Kay, Lotte, Dor, Omama. (b) Bournemouth, August 1933, first weeks in England. Kay, Lotte, Dor, Franz.

(a) (b)

11 The house at 10 Belbroughton Road, 1930s. (a) Front view. (b) Rear view, from the garden.

12 An open house in the garden, held on Sunday afternoons in the 1930s.

13 Lindemann and Schrödinger in the garden. From 1934 to 1936 the Schrödingers lived just around the corner at 24 Northmoor Road.

14 Walther and Emma Nernst in the garden with the Simon children, when Professor Nernst was awarded an honorary degree from the university, 1937.

15 In the early Oxford years, Sunday mornings were for bicycle rides. From left: Franz, Dor, Kay, Franz's nephew Thomas Frank, in front of the old Clarendon Laboratory.

16 The Second World War brought 4 years of family separation, broken by brief reunions during Franz's working visits to North America. This view shows Lotte and Franz in New York in 1942, probably during Franz's second visit of that year (September to November).

17 Another meeting in New York labelled 1943. If Franz looks preoccupied, this visit was for the crucial Manhattan–Tube Alloys barrier meetings of December 1943–January 1944.

18 Simon with Brigadier Charles Lindemann, Cherwell's brother, who was Air Attaché at the British Military Mission, Washington, probably 1944.

19 Southern France was also appreciated. This photograph is probably from a holiday on the Riviera in 1948.

20 Ties with the Mendel family, who welcomed Kay and Dor as evacuees to Toronto in July 1940, continued after Bruno Mendel and his wife returned to Europe in 1950. This photograph shows a reunion in Switzerland. From left: Kay Simon, Ruth Mendel, Herta Mendel, Franz, Lotte, Bruno Mendel.

21　Franz relaxing in the garden, Oxford, August 1953. Photograph by Kay Simmon.

22 Franz in the garden during a visit by New York cousins, August 1953. Photograph by Kay Simmon.

23 Franz with his first grandchild, Patricia, summer, 1956.

24 Nicholas Kurti, Franz's student and research partner for four decades, remained a close friend of the family to the end of his life. Here he is shown with Lotte on his eightieth birthday in 1988.

1

GROWING UP INTO A WORLD AT WAR

Despite a four-year difference in their ages, the central event in Franz's and Lotte's intellectual coming of age was the First World War. Nothing before it was even remotely comparable to its impact on their lives. Though they had never met—and would not meet until 1920—the year 1913 marked a turning point for both of them. For Franz, who had completed gymnasium-level schooling early in 1912 and begun university studies the same year, the summer of 1913 brought an interruption, a call-up for military service. For Lotte, aged 16, 1913 marked the end of a ten-year curriculum in a private school for girls.

For Franz and Lotte, the house at 10 Belbroughton Road, Oxford, contained all the relevant school reports, arranged chronologically and carefully preserved in embossed binders. These reports ran from 1899 to 1912 for Franz, and from 1903 to 1913 for Lotte. They show not only two sets of individual marks but also the divergent educational values for boys and for girls in this period.

For all 13 years, Franz attended the Kaiser Friedrich-Schule in Berlin-Charlottenburg, a state school, serious, demanding. His earliest reports begin strongly. He is typically ranked in the top half-dozen of the 40–45 pupils in his class. Later, he drifts downwards to the middle level. In the upper years, his marks in mathematics and physics are strong, but in a curriculum that includes five languages his overall showing is only adequate. In addition to German, his courses included 10 years of French, seven of Latin, five of Greek, and a couple of years of English towards the end.

By 1912 he had completed the gymnasium or upper secondary programme and proceeded to university. His choice had been to study physics and mathematics, but this field had run into initial opposition from his father, who was doubtful of the economic possibilities of an academic career. At this point a family friend, Leonor Michaelis, a research biochemist and physical chemist at a Berlin hospital, persuaded Franz's father to drop his opposition and allow the son to follow his inclination.

Like other German students, Franz could move between universities as his studies required. His first two terms were spent at the University of Munich in 1912–1913. Photographs of him in this year show a serious young man, always in suit and tie, in characteristic settings: seated formally in a chair; smiling, at a table in a restaurant; or standing in the doorway of a photography shop, apparently indulging what would become a serious lifelong hobby. The year in Munich was followed by a summer term in the University of Göttingen, to which he was drawn in part by the reputation of its mathematicians. Completion of this term was blocked by his military call-up.

Lotte's education was different in content and in purpose. In the private school that she attended for ten years, her studies included German, French, English, religion, world history, art history, geography, art, music, and gymnastics. There is some mathematics and natural science, but not the physics, chemistry, Latin, and Greek required in Franz's state school, and needed for admission to a university. In middle-class families, men were educated to enter a gainful profession, women to practise and enjoy the arts of civilization. In her later years, Lotte would criticize her schooling as undemanding and non-serious, but its primary purpose was not professional but cultural. Nevertheless, her carefully developed fluency in French and English would serve her well in the turbulent years ahead. For her parents, the notion that their only child should undertake university studies seemed unnecessary, even inappropriate. As her father often assured her, life is for enjoyment, not for work. In the next decade, these comfortable assumptions would be seriously challenged.

Franz's war

Called up after three terms of university work, Franz became an *Einjähriger*, fulfilling the army training required of every German male. As a science student in Munich, he was assigned to a local artillery unit. Towards the end of that year, his unit was ordered on some extra exercises and unexplained manoeuvres. In a letter to his parents he writes from Augsburg:

> Dear Parents,
> Finally, I have a chance to write you. The day before yesterday we had a night shooting drill. When we got back in the morning, all was quite peaceful. In Grafenwöhr one can learn nothing, and the army lets out nothing. The only things we know come from the *Münchner Neuesten*, which I get forwarded, but it comes a day late. At noon orders came to get ready at once to march . . . . By about 2 p.m., the horses were harnessed and all was ready to move away. Then it appeared no wagons were there, and it was put off again. From noon to 7 a.m. yesterday, we stood in the stable in full equipment, helmets and swords, horses saddled . . . . Finally we got away around 7 and arrived here about 9 p.m.
> People here do not think it will come to a war, but a mobilization is possible. We would then go for a few days to some little spot on the French frontier, and then back again to Augsburg.

Military efficiency seems much the same everywhere. The letter is dated July 31, 1914.

Franz's war years can be documented by combining four different sources. First, in one of the Simon family photograph albums, more than 20 pages are filled with scenes from his service years. These can be linked with other sources. The family papers include a packet of wartime letters and postcards sent to his parents in Berlin. Another family file contains official papers, his commission, certificates of decorations, and the like. Finally, and quite unexpectedly, a regimental history of his unit surfaced in the bookcase. Taken together, these four sources constitute almost a daily record of Franz's war experience, an experience that he never discussed with family members of our generation.

The opening months of the war brought rapid expansion of the army, reorganization, and completion of a continuous defence line along the frontier. By December 1914, Franz's previous artillery unit had been joined with other units to form a new regiment, the Royal Bavarian Reserve Field Artillery, Regiment no. 8, which became part of a reserve division of the Bavarian Army. From August 1914 onwards, Franz reminds his parents several times that mail for him must specify the *Bavarian* Army. Closer inspection of the official documents shows that his officer's commission and two of his four service decorations are issued in the name of the King of Bavaria. The regimental history also records occasional personal visits from this monarch.

The regimental volume resolves the longstanding puzzle of why the photograph album contains scenes from several different fronts. As a reserve artillery unit, Franz's regiment—and sometimes the entire reserve division—were moved from one theatre to another, providing extra support to first-line troops wherever needed. By combining his own war photographs with the official record of the regiment, one can chart a good deal of his military experience in remarkable detail.

In the closing months of 1914, Franz's previous unit had been part of the rapid German advance into northern France. The first war scenes in the photograph album show the heavily damaged village of Fay, just west of Péronne, where his battery had been stationed. In December, after this battery was transferred to the new regiment, it moved to the mountainous Vosges front in Alsace. There, the regiment faced strong French forces intent on recovering the territory France had lost to Germany in 1871. The fighting in the Münstertal soon became intense.

Late in April, a letter dated 22 April, and written from the Münstertal in an unfamiliar hand, was delivered to Franz's father in Berlin. It was found among Franz's own letters home, but a combination of unfamiliar script and faint writing in pencil make the text almost indecipherable. A few key words stand out: "... peak 830 near Metzeral ... April 20 afternoon around 5 o'clock ... your son ... Metzeral ... Münster ... Colmar ... hospital." The sender's name, however, is written clearly: Aarlo Merker.

Another shorter letter soon followed, postmarked from Colmar the next day. This one is in another, more legible hand, and is obviously dictated by Franz from the hospital. He announces a light shrapnel wound in his right elbow, nothing to get worried about, and sends his greetings.

On 30 April Franz sends a letter of his own, this time typewritten, but using no capitals. He writes about the affair from a hospital in Freiburg, where he had been sent for surgery:

> I was away from the battery for three days as artillery spotter on the famous peak 830 north of Steinabrück. With me as telephonists were two one-year men, Merker and Schmidt-Baumler. I will give you the details later orally.
>
> In any case the summit was almost captured, and the rest were made prisoners. Nothing happened to Merker, but Schmidt-Baumler has fallen and I have got a shrapnel wound.

Now the picture becomes clearer. Merker, the telephonist who was unscathed, wrote the first report to Franz's father. The regimental history, on its list of those fallen or missing, gives the name Wilhelm Schmidt, 6th Battery, with the date of 20 April 1915. He was born four months after Franz, in Augsburg, and was apparently married. The Simon album contains a photograph of him, labelled, and shows a tall, lanky lad in an untidy uniform. Some of the others labelled in the photograph album can also be matched to the regimental list of those *gefallen,* whereas many of the photographs, other than those of senior officers, do not carry names. These album labels are evidently Franz's way of memorializing those who did not get through the war.

In his letter of 30 April from Freiburg, Franz estimates his recovery period at five or six weeks, and reminds the family that civilian visits to this city are possible without military authorization. He also notes that the alternatives after recovery are immediate return to his regiment or a period of leave and later reassignment to some other unit. If given a choice, he would prefer the former. The photographic evidence shows that he did return quite soon to his unit, but another photograph in the album also testifies to a visit from his parents and both sisters to Freiburg during his hospital stay.

While he was still in hospital, his regiment, after a short break in Colmar, was ordered to the southern sector of the Eastern Front, where General Mackensen was following up a notable breakthrough in Galicia, which had been successfully invaded by the Russian forces in 1914 as far as the passes of the Carpathians. A joint German–Austrian counteroffensive the next year forced the Russians to retreat, but more German troops were needed to complete the campaign. In the first week of June, Franz's regiment completed the long rail journey via Strasbourg, Nuremberg, Dresden, and Breslau (now Wrocław in Poland) towards Cracow, Tarnow, and Jaroslaw. "During the journey," the regimental history states laconically, "the transports were greeted enthusiastically, especially in Saxony, and hailed as liberators in Galicia – mainly by the Jewish population."[1]

Franz, however, was not yet with them. One of his letters home is on the letterhead of a Munich hotel, dated 16 June 1915. In it he tells of his imminent departure by train later that day to rejoin his old unit, now serving as part of General Mackensen's Eleventh Army near Przemysl. "I travel alone," he announces confidently to his parents, "I am equipped with everything."

The photograph album has eight photographs from this Galician action, most of them with labels: a destroyed bridge; Galician children; vast piles of stores and equipment, labeled "booty collection point"; and the church at Radymno, a rail point north of Przemysl where the regiment and its equipment were loaded for the return to Germany. These photographs show that Franz was with his unit for the later stages of the short Galician engagement. By 27 June, a scant 3 weeks after its arrival, the regiment, having shared in the advance and completed its assignment, returned to its previous location in the Vosges sector of the Western Front.

On its return to Alsace, the regiment faced a strong French attack through the Münstertal, aimed at Colmar. Once this situation was stabilized, the regiment became part of a defensive line in the mountains until the summer of 1916. When the French and British forces launched a massive combined attack along the Somme on 1 July 1916, half of the regiment was moved to support the German defences northwest of Péronne. Franz's battery was the first to be sent to the area, arriving on 12 July, followed in August by the two other batteries of the regiment's Section II. From their arrival until late September, these batteries faced continuous heavy fighting. As reserve artillery units, they were moved around as necessary to the most critical points of their sector. In these few weeks, the official account records, this half of the regiment lost 207 officers and men through death, wounds, or illness, as well as major losses of horses and guns.[2]

Franz's album has a single page of photographs of this battle. Taken together, they illustrate the bleakness of the battlefield: a desolate, flat landscape, with a few stunted trees and distant bursting shells; fire bombs, closer at hand; the ruins of a building reduced to rubble; dead horses on a country road; a pile of shell cases; an artillery lookout station; a field kitchen.

In the summer of 1956, the BBC presented a special programme to mark the fortieth anniversary of the battle of the Somme. Franz wanted to listen and to concentrate on it, alone, without interruptions. The rest of the family were invited to move to the garden. They did so, though hardly aware of the significance of the event for Franz. His revisiting of the battle was a small ceremony, a private memorial to the comrades who had shared that ordeal in the summer of 1916.

The roll of honour in Franz's regimental history records the German losses for these weeks. For August and September alone the three batteries of the half-regiment show 20 deaths, more than double the total for the preceding 2 years.

After the slaughter on both sides at the Somme, the war dragged on. Following a period of mainly defensive action on the Western Front, Franz's unit was ordered again to the Eastern Front, where Romania had entered the war on the Russian side at the end of August 1916. Romanian troops immediately invaded Transylvania, capturing Kronstadt and Hermannstadt. German and Austrian forces counterattacked in October, retaking both cities and advancing to the passes of the Carpathian mountains.

Franz's regiment, arriving in late October, became part of this counteroffensive. Half of it was attached to a German–Austrian force that took the Red Tower pass south of Hermannstadt and then advanced eastward in hard fighting across the muddy plains of northern Wallachia. The other half, with the regimental staff, was assigned to other, more northerly Carpathian passes. From October 1916 to November 1917 the two halves operated in this divided fashion, sometimes far apart. But now a problem emerges. The war photographs in Franz's album do not correspond to the movements of his original 6th battery, and for this period we have no letters home, or other evidence, to explain why. From these photographs one has to assume that he may have been either with the regimental staff, which stayed with Section I, or with one of the three batteries of this section.

There are several dozen photographs with captions in the album for this period, scenes with snow-covered mountain vistas, soldiers waist-deep in snow, gun crews, artillery lookouts high above the valley, stout log shelters and defensive positions built into the hillsides, group photographs of officers. In these scenes from the Trotostal there is even a suggestion of relative comfort. But the guns are also present, laboriously dragged into position on the hillsides, and one photograph of dispirited men is labelled "Cossack prisoners."

Not recorded by Franz's camera, there had been strong Russian–Romanian resistance before these defensive positions in the passes could be won and consolidated. On the whole, however, Section I escaped the harder fighting and heavier losses of the Section II batteries as the latter fought their way eastward across the Wallachian plains south of the mountains. By January 1917, most of Romania, including Bucharest, was under German control.

Despite the outbreak of revolution in Russia in March 1917, the Provisional Government pursued the war actively. On 1 July, General Brusilov launched a major Russian offensive in Galicia. With most of Romania under German control, Franz's unit was moved from the Carpathians in mid-July for a counterattack on the new Russian line in Galicia. His photograph album contains a full page of scenes from this engagement. The majority are from Kalusz, the Galician town close to where the half-regiment arrived by train, unloaded, and set up a camp on 16 July. Eight views of the town portray devastated streets, burned out houses, a ransacked office, a dead horse outstretched grotesquely on a deserted street—all of these with a single caption, "Kalusz, destroyed by the Russians." Another scene shows a field encampment or bivouac of tents, guns, and soldiers. In the foreground, half a dozen women of the local population sit on the ground, their heads covered, and flanked by baskets of fruit or vegetables. Two more photographs show a ruined Russian trench. One has bodies in it, either two or three, covered, positioned unnaturally on a slope, possibly dismembered, and labelled "Dead Russians." Once seen, this picture lingers deep in the mind.

With Russian resolve weakened by the spreading revolution, the battlefront in Galicia and Bukovina was largely restored to the previous frontier by October. Franz's unit was then recalled to the Western Front, where it unloaded in pouring rain at Thourout in Belgian Flanders on 17 October. For the next several weeks it faced sharp fighting. By December the other half of the regiment, Section II, which had operated in other sectors of the Eastern Front for more than a year, also returned to Belgium to be reunited with Section I. There followed a badly needed period of rest, refresher courses, and extensive refurbishment of men, equipment, and horses. This continued into early 1918.

In the photograph album there is a page that illustrates these activities: views of shell craters and bleak, foggy fields; a destroyed factory or warehouse, still smouldering; a house intact in open countryside behind the lines, labelled "rest barrack"; four officers, including Franz, grouped around two machine guns, labelled "machine weapons course, Antwerp"; and sightseeing views of Antwerp and a deserted sea front at Ostend.

This period also saw extensive organizational changes, devised by the high command in anticipation of massive German offensives being planned for 1918. The regiment was detached from Bavarian Reserve Infantry Division 8 and reclassified as army artillery to be used wherever needed. It was enlarged by creation of a Section III, making three sections of three batteries each, each section with its munitions train. The new section was staffed two-thirds from the conscript class of 1899 and the rest by transfers of officers and seasoned troops from other units, especially from Field Artillery Regiment 9, the regiment's "twin" or parallel artillery unit on the organization chart of the Reserve Infantry Division.

These changes raise questions concerning Franz's exact whereabouts during 1918. Among his official documents are two, dated May 1918 and October 1918, which both identify him as a member of Regiment 9, not Regiment 8, at these dates. The first of these is a certificate, signed by the Commander, Bavarian Infantry Division 8, conferring, in the name of the Kaiser, the Iron Cross, First Class. References to Regiment 9 occur sporadically in the official history of Regiment 8, but its movements are not traced systematically. Further, Regiment 9 appears to have remained with its original infantry division after Regiment 8 was detached and reclassified. The sole reliable source for the remaining months of Franz's war is therefore the photograph album. Even this evidence is limited, probably owing to the intensity of the fighting and the swiftness of the final German retreat.

For these last months of the war two things can be said with some certainty. The album contains a group of four photographs with the caption "Offensive Reims July 1918." As documentary photography, they are hardly memorable: a desolate crossroads, littered with destroyed equipment; a crater-ridden field, with a distant view of a team of horses, untended but still tethered to a gun carriage or wagon; a close-up of a heavy artillery piece, apparently destroyed. These photographs testify to Franz's presence at the front through the turning point in July, when the last major German offensive was stopped and reversed by an Allied counteroffensive, the point that marked the beginning of German withdrawal from France.

The other significant event from 1918, known from family sources, is that Franz was wounded again at the very end of the war, this time by shrapnel in the leg. The event is corroborated by one more photograph on the last wartime page of the album. It features a single enlarged picture of several nurses and two officers—presumably medical officers—posed formally in front of a substantial building, with a caption reading "Lazarett Hanau 1918." This military hospital in Hanau, near Frankfurt, is almost certainly where Franz was initially treated for this second wound in the weeks after the armistice. Once back home in Berlin, he would spend the early months of 1919 in and out of hospital for operations to remove metal fragments.

Lotte's war

In her later years, there were two things that Lotte looked back upon with pride. The first was her work during the war as a laboratory assistant. The second was a lengthy tour of Italy in 1921, when travel abroad again became possible. After she was 90, when her short-term memory became unreliable, she could recall these formative events with undiminished clarity and describe them to any visitor. When the war began, Lotte was too young for the nursing service. She enrolled instead in a training course as a laboratory assistant. On its completion she was taken on in a biological laboratory making anti-cholera vaccines, intended as a defence against possible epidemics among the troops serving on the Eastern Front. She also remembered clearly how the professor in charge warned the assistants of the dangers of handling live bacteria. "If you make a mistake, you die," he would say, always standing behind them at a safe distance.

Documents found in the house in Oxford add precision to Lotte's account. A letter describing the content of the training course for *Laborantinnen* and Lotte's successful completion of it is dated 29 January 1915. She had then gone to work at the Charité, the university hospital founded by Frederick I of Prussia in 1710. With the first letter is a laudatory reference from Professor Morgenroth, head of the Bacteriology Section at the Institute of Pathology, attesting to her valuable service there from February 1915 to June 1917.

There is more. Amid a collection of scientific offprints from Franz's students and colleagues on physics and physical chemistry, there are five more authored or coauthored by the same Professor Julius Morgenroth, dating from 1915 to 1920. These stand out oddly. They also show that while Lotte might remember most vividly the dangerous work on cholera vaccines—as who would not?—these reports indicate a wider range of research questions focusing on "experimental chemotherapy," investigating the use of various disinfective agents against several types of bacilli. In an offprint of 1917 describing experiments against strains of *Staphylococcus*, a footnote acknowledges the author's strong support from three named laboratory assistants, among them Fräulein L. Münchhausen. Morgenroth himself had begun his career in chemotherapeutic research as an assistant to the renowned Paul Ehrlich.

In time, an issue of salary arose. Lotte had entered the workforce in 1914 as a volunteer, for patriotic reasons, since her parents were relatively well off. After a few months, in her telling, her father pointed out that she was displacing a paid worker, and in fairness she also should be remunerated. She raised the issue at the laboratory, and a salary was agreed to. This part of the story is corroborated by the discovery of her original social security card, which shows social insurance premiums of 4.80 marks per month from March 1916 to May 1917. Each month lists Morgenroth as her employer. A table on the back of the card indicates a salary between 70 and 95 marks per month.

In her later years, Lotte remained proud of this wartime work experience. Though she could hardly realize it at the time, she was entering an adult world that had reached a turning point. The pre-war ideal of a life of cultivated leisure would no longer be respectable, or even available, for families like hers. In the war years, however, Lotte was able to combine working in the laboratory by day with living at home and access to the rich cultural life in music and the theatre that her education envisaged. Among her papers is a list of remembered musical events from the war period, probably jotted down in preparation for an interview or broadcast. Her list begins with a memorial concert to mark the death of Mahler in 1911. She mentions the 1914 Mozart Opera Festival, noting it for her first hearing of the *Marriage of Figaro*, conducted by Richard Strauss. The list goes on to include other conductors: Artur Nikisch, Bruno Walter, Wilhelm Furtwängler; pianists: Eugen d'Albert, Artur Schnabel, Emil Sauer; and a concert of Schubert lieder. She also lists her own participation in the Philharmonic Choir, directed by Siegfried Ochs, a connection that lasted from 1915 until 1929. Her collection of music in the book cases included choral scores from this period, as well as others later on from the Oxford Bach Choir. This choral activity over half a century provided a link between her life in Oxford and that left behind in Berlin.

Concerning theatre, she had developed as a teenager an infatuation with Shakespearian drama, performed in German on the Berlin stage. This interest, she explained, arose first from seeing Shakespeare as theatre. Reading the plays came later. Again the bookshelves yielded supporting evidence. Lotte's bookshelves contained several volumes of Shakespeare, all in German translation, and all dating from this formative period.

The military defeat of Germany in 1918 appears to have ushered in a cold, grey period for Berliners. There are more deaths than usual in the family trees for those years, and memories of shortages of food. Housing was difficult. Even for those who owned a house or flat, space was controlled, on the basis of one person per room, plus an extra room for academics or others requiring an office.

For Lotte there are few records from these dark years, but her situation brightened in the spring of 1921, when she was given 1,000 marks by her parents for travel to Italy. On 8 April, accompanied by an acquaintance of similar age and apparently from the same social circle, she set out from Berlin by train. By careful budgeting, the two travellers were able to stretch their tour to almost three months, taking in Verona, Vicenza, Padua, six days in Venice, Ravenna, Bologna, 21 days in Rome, six days in Naples, Salerno, 15 days on the island of Capri, Siena, Pisa, 17 days in Florence, and finally Mantua, from which town, on 29 June, Lotte sent her parents a last postcard before their return.

Unlike the preceding years, the tour through Italy is closely and luminously documented in three separate sources. First, she sent reassuring postcards, posted almost daily, home to her parents. Second, she kept a formal journal, in imitation leather cover, to record her impressions of what she was seeing. Finally, there is a separate picture album, filled mainly with postcards of places visited and things seen, together with a few photographs of the travellers. These items turned up in three different corners of the house, but when put together they capture Lotte's outlook and sensibility at that point in vivid detail. Her travelling companion, another Lotte, figures surprisingly little in these sources; the two may not have shared the same enthusiasms.

The sources leave no doubt of the serious intent of the trip. This Italian interlude is a scaled-down version of the classic Grand Tour. Its purpose is not scholarly study in an academic sense but aesthetic appreciation, a maturing of the senses, a rounding off of an academic education in the arts. The tone is bright, optimistic, at times ecstatic. "Really in

Italy!" the diary exclaims in Verona, and two weeks later the same exultation proclaims "In Rome!" This tour obviously represents a long-postponed realization of a cherished dream.

A closer look at the records indicates the range and emphasis of Lotte's artistic interests. The photograph album in particular emphasizes churches, both external architecture and internal sculpture, and other art. The diary, in German, lists visits to churches, to galleries and museums (some more than once), to the excavations at Pompeii and Ostia. She has an eye for architecture, and for noting architectural detail under changing conditions of light. Even the weather, and the colour of the Italian sky, are carefully recorded. In the evenings she attends a few operas and orchestral concerts. In her recurring emphasis on church interiors and Christian art, Lotte gave a clear priority to Renaissance treasures over those from antiquity.

Her dedication to art and culture on this tour slipped only once— during the visit to Capri. Here for a while nature superseded art, and her lyricism found a new orientation and a new height. From 24 May to 7 June the travellers spent two idyllic weeks on Capri, staying longer than they planned. A journal entry on 24 May sets the tone:

> Capri, indescribably beautiful, paradise on earth, wine, lemons, flowers in abundance. Steep cliffs plunge into the blue sea. Heavenly days. Bathing in the Baths of Tiberius. Excursion to the Blue Grotto, wonderful swimming, silvery bodies. Procession in the enchanting little town. Small children wearing charming shawls. Very much the land of children. Fishing trip, balmy air, starry heavens, boats with lights, beautiful illuminations.

The next two pages are left blank, doubtless intended to be filled in later. In the diary, some aspects of this "indescribable" enchantment of Capri remain undescribed, apparently beyond Lotte's powers.

The daily postcards sent home fill in more prosaic details of the days on Capri, details missing in the diary: excursions by boat or on land, excellent meals, evening promenades, processions in the town. The visitors found the local population *simpatico* and rejoiced in learning to speak some Italian. They became friends with an old fisherman, who took them on excursions to local sites and appointed himself their protector against his younger countrymen. They alternated between lazy, quiet days and excursions. The superlatives continue, and they arranged to prolong their stay beyond the original schedule. Lotte ends a long letter home by regretfully admitting a need for sleep, but "this time in Italy is so unbelievably beautiful that I want to experience every minute to the very fullest." The fortnight on Capri was a peak experience that she would carry with her to the end of her life.

Even paradise, however, can have its limits. By the end of the second week the postcards home switch to their plans for the rest of the tour. Still ahead were the treasures of Florence, and their itinerary had other towns that were almost obligatory. By 8 June the travellers returned to the serious study of art and culture. The summer was approaching, and as a northerner Lotte records the last days of the trip as oppressively hot.

Her Italian journey, with its escape from postwar hardships and its first exciting experience of a country that the war had made inaccessible to Germans, marked the completion of a formative stage of her life. Just over a year later, in 1922, she would marry. By Lotte's oft-repeated account, she first met Franz Simon at a sailing party on the Wannsee, which both had attended with other partners, in the summer of 1920. This date is also supported, though cryptically, by a few typescript pages of improvised verse in mixed German and

Italian, decorated with original art work, that could have originated only from her travel companion to Italy, the other Lotte.

These verses have a title page, "A Watery Story" (*Eine wassrige Geschichte*), and a date, 13 August 1922. At the bottom of this page, two events are linked:

IN MEMORIAM

| Wannsee 1920 | | Capri 1921 |

L.E.

The four pages of loosely improvised verse that follow have many insider references to the carefree days in Capri, and a few to the Wannsee sailing party. The writer evidently was familiar with both. She makes clear that our Lotte is making a fundamental choice, and ends with a warm congratulation:

> Tant' belle cose
> Wünsch von Herzen ich Dir ganz
> Botton' di rose
> Dir und Deinem lieben Franz.

For a long time, the identity of the enigmatic "L.E." remained unknown, but one day *Dor's sister* telephoned with a discovery. The pictorial record of Lotte's Italian journey, which remained with the other photograph albums in London, had a lone photograph of Lotte's fellow voyager with a caption "Lotte Eisner." Once named, she could be easily identified as the distinguished film critic and film historian of the German cinema. The clinching point is in Eisner's memoirs. She describes the same three-month Italian trip, with its idyllic fourteen-day stop in Capri, as the fulfilment of a cherished dream of the south and the sun: "I was with a Berlin friend," she notes, "a relative of my future sister-in-law Paula, Charlotte Münchhausen."[3]

In March 1933 Eisner, already known in her field, left Germany for Paris, where she continued her career. In May 1940 she was arrested and sent to the south of France, where she was interned, with several thousand other Jews, in the Vichy government's sprawling, primitive prison camp at Gurs, near Pau. At some point she was allowed to escape from this camp by a sympathetic French officer,[4] and to link with relatives in Montpellier. From 1942 to the war's end she lived in this region with new identity papers as Louise Escoffier.[5]

2

BERLIN 1919–1930

For Franz and Lotte, the 1920s were a decade for going through the stages that many young people everywhere experience as they evolve into mature adults and citizens. Simple stages, normal, routine, mundane. Completing an education. Starting a career. Finding a life partner. Embarking on marriage. Setting up a household. Having children. Visits to and from relatives. Occasional holidays. At first glance, this is so ordinary as to be scarcely worth listing. The context in their case, however, is Berlin in the 1920s. The place and the period were far from ordinary.

The immediate postwar years are only sketchily documented: a few letters, a few uncaptioned photographs, to illustrate their lives at this point. One photograph shows Franz with his younger sister Ebeth on holiday, on a sunny southern balcony in Lugano in southern Switzerland, in 1919. Both look emaciated, undernourished, and tired, which ties in with other evidence of physical shortages and mental let-down after the military collapse.

There was also a better side. Four years of war had ended. The Simon family was still quite well off. The sun shone in the summer, and the Wannsee and other waterways were ideal for boating. The album has photographs of sailing parties on the lakes, and one photograph shows a small inboard motor boat that the family used at this time, a boat named for the sound of its single-cylinder engine: *Muckepicke*. The new German Republic had a difficult birth in these years, but the family history reveals no hint of the violence and instability that marked Berlin and other German cities. The new constitution of 1919 promised greater equality for all its citizens than anything achieved by Imperial Germany.

In 1919, Franz resumed his studies. He had previously completed only one year of university work in 1912–13, and he was now 26 years old. In very short order he became a serious and dedicated student of physics and physical chemistry at the University of Berlin. By December 1921 he had completed a doctoral thesis in low-temperature physics under Walther Nernst, and was inclining, despite his father's earlier doubts, towards an academic career.

He could hardly have found a better setting for studying science. Berlin University had become pre-eminent in physics in the decade before the First World War. Now, in the 1920s, the university and the nearby state scientific institutes were consolidating and enlarging upon prewar pathbreaking work by Planck, Einstein, and Nernst, until the local constellation of physicists and chemists included more than a dozen figures of world renown. The seminars and colloquia generated by the interaction of this group were making Berlin the foremost centre in the world for advanced studies in the physical sciences.

Franz thrived under these stimuli, rapidly making up for the time lost to the war. By 1923 he had published six papers, and by the end of the decade his publications would number more than 40. In 1924 he obtained the *venia legendi*, which gave him the right to teach. He acquired research students of his own, developing an experimental niche for himself and his students in the physics of very low temperatures. Beyond this experimental area, he was also building a comprehensive understanding of Nernst's theoretical work on thermodynamics.

Along with professional development came economic challenges. As noted earlier, Franz and Lotte married in August 1922, but they continued living in the large flat with Lotte's parents in Pariserstrasse throughout the 1920s. Housing was scarce, and controlled. Franz's initial income as assistant to Nernst was scarcely enough to live on independently. They did, however, set up a separate household budget for themselves, and these household records survived intact among Lotte's papers in the Oxford house. There are 40 large accounting-style pages, removed from their original binder and preserved as loose pages. They cover 30 months, from October 1922—after Franz and Lotte returned from a honeymoon in Heligoland—to March 1925. This period spans the acute phase of the great German inflation and a little beyond, showing also the early months of the new currency, the *Rentenmark*. As such, they are of interest not only as a family record but also as vivid testimony to the financial crisis that was ravaging the German economy and destroying the social order.

Entries in the register are written by both spouses, but mainly by Lotte, with income on the left-hand page and expenditures on the right. The first month in this record, October 1922, shows a wedding present from Franz's brother-in-law, another from his uncle Justus, and a very sizeable grant for research apparatus of just under 100,000 marks. Other income in this period includes modest supplements from Franz's parents. But one key discovery from these pages is that Franz's earned income or stipend, though small, is rising with each month. It is indexed and, like other public-sector wages, it includes a retroactive bonus for the currency depreciation during the preceding month. With this bonus added, his earned stipend for October was almost 28,000 marks, for November 41,000, and for December—the first time it is listed as a *Dozent*'s stipend—71,000.

Even in these opening pages, inflation is already alarming. Lotte's three-month tour to Italy in 1921 had been managed on just 1,000 marks. Now this amount was just enough to buy the "coffee machine" that Lotte bought on 17 October 1922, but would not be enough for the "soap and chocolate" that she bought 10 days later for 1,400 marks. She was fortunate to have travelled when she did, while foreign travel was still possible.

Franz's father offered prolonged resistance to the very idea of inflation. "A mark is a mark," he would insist angrily, echoing the official line. To think otherwise was to be unpatriotic. He refused to be persuaded by his son and daughter, both of whom had been in Switzerland in 1919 and were thus well aware of the deteriorating mark.

In subsequent months the rate of inflation increased exponentially. Through 1923, these records are not easy to follow. In January, Franz's stipend, in two half-month instalments, appears as 121,000 marks. For February, it is more than doubled, and by April it is 418,000 marks. By July, three payments are each in seven figures, for a total of more than 6,000,000 marks in that month. In August, with payments reaching eight figures, the accounts begin to be recorded in thousands of marks by dropping the last three zeros, at first informally, and then in September formally. On 1 October the unit shifts again, to be reckoned in millions. By the middle of the same month, with printing presses roaring, the unit shifts again, from *Millionen* to *Milliarden*. For October the stipend payments were made nine times, every two or three working days.

The final explosion came in November 1923. With inflation increasing twelvefold from 1 November to 16 November, the record shifts again on 18 November from *Milliarden* to

Billionen, that is, from American-style billions to trillions. The eight payments recorded for November add up to just over 50 trillion marks. But now Franz has begun to list in parentheses the equivalent dollar value of these eight payments at the moment they were received. These equivalent amounts total just under 36 dollars for the month.

The last three payments for November 1923 show a modest supplement payable in the new replacement currency, the *Rentenmark*. By December, almost all entries, both income and expenditures, are appearing in the new units. These remain relatively stable through 1924 and early 1925.

If the income side of these records tells us of the general impact of inflation on society, the expenditure side is more revealing of how Franz and Lotte lived during these months. To trace the prices of a few basic needs, one can select bread, eggs, and beer as more or less standardized products. In October 1922, the first month of the records, typical purchases show bread at 100 marks, eggs at 105 marks, and beer at 122 marks. Bread prices then rise to 310 marks on 3 January 1923, 1100 by 5 April, and 7,000 by 29 June. Bread then seems to be missing altogether during the summer, although a *Torte* at 60,000 marks appears on 2 August. They also buy rice, at 65,000 marks. When bread purchases reappear in the autumn, the price shown is 1,040,000 marks on 20 September, rising to 240,000,000 marks on 13 October.

Egg prices follow a similar curve, rising from 105 marks in October 1922 to 500 marks in January 1923, 2560 marks in mid-April, and 6,000 marks by mid-June. No quantities are specified. Then on 13 July Lotte purchased 20 eggs for 90,000 marks, and on 30 July another 41 eggs for 369,000 marks. The price has doubled in the 17 days. More than that, eggs had become a better storehouse of value than marks, an alternative currency to paper money. When eggs were purchased again on 2 November, the price paid was 150,000,000,000 marks.

Similarly, these accounts show the prices paid for beer as 122 marks in October 1922, 2,705 marks in February 1923, and 10,000 marks on 13 July. After this date, beer seems to disappear altogether from the grocery list. After all, one can drink something else.

Living with a fast-depleting currency had other consequences. With the mark losing from 50 to 90 per cent of its purchasing power every month, the only logical strategy soon became obvious: to buy something—anything—as soon as money was received. As inflation soared, salary payments increased from twice a month to twice a week. Purchasing anything at all was better than holding shrinking paper money.

One can see this dilemma clearly in the household accounts from their first entries in October 1922. For Lotte, shoes were initially favoured, but there is a practical limit to the number of pairs that one can use, or even store. For Franz, October saw a sizeable purchase of photographic paper. He also maintained throughout the inflation period his subscriptions to professional scientific associations and scientific periodicals, though their costs soared like everything else. Lotte's interests in music and theatre, already noted previously, continued. Some account entries are listed only as an opera or a concert, but others are specific. The accounts show a concert by Fritz Kreisler, two by the violinist Arnold Rosé (or by his quartet), two operas (*Don Giovanni* and *The Barber of Seville*), and a program of modern dance by Mary Wigman. The ticket prices rise from about 300 marks or less in October to 11,000 for the Wigman performance on 1 June 1923.

By summer, other expedients were needed to keep abreast of the presses pouring out more and more new money. On 8 June, Lotte purchases a book on art history, not further identified, for 36,720 marks. On July 5, she records two volumes on Goethe and one on Stendhal, for a total of 225,000 marks. Just before the end of June, she buys two etchings, also unidentified, for a total of 300,000 marks. Some of the volumes on art history and literature that have turned up on the bookcases in Oxford were originally acquired as a substitute for currency, and the same holds equally for some of the etchings.

The height of the inflation brought one more anomaly. In the autumn of 1922 Franz's inadequate income as an assistant was occasionally supplemented by small gifts from either parent. But his regular stipend, though small, was indexed like other state wages. As inflation soared in 1923, he became better off than anyone who lacked an indexed income. In this world turned suddenly upside down, a considerable sum was advanced to the once wealthy "Papa."

By December of 1923, the final bout of madness subsides, and expenditures are recorded in the new *Rentenmark*. Bread is at 0.27 marks, beer at 0.40, eggs at 0.90. Franz's stipend, paid in two instalments, shows as a steady 220 new marks per month. The modest parental supplements to his income are resumed, and later on some of this will be flagged as repayment of loans. In 1924 holders of some paper securities were partially compensated by state ordinances that revalued these issues at a small percentage of their original gold value. Apapa Ernst Simon would die just over a year later. The loss of a fortune originally worth millions of pre-inflation marks may well have hastened his death.

Cultural life continued. There are expenditures shown in the accounts for theatre in December (1.90) and for the opera in February 1924 (16.00 marks, apparently for four tickets). Best of all, the first week in January shows payments for a visa, a *Schlafwagen* reservation, and rail tickets to St Moritz (in total 77.00 marks). After the long nightmare, Germans once again had the means to travel abroad. Their Swiss visit was the first of several forays abroad for Franz and Lotte in 1924. Others included visits to the Tyrol, Vorarlberg, and Rome. Inflation had been a form of imprisonment. At the time, probably few understood the extent to which inflation had also devastated the country's social structure, and especially its middle class.

The household accounts list expenditures to December 1924, and income to March 1925. After this a gap occurs in the records. In October 1926, Lotte began to keep a diary or journal like the one she used for the Italian tour in 1921. It has the same size and format, with a different cover. She would keep several small diaries during her long life, most of them spurred by travel. This one records a voyage of a different kind: motherhood. Its first entry begins when her infant daughter Kathrin is 10 months old. ". . . Baby has two teeth," she writes, "turns herself over on her stomach, but not yet back again." Subsequent entries make the purpose clearer. Lotte is keeping a record mainly of child development, which will run until the summer of 1929. The entries are sporadic, and they fill about 60 pages.

This diary shows clearly that Lotte—and Franz as well—soon discovered what most new parents discover: that infants of a certain age become interesting, and sometimes astonishing, for their capacity to assimilate new knowledge rapidly. Lotte first records the evolution of physical movement—ability to turn over, sit up, crawl, stand, walk, and so on—but soon devotes more attention to the child's sensory and thinking experiences: awareness of sounds, making sounds, saying syllables, attaching meanings to specific syllables, and the development of language. Being musical herself, she notes carefully the children's songs that Kathrin knows at various ages, and later Dor's songs as well.

When Dor was born in April 1928, the domestic scene became more complex. Much of the interaction is triangular. With a gap of just over two years between the children, each is a major influence on the life of the other. Unlike her sister's case, Dor's life would be observed and recorded from her earliest days. In this diary is a postcard written by Franz to his mother, then holidaying in Rome, just four days after the baby was born. His ready sense of humour could always find the lighter side of any situation short of the catastrophic:

> Here everything goes well again, and Lotte should be up by Sunday evening. *Das Kind* – still without a name, by the way – is very voracious, and nurses well. Throughout the day she is quiet and gathers strength, so as to be able to roar all night long.

After a few months, Lotte resumes the diary, making retroactive entries back to May to note her daughters' more complex pattern of interaction.

If Lotte had a diary to keep the record, Franz had his camera. As his earlier life shows, he was already an accomplished and dedicated photographer. The arrival of children opened up a new dimension. The children became preferred subjects and models for hundreds of photographs over the next three decades. A good many of these photographs he would also develop, print, crop, and enlarge, using his own equipment and dark room, until he was satisfied with the result. For these Berlin years, Lotte's diary and Franz's photography complement each other to create a combined portrait of family life.

The diary is more than a record of the children. It provides incidental sidelights on visits with relatives, birthdays, and holidays. There were frequent visits to the large house in Landhausstrasse to see the widowed Anna Simon and the family of Franz's older sister Mimi Frank; visits to Hamburg in 1927 and 1928 to the family of Omama's younger brother, Carl Lipmann; as well as a return visit by Tante Ellie Lipmann, Lotte's aunt, and the children to Berlin. There are visits from Omama's sister, Tante Clara, and her husband Onkel Jacob Aberle to the flat in Pariserstrasse. Not every Tante or Onkel is a real relative. There is, for example, Onkel Herr Lehrer, Lotte's music teacher. There is news by letter from Franz's favourite uncle, Justus Simon in Hanover, who is feeling quite well again after an illness.

Holiday seasons are rather special. More visitors come. At Christmas, which coincides with Kathrin's birthday, a small Christmas tree decorates the balcony of the Pariserstrasse flat. In April 1929, on Dor's first birthday, Kathrin asks, "Where is the Christmas tree?" At Easter, at Landhausstrasse, there is a joyful hunt for Easter eggs. To all appearances, both households adjust smoothly to the social customs of their milieu, though neither is observant religiously.

From time to time the diary shows a darker side. It chronicles the usual stream of childhood illnesses, fevers, and indispositions, a reminder that parenting is not an unmitigated joy. The entries for the autumn and winter of 1928–29 seem to have more than their share of these illnesses. By summer, Lotte felt the need for a change. The family spent the month of June 1929 at the small resort town of Heringsdorf on the Baltic, and in the photograph collection dozens of scenes from this holiday show that Franz was present for at least part of the month.

The camera record from Heringsdorf portrays a summer idyll: scenes of sunshine, the sea, vast stretches of almost empty beaches; of parks, statues, and buildings in the town; of Lotte and the two children relaxing in sand and sunshine; of tots playing free and naked—save for small, round sunhats—at the edge of the sea. Cameras, however, do not always tell the complete story. The diary describes the month with more nuances. Along with high points, it mentions the children's various minor ailments, and also notes: "weather mostly cold and rainy," a reminder that the Baltic is not the Mediterranean.

The pages on Heringsdorf are the last regular entry in this diary of the children's early years. After a few blank pages, a one-page coda, dated 1931 and 1932, lists a few of the children's early sayings. Six-year-old Kathrin asks "How far can one count? What is the last number called?" The rest of the volume is blank.

In the larger context, the stage was being set for changes. The first postwar decade to 1929, which witnessed Germany's first halting experiment in democracy, was ending on a positive note. Born in conditions of military defeat, civil disorder, and destructive hyperinflation, the Weimar regime was showing modest evidence of stability and economic improvement. In the summer of 1929, no one yet imagined that worse conditions might lie ahead.

Through bad times and better times, Franz's professional career progressed steadily. His experimental work and publications were winning increasing recognition in Germany and abroad. Step by step, he moved up the academic ladder: special assistant to Nernst in 1922, regular assistant in 1923, *Privatdozent* in 1924, senior assistant in 1927. In 1928 he gained the title of professor as an *extraordinarius*, with a modest salary increase, and in 1929 an additional mandate, with salary supplement, to lecture on thermodynamics. This last step came after he had considered, but declined to stand for, a chair in the German University of Prague, where funding for research would have been inadequate. The time for a more promising offer was close at hand. That call would come in December 1930, with an invitation from the Technische Hochschule in Breslau to a permanent chair and directorship of the Institute for Physical Chemistry. This was an offer that he could hardly refuse.

Among Franz's documents in the house is a programme for the 34th annual meeting of the Deutsche Bunsengesellschaft, the society for experimental physical chemistry, which met in Berlin on 9–12 May 1929. The list of participants includes "Prof. Dr. Franz Simon" and "Frau Simon," along with several other familiar names: Fritz Haber, Otto Hahn, Hans von Halban, Lise Meitner, Walther Nernst, Max Planck, Michael Polanyi, Paul Rosbaud. This was an important conference. Founded on the results of a highly productive decade, it stands out as a landmark in the careers of some of its younger participants. At the opening session, Franz delivered the first scientific paper, on the melting curve of helium. By this date his research group was widely known for its success in liquefying helium, and also for developing a compact, inexpensive apparatus for low-temperature research generally. His scientific papers include correspondence concerning German patents for this apparatus.

Before leaving Berlin, Franz and Lotte would attend another conference that would leave a deep impression on them. In 1930 they were invited to a scientific conference held at Odessa in the USSR from 19 to 24 August. This was primarily a meeting of Soviet physicists, but selected foreign scientists were invited to attend and to visit the USSR for up to a month at state expense. After the formal meetings, all participants were given a tour by ship on the Black Sea, along the Crimean coast from Odessa to Batum. Franz and Lotte then went on to visit colleagues and laboratories in Moscow and Leningrad. They set out from Berlin well prepared. The bookshelves in Oxford revealed a thick 900-page guidebook, *Führer durch die Sowjetunion*, prepared by the Soviet authorities and published in Berlin in 1929. The text and maps are in Baedeker style and format, in a red binding with a hammer-and-sickle crest on the front cover.

Other mementoes from this trip turn up at several points through the house: unused postcards from Odessa, Moscow, and Leningrad; delegates' registration cards; tourist brochures for Yalta and the Crimea, all of these in Cyrillic print or script, plus numerous photographs from Franz's camera. The photographs best portray the human dynamics. When the conference was not in session, Soviet and foreign participants relaxed and talked together at the Lusanovka beach. Other pictures show the delegates on deck and in various ports on the excursion to Batum. Franz's photographs capture these moments in an informal, naturalistic way. In wider context, this conference marked a renewal of contact between German and Soviet scientists.

Before leaving Russia, Franz gave lectures in Leningrad on low-temperature physics. They returned via Finland and Sweden. A photograph from the return trip shows the harbour and market in Helsinki, a city which Lotte remembered fondly for her first encounter with plentiful food after a month of meagre meals in the USSR. To the end of her life, she called it Helsingfors, and thought of it as the Swedish-speaking city that it still appeared to be during Finland's early years of independence. By 17 September, they were in Stockholm. Among the memorabilia I find a small brochure and map of the central city in German, along with some dated receipts as evidence of their visit.

For Franz and Lotte, the month in the USSR remained a major experience in their lives, one that left powerful but mixed impressions. It was a bold adventure, unusual, undertaken when the Soviet Union was still all but unknown to the outside world. For a full month of travel and discussion with fellow scientists, they could observe the climate of a dictatorship. They were appalled by the poverty and shortages of food. On the positive side, they were impressed by Soviet priorities for science and scientists at a time when financial stringency in the Western countries was slashing science budgets. They returned from the USSR with one persistent idea: that if conditions in Germany should ever become unbearable, a research career in the Soviet Union in Franz's own field would be a continuing option.

3

BRESLAU 1931–1933

Franz's acceptance of the Breslau chair and directorship of the Physical Chemistry Laboratory established him firmly as a leading figure in low-temperature research. His predecessor in the chair had been Arnold Eucken, a well-known physical chemist who moved on to a chair in Göttingen. As early as 1925, Franz had dared to question some of Eucken's theoretical work on Nernst's heat theorem, work that he had extrapolated to lower temperatures without the experimental data obtainable by newer techniques of cooling.

In Breslau there was money for the laboratory and for new furniture for the Director's office. For equipping the laboratory, some of the low-temperature apparatus was bought from personal grants, and some was of Franz's own design, for which he held patents. These could move with him from Berlin. The budget allowed for further equipment, though a major advantage of his low-temperature liquefiers was relative simplicity and lower cost compared with earlier methods of cooling. Some of Franz's current doctoral students transferred from Berlin with him.

On the domestic side, living arrangements in Breslau were greatly improved. For the first time in their marriage, Franz and Lotte had an independent household. For their growing family, the rented house in Morgenzeile had its own deep garden and an ornamental pool. Franz's salary had more than doubled from his years as an *extraordinarius* in Berlin. Breslau was not too far from Berlin, and good rail connections made visits easy. Both the Simon and the Lipmann families had relatives in Breslau.

The Oxford house had many reminders of the Breslau period. Most prominently, there are numerous photographs in the family albums: the house and garden from many angles, children, visitors, grandparents, and other relatives. There are letters to or from Morgenzeile 15, and congratulatory notes from colleagues on the new post. Among Franz's formal papers, there is correspondence from the ministry concerning the official duties of the chair, salary, family allowances, and so on. Other photograph files show the laboratory, its workshop, its main apparatus, and a luxuriously furnished office for the director.

Other Breslau material shows up from less accessible places, including files of yellowed press clippings on the new professor and the type of low-temperature research that will be undertaken there. Much of this coverage comes from local or regional papers. These reveal an element of local Silesian pride that research of international importance will be done in a local institution. After months of preparation, the new low-temperature laboratory was formally opened on 1 December 1931. Against heavy odds, it was completed in the depths of the Depression. The press coverage of the event emphasizes the relatively low cost and the multilevel local support that had made the project possible. These fading clippings from the early months in Breslau, along with similar clippings from the Berlin period, have been relegated to near-oblivion in the remote attic storage area of the Oxford house. In the business files, however, is an item that Franz had retained. At mid-morning on opening day, the Heylandt firm which had built apparatus for the new laboratory, and doubtless hoped for future orders, sent a telegram: "Cordial best wishes on today's opening of the new *Kältelaboratorium*." Franz's star seemed clearly in the ascendant.

The path ahead would prove less than straightforward. Even as the new laboratory opened, Franz was about to begin a leave of absence for the first term of 1932. Early in 1931, before taking up the Breslau chair, he had been invited to spend a term as a visiting professor at the University of California at Berkeley. Among his papers is a formal letter of appointment, dated August 1931. It designates him "Lecturer in Chemistry and Physics" from 1 January 1932 for six months, at a salary of $4,000. The invitation had been suggested by Professor Joel Hildebrand, who had studied in Berlin with Nernst in 1906 and had met Franz in 1930 on a return visit to his former institution.

The opportunity to visit one of the foremost departments in North America was too good to be missed. Yet it posed one serious problem. On the salary offered, moving the entire family to California was not feasible. On the other hand, for Franz it was unthinkable to leave them in Germany in such unstable political conditions. As Lotte later reminded us, California in 1932 appeared half a world distant from Germany—a week by steamship, then most of a second week by rail across the continent. In uncertain times, much could happen in a fortnight.

By 1932, conditions in Germany were clearly deteriorating and becoming potentially explosive. The number of unemployed doubled in two years to about six million. Parliamentary government was weakening and collapsing into deadlock as political parties of the centre lost ground to fast-growing militant forces of the left and right. With parliament paralysed, governments acted increasingly through emergency presidential powers that bypassed the legislature.

Further, Breslau was not Berlin. In Silesia the academic atmosphere was more provincial, more closed, more traditional. Even before the crisis, the city had retained a streak of older style antisemitism, something that the local Jewish population had had to learn to live with. In pre-Hitler times, this was, in Rudolf Peierls' lapidary phrase, "a bearable handicap."[1] The atmosphere in Breslau differed sharply from that which Franz and Lotte had known in Berlin.

As the Depression deepened and disorder grew, the quasi-military forces of the radical right and left infringed further and further upon the weakened legal order of the German state, an order that soon became powerless to prevent the growth of a more strident, violent antisemitism. Parades and demonstrations became the dominant form of politics. The house in Morgenzeile was just across the road from a sports stadium, where much of this activity took place. To the dismay of their parents, Kathrin and Dorothea, like other children, were fascinated by the bands, the marching, and the parades that filed past the house. At the ages of five and three, they could hardly understand that the slogans of irrational hatred fuelling these marchers were aimed at them.

To meet this dilemma, the family had to be relocated. Not for the first time, Switzerland became a place of refuge. In early November 1931, Lotte and the children, with their housekeeper, moved to a house in a hillside village above Locarno. Southern Switzerland was attractive for its mild climate, and Locarno may have been selected because Lotte's uncle, Carl Münchhausen, was living in retirement nearby. When preparations were complete in

Breslau, Franz headed for Locarno for a few days with the family before sailing for America. The photograph collection shows several outdoor scenes of Franz with the children. They were clearly intended to ease the pain of a half-year of separation distant by 10 time zones.

Discovering America

On departure from Locarno, Franz proceeded via Paris to board his ship, the *Albert Ballin*, at Cherbourg on 17 December. The sea passage in winter was not a smooth one, and he spent two days in bed, seasick. When he did get to the dining room, he found himself at the captain's table, in company with an opera singer, a banker, and a professor of economics. The last two, both Americans, were returning from protracted meetings of the Wiggin Committee in Basle, which was trying to rescue the international monetary system after Britain had been forced off the gold standard in September. Other countries would follow Britain, they predicted, beginning with Germany. From them Franz obtained a pessimistic forecast for Europe's monetary future and their estimated odds of seven to three favouring renewed German inflation. The ship, carrying about 10 per cent of its passenger capacity, reached New York on 24 December.

From the first arrival of the family in Switzerland in November, Franz and Lotte stayed in touch by post and telegram. Their lives for the next few months therefore became documented almost on a daily basis. After Franz sailed for New York, mail delivery became irregular, dependent on ship sailings. For Franz, a day without a letter became a source of anxiety, almost a minor crisis. In these 300 or more pages of letters and telegrams there is an entire book in itself, but that is for another time and place.

The American interlude can be viewed from several angles. First, Franz was a European traveller, like thousands before him, seeing and reacting to America for the first time. Second, he was an academic visitor, bringing his own expertise but also assessing American universities in his own and related disciplines. Third, there were cousins, friends, and colleagues to visit. Finally, he remained a young German professor with a newly launched scientific laboratory to direct, at a critical time of budgetary cutbacks and political upheaval. His letters illustrate all these dimensions.

His itinerary was complex. After arrival in New York on 24 December, he went on after a day or two to the winter meetings of the American Association for the Advancement of Science (AAAS) in New Orleans. The meetings, held from 28 December to 2 January, attracted over 1,300 papers and 1,447 registered participants distributed among 15 different discipline groups.[2] One of the keynote events, the Gibbs Lecture, was given by Percy Bridgman of Harvard University, who became one of Franz's closest friends. From New Orleans he headed westward by train towards Los Angeles and Berkeley. When the term was finished in mid-April, he would complete a circuit through Seattle, Vancouver, Chicago, New York, and Washington.

For a first-time visitor writing to a spouse equally unfamiliar with North America, the first priority was to describe the physical and social landscape. Moving westward from New Orleans, his interest grows as the train traverses vast empty spaces and desert scenery, marked first by cacti and later by palm trees. In a small side trip, he views the Grand Canyon and descends into it by mule along a narrow path. Once settled in Berkeley, there will be further descriptions of redwood forests, rocky coastlines, camping in the Yosemite Valley, and evening walks on the hillside above the university to view magnificent sunsets across San Francisco Bay.

The return journey from Berkeley was planned similarly to include several major scenic highlights: a stage by autobus from Eureka to Grant's Pass through magnificent coastal forests; by ship from Seattle through the Gulf Islands to Victoria and then Vancouver; a daylight crossing through the Rockies; an extra day to visit Niagara Falls. Most of these experiences of a first-time European visitor are recorded both in letters and in scores of photographs. A selection from the photographs fills an entire volume among the family albums. It includes scenes of nature, friends, and colleagues—mementoes of a pristine region now changed almost beyond recognition by generations of population growth. California at this time had just under six million inhabitants.

Beyond the natural landscape, the early letters include first impressions of the American social scene. He notes the size and social status of the Jewish population of New York, and also of other cities when he travels across the South. He describes the position of the black Americans, noting the segregated waiting rooms and separate rail cars. He is fascinated by the complex ethnic diversity of the Southwest: Indian, Hispanic, black, white, and its resulting mixtures (*ein sonderbares Menschengemisch*).

His first comments on American Jews, perhaps unduly influenced by his New York informants, are dismissive of their immigrant experience in New York: "The second generation especially must be quite crazy (*ganz doll*), lawbreakers, smugglers, prizefighters." In the South he also finds many Jews, concentrated particularly in the apparel sector. He also notes a more reputable group of Jewish professionals and major merchants, "which is however not very large."

Notwithstanding scenic wonders and social observations, the scientific side of the journey is never far distant. To the extent that his timetable allowed, Franz made a regular practice of visiting major university science departments and scientific institutions along the way, whether or not he had a personal contact there. Heading west from New Orleans, he stopped over in Houston to see the Rice Institute. He stopped again in Pasadena to visit the California Institute of Technology, where he saw his colleague from Berlin, Albert Einstein. He writes a quick note on 7 January:

> Saw the Institute [Cal Tech] and visited Einstein. Had a very interesting talk with him about political questions. He remains optimistic still, but then he always thinks only the best of everyone. In the evening with Pauling, a very good physical chemist appointed here, very nice.

Several weeks later, during the Berkeley mid-term break, he would return to Pasadena and visit Cal Tech again.

The stopover at Cal Tech was more than incidental. Einstein, probably the best known scientist in the world at this time, had been seriously wooed to visit Cal Tech by Robert Millikan and to visit Oxford by Frederick Lindemann. Both had proposed short-term annual visits, without specific teaching obligations. Einstein, who was facing mounting attacks in Germany for his Jewish background and leftist, pacifist, and internationalist political views, visited both universities in 1931–2 and would visit both again on the same basis in 1932–3, though still nominally in Berlin.

Franz, who had seen Millikan at the New Orleans meetings, would have known of Einstein's visit to Cal Tech and clearly welcomed the chance to discuss the political situation with him in private. In 1932, Einstein's assessment as reported by Franz may well have been reassuring, but it would change sharply a year later when Hitler became chancellor. Within weeks, Einstein would renounce his German citizenship, resign from the Prussian Academy of Sciences, and decide against returning to Germany. His well-publicized example would not be lost on younger colleagues.

On 8 January 1932, Franz arrived in Berkeley. Once settled there, he had opportunities for a closer look at American mores. He is impressed by the easygoing friendliness and hospitality; the efficiency of supermarkets; the ubiquitous role of automobiles; drugstores scarcely recognizable as pharmacies; the prevalence of divorce; wide differences from Europe in gender relations. From differences in table manners to undergraduate antics and dress, few topics go unmentioned in his letters home. By this date Berkeley had a football stadium capable of seating 75,000 spectators, and the noisy Axe Rally held in the stadium against rival Stanford served as a vivid introduction to student life.

He samples American newspapers and radio, but finds news about Europe inadequate in both. In January, the California papers are paying more attention to the Japanese invasion of Manchuria, and there is wide discussion whether the public should boycott Japanese goods. By early March, the papers are full of the kidnapping of the Lindbergh baby. He does hear the result of the German Presidential election of March 13 almost instantly, but for detail he seems to have been dependent on material forwarded by Lotte. American radio, with its heavy emphasis on advertising, is a source of culture shock.

The day after his arrival, Franz delivers his first lecture. The teaching load proves taxing. With an hour-long lecture four days a week for 14 weeks, alternating between chemistry and physics, this is far more demanding linguistically than presenting a brief report at a scientific congress. Further, the lectures are attended by most graduate students and many senior faculty colleagues. Expectations are high. The lectureship was modelled on inviting a distinguished visitor for one term each year. Among these, Franz notes, was his predecessor, the Dutch theoretician Hendrik Kramers, who had been preceded in turn by Otto Stern from Hamburg, James Franck from Göttingen, and Abram Joffé from Leningrad.

Franz's early letters mention some language difficulties. Initially he found difficulties in comprehending colloquial American English. Further, the challenge of lecturing in English for the first time revealed gaps in scientific terminology. "If I don't know a word," he reports, "I say it in German and then someone translates it."

Alongside the lecture courses, Franz continued with his low-temperature research. He was given working space in Professor William Giauque's laboratory, a postdoctoral assistant—a *Privatdozent* from Sweden—and a mechanic in the laboratory was available to help in building apparatus. By 3 March, he reports that he has run his first experiment:

> It went well, even if nothing was correctly prepared beforehand. The day before I was at Stanford and had told Ahlberg that everything must be ready. When I arrived an hour early about half was missing. In experimental work here they are really not so much on top of things.

On 27 March, he mentions a further experiment:

> Recently here, with a small homemade apparatus I have liquefied helium and have produced and reached the lowest temperature in America.

His one-sentence announcement gives Lotte no details, but there is a hint of triumph in breaking a scientific record. Franz's publications list includes an article, jointly authored with Ahlberg, on this apparatus built during the weeks in Berkeley.

His living arrangements on campus were simple and convenient. The Faculty Club, close to the Chemistry Department, housed and fed most of the bachelors and young single faculty, and was also a midday meeting point for faculty living off campus. With three of these club residents Franz soon discovered common interests and close friendship. One was a young physicist, Ernest Lawrence, "a very nice and well qualified man." They frequently walked together in stimulating discussion in the evenings, and were soon on first-name terms, a practice that Franz found quite odd. He also talked often with an "old professor of French," Richard Holbrook, and "a nice Englishman, 60 years old, a professor of Latin," who was also visiting Berkeley for the term. In later letters this unnamed Englishman is identified as Cyril Bailey, a Fellow of Balliol College, Oxford. "We often talk together," Franz reports. "He says 'We are both exiles'."

These four residents of the Faculty Club became fast friends, sharing weekend forays to San Francisco and other points nearby, but Franz also had social invitations from colleagues living off campus and from visiting German academics. He describes to Lotte many of these occasions in detail, with attention to the style of house and standard of living. These people live in small wooden houses, he reports, and they have no live-in domestic servants. On the other hand, they all have automobiles, and most of them have a summer cottage or cabin as well, which may be three or four hours away by car. "The standard of living of professors here is certainly the equivalent of ours . . . . They live not luxuriously, but very well." As early as January, the American living standard becomes a topic of careful attention. In a good university, he notes earlier, the starting salary for a full professor is about $4,000. It can rise to only about $6,000, or up to $10,000 in exceptional cases.

Towards the end of term, his midday and evening social calendar filled up completely, to the point that he could not report on it fully in letters home. In return for this hospitality he arranged a farewell dinner for these colleagues and friends, catered by the Faculty Club, to be held on the eve of his departure. He chose a menu of German food—spare ribs Kassel style, sauerkraut, erbsenpuree. "They made it really well and people enjoyed it very much. The whole thing was a great success." The drinks were supplied by the club's "unofficial" bootlegger. Franz spoke briefly, and Holbrook presented him with a book of photographs showing scenic highlights of California: mountains, forests, beach and coastal scenes, cities, early Spanish missions. Cyril Bailey composed a fine italic dedication slip in Latin, which was signed by the 21 guests present and attached at the front of the book. This volume was found in one of the Oxford bookcases, with its original inscription intact.

Before leaving, Franz received another present from the mechanic in the laboratory. He describes it in a letter home:

> He has taken a fancy to me and has made for my departure a little bear–the University of California heraldic animal–as a paperweight, and gilded it with California gold.

This bear is a solid casting in metal on a slightly cambered metal base, surprisingly heavy for its 6-centimetre length. It was a fixture of the house and of Franz's desk for over six decades. The gilding has worn thin in a few spots from long use, but henceforth its California origin will give it a special place as a memento of Franz's voyage of discovery to America.

On his return journey eastwards, he writes longer letters describing his final days in Berkeley and the dramatic scenery along the return route. One very long letter is composed as the train traverses the drought-ridden prairie of southern Saskatchewan—"other than the name, nothing interesting here." In it he looks at his American experience in perspective. By this point he knows the country rather well, and along the way he notes a few first impressions of Canada and Canadians to serve as a comparison, a Canada that in his eyes seemed still a British colony.

In this retrospective exercise, he is quick to find paradox and irony. In America, he notes, it is the rich who ride in trains and sleep in pullman cars. The poor travel in their own vehicles, and sleep in tourist cabins. A Ford in working condition (called a "Flivver," he notes) costs $30, or with a little bargaining even less. In alcohol matters, the delivery service of the bootleggers under American prohibition is both faster and less expensive than the cumbersome permit system for government liquor stores in Vancouver, where drinking is forbidden in restaurants and confined to private homes.

Beyond the bizarre and paradoxical, he also has important positive findings. Earlier, he had remarked that class conflicts scarcely exist in America. "There is also no trace of class hatred. If something goes badly for someone, he doesn't blame it on others but treats it as his own bad luck. But who knows how long that will last!"

After the plains of Saskatchewan, the return itinerary becomes more hectic and letters home are fewer and shorter. From the last two letters in the file we learn that he has been to Chicago, visited the university and physics colleagues, and seen Lotte's cousin Ernst Koppel, who arranged a theatre party for several mutual acquaintances.

He makes time for a visit to Niagara Falls, an event documented by photographs from several angles. He then moves on to New York City, where he reports a meeting with his Berlin colleague Leo Szilard, "who has been in industry here since Christmas," and a large party hosted by Professor Leonor Michaelis. The latter had held the title of professor in Berlin as an *extraordinarius* since 1908, but left Germany in 1922 after failing to secure a regular chair. By this date he was in New York at the Rockefeller Institute of Medical Research.

After New York came Washington and Baltimore. Franz had written earlier of speaking commitments in both places. In Washington he was an invited foreign guest at the American Physical Society banquet on April 29. The handwritten invitation, sent to California by Fred Mohler of the Bureau of Standards, mentions that "the toastmaster will call on you for a few remarks." Among Franz's papers there are a few rough notes that appear to be linked to this dinner. They are routine, but one term stands out. He will speak, the notes begin, from "a German" point of view. Accidental or not, the time was close at hand when he would avoid using any such term. While in Washington, he made a routine visit to the German Embassy, and was surprised the next day to receive two invitations: to lunch with the ambassador and his wife, and to dine with the first counsellor.

After Washington, campus and laboratory visits continued, first to the University of Virginia, to which he was invited by Günther von Elbe, another German colleague and collaborator from Berlin days, and then to his prearranged lecture in Baltimore. An earlier travel plan in March envisaged further visits over the next two weeks to Princeton, Harvard, the General Electric laboratories in Schenectady, Pittsburgh, Yale, and Columbia. In the absence of further letters home, it is not known which of these visits were carried out. His return passage to Europe was again on the *Albert Ballin*, sailing 18 May.

Franz's first voyage to America can be placed in wider context. Over five busy months, he had come to know the American academic world in some depth, and had adapted to it successfully. He could lecture in English, even to his peers. He had run successful experiments in the Berkeley laboratory. He had looked up earlier acquaintances, and had made new friendships across the continent.

When read carefully, his letters to Lotte reveal the gradual shaping of an idea, similar to the one that surfaced during their visit to the Soviet Union. In April 1932, even as Franz was preparing to return, his former student and coauthor on five publications, Martin Ruhemann, was about to move to the USSR for a research position in Kharkov. His wife, Barbara Ruhemann, who had also studied and copublished with Simon, went with her spouse. If conditions in Germany worsened, there were other alternatives than the Soviet Union to staying in Germany.

Despite Franz's intensive immersion in American society and scientific life, thoughts of Germany were often close to the surface. In Houston, on the way west, a letter records how odd it seemed—"in so foreign a setting"—to listen to President Hindenburg's New Year message for 1932, relayed from New York by a local radio station. In Berkeley, a rebroadcast of German music from Berlin, with lieder sung by Elena Gerhardt, touched a sensitive nerve and provoked a surge of homesickness. He remains continuously interested in German politics and American views of Germany, and also in the adjustment of German immigrants and their children to American society.

As director of the laboratory in Breslau, he was in touch regularly with his acting director, Professor Rudolf Suhrmann, on budgetary and policy questions, and also for news of colleagues. On 22 January, he reports letters and a New Year's card "signed by the entire Institute including all the Nazis!" As the Depression worsened, budgetary cutbacks and how to keep programmes going became a regular preoccupation in reports home. He also received occasional reports from his doctoral students living in the house in Morgenzeile.

From the distance of California, Franz could also view his own country in broader perspective. At the end of January he reports to Lotte in some detail on a "truly interesting" lecture by a Professor Schlick on "contemporary philosophy in Germany":

> He could not give much factual material, but he described the situation as a whole very well. According to his account, the mystical tendency now dominates in the universities. People are concerned almost solely with metaphysics, therefore purely speculatively and without connection to the natural sciences, which are in a certain sense disdained. One believes that wisdom (*die Weisheiten*) can be drawn out of the depths of the mind (*Gemüt*). This tendency is to be seen everywhere. I did not know that it is already so widespread also in the schools. The leader of this tendency is Heidegger. Schlick read some very comical passages from his books. Of course, he comes from the other tendency. This second tendency is concentrated mainly in Vienna; in Berlin, Reichenbach is still an adherent (you remember – the one I was with in Arosa).

Moritz Schlick, who had earlier studied physics under Planck, was Professor of Philosophy in Vienna from 1922. In 1936 he was murdered by one of his students. Hans Reichenbach became Professor of the Philosophy of Physics in Berlin in 1926.

At another level, both Franz and Lotte were watching day by day the deteriorating political situation in Germany. Lotte was forwarding clippings and articles from Swiss publications to keep Franz better informed, and both discuss new developments frequently. One special concern was results from the *Land* or provincial elections. In mid-November, with the family already removed to Locarno but Franz still in Breslau, he notes the ominous 40 per cent vote for the Nazis in Hesse, mainly through gains from the liberal parties.

In Berkeley, he awaited with increasing apprehension the results from the *Landtag* election in Prussia, scheduled for early May. The possibilities for violence, both during the campaign and when the new *Landtag* was scheduled to meet in Berlin on 24 May, were serious enough for him to consider delaying the family's move back to Germany. From early April the letters canvass several scenarios for the return and reunion. With no agreed plan in sight, the decision was deferred.

When he left Berkeley on the journey back to the east coast, the matter remained unresolved. After six months in Locarno, Lotte had grown restive and longed for a return to Berlin. She had not been there, she pointed out, for more than a year. A worried correspondence ensued, ending with a flurry of telegrams at the end of April. Against Franz's deep misgivings, Lotte returned with the children to her parents' flat in Pariserstrasse early in May. She then went to meet him at the ship's port of arrival in Cuxhaven towards the end of the month.

For the time being, Franz's dark fears remained unrealized, but on 30 May Hindenburg forced Chancellor Brüning to resign, to be replaced by a right-wing ministry headed by Franz von Papen. Two weeks later, the new chancellor removed a ban on Nazi storm troops imposed by Brüning. Seven weeks later, amid growing street violence, he removed the Socialist-led *Land* government of Prussia and placed Berlin under martial law. Though few yet realized it, the right-wing revolution had begun. Franz's forebodings were about to be proved justified.

Portents of the storm

One of the hardest acts of academic life is returning to a normal routine after a leave of absence, especially when that leave has opened up new vistas. Franz's return to Breslau for the 1932–3 session was no exception. He found himself with added administrative duties as Dean of the Faculty of Chemistry and Mining. Lotte was away in Berlin for two extended periods during the year, in September–October 1932 and in January 1933. As they did during the American visit, they again exchanged letters almost daily. In these, Franz gives a double picture of family life in Lotte's absence and academic politics at a critical time.

The Depression was forcing sharper cutbacks everywhere, but the main academic issue of the year was a proposal to merge the Technische Hochschule with Breslau University. The question had surfaced more than once in Franz's letters from California, but it became acute in January 1933 when the ministry decided to proceed with the merger, to take effect on 1 April. In a letter of 6 January, Franz describes the uproar that ensued when the ministry decision was announced in the Hochschule senate. In several letters spaced over four weeks, he details successive day-long meetings, bitter infighting, increasingly intemperate language, exaggerated fears, inaccurate press coverage, and student demonstrations. Several of these letters are marked strictly for Lotte's eyes only. In them, Franz writes openly of the inner politics, unleashing his frustrations with colleagues and a growing despair for any peaceable outcome.

The debate on the ministry plan went badly from the start. The Rector, Bernhard Neumann, "described everything in the blackest colors" and soon joined those in blanket opposition.

Franz considered the plan to be workable, arguing in senate that it "coincided about 95 per cent" with the two institutions' own proposals, with room for negotiation on the remainder. As a dean, he met at length with his own faculty and with leaders of its student organizations to explain the plan and calm their fears, seemingly with some success. He consulted with the Dean of Philosophy in the university to work out compromises on the more difficult areas of implementation. He then discussed this plan with the previous year's Rector of the Hochschule, Erich Waetzmann, a physicist who held concurrently an honorary chair in the university.

After two days of wearying debate, Franz summarizes the situation in a letter of January 13:

> Yesterday again a whole-day session, in part not very enjoyable, since the Mechanical Faculty has also totally lost its senses. Overall, the tone is one of extreme irritation, except in our faculty. The same in the university, not because of this project, but because of Cohn. Tomorrow is the final decision of the large senate, which apparently will be reasonable, whereupon the student organizations will go out again.
>
> For me personally it is not pleasant to be forced now to speak out against something. But since I am dean of the largest faculty – we have 40 per cent of the *dozents* of the whole school – and since our faculty is the only one that has remained quiet, and since unfortunately Waetzmann no longer holds any office, it cannot be helped. I now see the point coming when someone, if objective arguments are lacking, will say, he's a Jew. Today there is a large student protest gathering according to Nazi principles. They have got themselves a speaker from Berlin, who will enlighten us on what has to be done here. The senate is invited to be there. Then we have a senate meeting. It is going to be a hot day!
>
> So now I must be off. To gather strength I shall go through the park, which still looks wonderful.

The Cohn case to which Franz refers was at a critical point in the university even as he wrote. As reported by *The New York Times*, Ernst Cohn, appointed Professor of Law in November 1932, had been immediately targeted by Nazi students. After initial support and then abandonment by the university's Rector, the university senate on 14 January voted to back Cohn and to provide for resumption of his lectures under police guard. The student rioting continued for several days. On 1 February, 2 days after Hitler became Chancellor, police protection was withdrawn at the demand of Nazi students, and Cohn ceased lecturing after the Rector informed him that he could no longer guarantee public order or the safety of his student audience. On 13 April, Cohn was among 13 Jewish and three Marxist professors in the first wave of dismissals by the new Prussian Minister of Education, Bernhard Rust.

While the administrative impasse in the Hochschule continued, the letters turn to other themes to relieve the tension: research activities, family, social life. Early in the month, at a regional meeting of the German Physical Society held in Breslau, he had spoken about his team's ongoing low-temperature findings, ending with a spectacular demonstration and a tour of the low-temperature laboratory. "The demonstrations," he writes, "really went very well and were very impressive," eliciting a congratulatory letter from Nernst in Berlin, who heard about them at second hand. This regional meeting seemed to bear out the optimism of 13 months earlier, when the laboratory had opened officially.

Other letters mention further projects: an article written with Nicholas Kurti (22 January); "two shorter works written and sent off" (26 January); a joint work written with Kurt Mendelssohn (29 January). Franz's list of offprints shows that no fewer than five items were successfully submitted for publication during January and February 1933.

On the lighter side, his letters report on the children, their sayings or questions, and various social events. On 22 January Franz mentions an evening at a social club cofounded by Lotte's "Onkel Ernst," the physician Ernst Lipmann, who had died in 1931. Earlier, with Lotte away in Berlin and Franz's mother staying with them in Breslau, they received a formal visit from three widowed elderly Simon aunts and a cousin—Cassilde Preuss, Selma Lindenberg, and Clotilde Simon (widow of Franz's uncle Isidor Simon) and her daughter. It was an occasion that Franz clearly did not relish. "Oh Freude über Freude!" he writes.

The feeling was reciprocated. When they departed 2 days later, Tante Selma remarked frankly, "If your mother were not here, I would not have come."

At the end of January, the merger issue continued, with nothing resolved. On 29 January, Franz reports an additional complication. By this time, the other faculties of the Hochschule, "behind our backs," have united on a plan to reduce the amount of chemistry required by engineers. This action has in turn enraged "our chemists," and the merger negotiations can no longer be settled peaceably.

On 31 January, with the issue still deadlocked, Franz sends a few lines in haste:

> We are still sitting, while the students are already rejoicing that the new government will roll back the merger. It's quite possible it will do so. I was just reading that a Nazi schoolmaster will become Minister of Education.

With Lotte's return from Berlin a couple of days later, these letters end, and the merger project fades from view. But Franz's information proved correct. The new Minister of Education was Bernhard Rust, a Nazi party member from 1922, who lost his job as a schoolmaster in 1930 but was then elected to the Reichstag.

The new government formed on 30 January, a Nazi–Nationalist coalition with Hitler as Chancellor, which soon revealed a much larger agenda. It moved with ruthless speed against its declared enemies: leftists, liberals, pacifists, Jews. On 1 February, President Hindenburg granted Hitler the dissolution of the Reichstag that he had refused three days earlier to the preceding Chancellor. New elections were called for 5 March, for the third time in eight months. On 27 February, the Reichstag building itself was set ablaze, for which the Nazis promptly blamed the Communists and banned their party.

On 5 March, a violence-ridden election gave the government coalition a slim parliamentary majority, which widened when parliament met because of the absence of the outlawed Communists and some Social Democrats. On 23 March, the government secured passage of a far-reaching Enabling Act, passed by the requisite two-thirds majority. The act granted the government full dictatorial powers for four years. By this date the first concentration camps—at Oranienburg and Dachau—were ready to receive their first inmates, primarily leftist enemies of the regime.

Three days after receiving full powers, Hitler began to address the Jewish question. Since early March, incidents of local violence, boycotts, intimidation, and forced resignations of judges and lawyers had occurred sporadically and with increasing intensity in several cities. On 13 March, in Breslau, Jewish lawyers and judges were expelled from the courts. Such incidents had been noted and condemned in the international press. To control the situation, Hitler decreed a countrywide boycott of all Jewish shops and businesses for 1 April, to be enforced by Nazi brownshirts and thugs. The boycott took place as planned, was deemed a success, and was called off by the government after three days. From this event Hitler had learned how to use the Jewish population as hostages, by forcing Jewish organizations to speak out internationally for an end to anti-German stories in the foreign press.

On 4 April, Hindenburg, as president and a field marshal, wrote to Hitler to protest against a wave of arbitrary dismissals of Jewish war veterans from the judiciary and the public service as a stain on the honour of the German army. In reply, Hitler promised that the matter would be handled not arbitrarily but by a law, which would take into account the President's concerns.

On 7 April, the government brought in the Law for the Restoration of the Professional Civil Service, the first of some 400 laws and decrees against persons of Jewish origin. The law provided for the elimination of Jews and political opponents of the regime from the public service. It extended to the judiciary, and to university teachers. A second law of the same date barred "non-Aryans" from practising law, and from future admission to the bar. In deference to Hindenburg, both Acts provided exemptions for Jewish veterans with front-line service and for relatives of those killed in action. In legal terms, Franz was exempt from dismissal under the Hindenburg rule.

Almost immediately, another former student of Nernst, Professor Frederick Lindemann, director of the Clarendon Laboratory in Oxford, correctly foresaw that any crisis may also spell opportunity, and that any exodus of scientists from Germany could bring important benefits for the countries receiving them. Einstein's visits to Oxford and Cal Tech in 1931 and 1932 had reinforced this point. With this in view, Lindemann, who kept in close touch with scientists in Germany, set out from Oxford to visit German universities during the Easter break of 1933. Travelling in his Rolls Royce with a chauffeur, he visited Göttingen, Berlin, and possibly other universities.

Behind Lindemann's visit was a prepared plan. He had grant money to offer to the ablest of the displaced scientists to continue their work in Britain. Though his mandate was a broad one, ranging across physics and chemistry in a wide sense, he had one priority for his own laboratory in Oxford: to build on its recent beginnings in low-temperature physics. The Clarendon had already obtained a small hydrogen liquefier from Franz's Berlin laboratory in the spring of 1931, and in 1932 it had ordered a more advanced apparatus to produce liquid helium. This one, constructed in the Breslau laboratory, had already been set up in the Clarendon by Kurt Mendelssohn by January 1933. Nernst, aware of this low-temperature interest, invited Franz to Berlin to discuss suitable candidates for these grants with Lindemann directly.

As Lotte described the meeting in a BBC broadcast many years later, she and Franz arrived in Berlin and proceeded directly to Nernst's house, where she met Lindemann for the first time. They were then told of a scheme to provide two-year grants in Britain for displaced younger scientists of exceptional promise:

> And they discussed this at great length – who, names and so on – and we were saying goodbye. Under his breath my husband said "Oh, by the way, I'm not going to stay either."
>
> And Lindemann was quite shocked and said, "But you don't have to leave in your position."
>
> And my husband said, "Well, I don't have to but I want to." And the next step was that a few weeks later came a very formal little letter from the Clarendon Laboratory saying would my husband care to give a lecture [in Oxford] on low temperature work.[3]

There followed a cautious negotiation by post, using coded language, for the postal system was no longer secure. In late May Franz visited Oxford, ostensibly to give a lecture and technical advice on the newly installed apparatus, but also to explore moving. Finally Lotte, who had never been to England, visited Oxford for 3 days in June. The decision was subject to her approval. She arrived in glorious sunshine, to see the city bedecked with flags for a royal visit from the Prince of Wales, who came to lay a hospital cornerstone. In contrast with the ugly violence surging through Germany, the scene left an indelible impression on her, and the move was on.

Documentary evidence to back up Lotte's recollection of the 1933 Berlin meeting can be found in the Cherwell Papers held at Nuffield College in Oxford. In one file there is a collection of more than a dozen visiting cards, all undated, but easily recognizable as the result of Lindemann's visits to Göttingen and Berlin in 1933.[4] Those from senior German colleagues list further names of younger scientists. For each of these names, Lindemann has neatly noted the basic information that he wanted: current address, fields of specialization, experience in Germany, contacts in Britain, age, dependants. Beside each name is also a cryptic figure—3, 3.5, 4, 5.5—presumably a tentative grant amount.

This collection has two cards from Franz Simon, both with his full academic address. On the first, Lindemann notes two candidates they discussed: Dr N. Kurti and Dr R. Kaischew. On the second, Lindemann has written "Letter before 20 June. Notice before 1 July. Insurance. Kurti." Was the second card exchanged in Berlin, or later? No evidence on that point.

The other cards have a few recognizable names. Two cards are from James Franck in Göttingen. On one, his own name is crossed out and on the verso side is written the name and address of his assistant, Heinrich Kuhn, plus Lindemann's notes about him. Kuhn, interviewed in Franck's presence, would soon be recruited to Oxford.[5] Among others recommended for grants are Dr Eduard Teller at Göttingen and Dr Bethe at the Institute of Theoretical Physics in Munich.

Also in the Cherwell Papers is a memorandum written by Lindemann himself outlining the main features of the grant project, which is to be financed by Imperial Chemical Industries (ICI).[6] Dated 24 May 1933, it asks for confirmation of the scheme by the ICI chairman, Sir Harry McGowan, "in order to prevent the possibility of any misunderstanding" over what has apparently been an unwritten agreement up to this point. The confirmation from ICI, dated June 6, also contains the first list of 13 physicists and chemists, including five professors, approved for ICI grants on Lindemann's recommendation.

The Lindemann memorandum and this first list of approvals show the intended dimensions of the scheme. The planned stipends were to range from £300 annually for young, single researchers to higher amounts for distinguished senior scholars, "rising to a maximum of, say, nine hundred pounds for a Nobel Prize man." Recipients were to work in England, in a university setting, on topics in their own fields of concentration, though they were to be available to ICI on a consultative basis if called upon. The list of approvals of 6 June includes "Prof. Simon" from Breslau, working on "Liquid Helium & Other Gases," with a stipend of £800 per annum.

A further paragraph of the memorandum authorizes Lindemann to go to Germany to interview and negotiate with likely candidates, as he had already done. By mid-summer he would be back in Germany, travelling with his car and chauffeur on a further recruiting round. The grants approved in June amounted to less than half of the £15,000 available annually. The memorandum itself has a characteristic Lindemann flavour, combining timely academic assistance with a keen eye for Britain's long-range national interest in scientific development.

Within days of Hitler's draconian law to "reform" the public service, the student organizations in Breslau launched an action of their own. On April 26, they addressed a letter to the Rector:

> Your Magnificence,
>
> The N.S.D.St.B. in the student body of the Technische Hochschule respectfully requests Your Magnificence to try to have the following professors and assistants take their leave immediately . . . [they] have generally spoken against the Aryan principle, and on this account it is not desirable for a German student to let him be taught by an unreliable (*artfremden*) teacher. Moreover, in the light of their political past, these men offer no guarantee to work in the interest of the national government.

The list appended begins "Herr Professor Dr. Simon," followed by two other professors, two *Privatdozenten*, and two assistants. The letter ends "*Mit studentischem Gruss und Heil Hitler!*" and is signed by Hans Bendix, leader of the universities section of the Nazi Student Federation, and Willi Eymer, leader of the Hochschule student body.

The Rector, Bernhard Neumann, who had done so little in the merger debate, remained in character now. He simply relayed the text of the letter to those targeted. "I bring to your attention a letter addressed to the Rector," he wrote, adding no further comment whatsoever. If there was one lasting resentment that Franz carried permanently from his academic life in Germany, it was a deep anger over the spinelessness of his Breslau colleagues. Franz's experience was far from unique. Stephen Remy's detailed study of Heidelberg University during the Nazi regime finds that "the dominant response of the professorate to the loss of their Jewish colleagues was almost total silence. In this respect Heidelberg was not exceptional among German universities."[7]

While Franz was preparing to visit England in May, the students moved on to a wave of book burning, targeting works of Jewish authors, living or dead, German or foreign. After four weeks of preparation, it culminated on 10 May, when some 20,000 volumes were collected and set ablaze in Berlin's Unter den Linden, in a savage ceremony presided over by Joseph Goebbels, and executed with enthusiasm by the students themselves. Similar bonfires were staged in other cities.

Among Franz's papers is a manifesto entitled "Against the unGerman spirit," issued in the name of the Deutsche Studentenschaft. It is typewritten on a single page, on flimsy paper, undated, and separated from any related document, but it mirrors the mindset of the violent anti-intellectualism of the period. Some of its language is close to that of Goebbels's speech of 10 May. In 12 numbered paragraphs, it develops a rationale for destruction. In 50 lines of repetitive text, a few key ideas stand out:

> Language and literature are rooted in the nation (*volk*) . . . . Today there is a gap between German nationhood and its literature. This situation is an outrage. Cleansing of language and literature rests on you [students] . . . . Our most dangerous enemy is the Jew, and anyone controlled by him. The Jew can only think Jewish. If he writes German, he lies. A German who writes German, but thinks unGerman, is a traitor.

In the final six paragraphs the manifesto moves to a plan of action:

> We want the Jew to be outlawed as a foreigner, and we want the [German] nationality [*das Volkstum*] to be taken seriously. We therefore demand from the censorship:
>
> – Jewish works to appear in the Hebrew language. If they appear in German, they are to be marked as translations.

- Severest penalties against misuse of German writing.
- German literature only at the disposal of Germans.
- The unGerman spirit to be removed from public libraries . . . .

We demand from German students the will and the ability to maintain the purity of the German language . . . . [and] to overcome Jewish intellectualism, and the liberal symptoms of decline associated with it, in German spiritual life. We demand the selection of students and professors according to the certainty of their thinking in accordance with the German spirit . . . .

On a single page, this manifesto develops an ideological foundation for the wave of book burning, and the destruction of German as a major international language. The target is wider than Judaism. It is the entire Western liberal tradition. For academics or scientists of international reputation and outlook, Germany is regressing towards barbarism.

After their two visits to England—by Franz in late May and by Lotte in June—their decision was made. Among Franz's key documents is a copy for the Rector of the resignation letter he sent to the Prussian Minister of Education on 24 June:

I announce to the Minister most respectfully that I have been offered a research professorship at the Clarendon Laboratory of the University of Oxford. I intend to accept this offer, and I respectfully request my release from the Prussian public service as of October 1 of this year.

Even the Rector is addressed with all the formal academic courtesies: "Your Magnificence . . . Expression of my highest consideration," etc.

The ministry replied on 13 July, acknowledging the letter of 24 June and granting a formal certificate and release from academic duties from the requested date. "At the same time," the letter concludes, "I express to you my recognition for your academic activity in this Hochschule." *Meine Anerkennung*. It is almost an expression of thanks—almost, but not quite. This veneer of formal courtesy covers an ugly reality. Is it to everyone's advantage to pretend that this transfer is just a change of jobs? This is the department headed by the same Minister Rust who dismissed Professor Ernst Cohn and 15 other Jewish or Marxist academics on 13 April, and whose main achievement as minister was to purge the Prussian education system of Jews, leftists, and all influences inconsistent with Nazi principles.

The Simon family's move to England, although a major upheaval in their lives, met with no special obstacles. Permission to leave came easily enough. Possessions and papers were subject to close scrutiny for anything subversive of the regime. The minor official assigned to this inspection proved compliant when supplied with beer and a few light bulbs. At this time, those leaving could take along all household furniture and books. Franz also arranged to exchange his office furniture, which he owned personally, for two helium liquefiers and other laboratory equipment. A more serious obstacle was the high British tariff on imported scientific equipment, but after intervention by Lindemann to explain its significance, duties were waived.

At the end of July, the rest of the family left for England, arriving on 4 August. They travelled on the *Albert Ballin*, the same ship that had carried Franz to New York. He remained in Breslau until 11 August to supervise the packing of furniture and complete other arrangements. In this interval, he writes three letters to Lotte that tell of his last days in Breslau. There is still much to be done: legal formalities, financial forms, banking arrangements, laboratory apparatus, even the dentist. On top of this is the social side: friends calling, sending their greetings; on 9 August, a farewell party of the faculty—"very nice"—with talks by three colleagues; on 11 August, "the whole institute, plus a few people from neighboring institutes," gather at the railway station to see him off. "The Institute gave me a silver cigarette case, with something engraved inside." This box remained in the house, still gleaming, the cedar lining still aromatic, a fixture of the living room in Oxford for decades. In time, it became a double-edged reminder of the exodus from Germany and of Franz's long years as a smoker—years that undoubtedly shortened his life.

4

OXFORD 1933–1939

If arrival in England had brought personal safety, the family now faced new problems. The first priority was to arrange basic shelter and sustenance in a new land. Then there were close relatives still living in Germany: parents, and both of Franz's sisters and their immediate families. As long as any of these were there, the Nazi threat remained. There would be workplace problems to resolve, cramped laboratory space and less advanced equipment, problems only partly resolved by equipment from Breslau. Finally, there would be a new social and academic setting to adjust to. To this end, Lotte was adept at languages and Franz, after his American experience, was quietly confident of his growing international reputation as a scientist. Both were eager travellers, naturally gregarious, but Lotte knew very little of England. "We will be people of consequence here," he assured her. Even for someone internationally known, a new start in a new country in a new language can be a formidable challenge.

Getting settled

When Lotte and her group arrived at Southampton on 4 August, 1933, they discovered a country on the move for the bank holiday weekend. They found space in a hotel in Bournemouth, where Franz would later join them. This became their headquarters while first Lotte, and later both, explored housing in Oxford. Because times were bad, many houses were available. By Lotte's account for the BBC, they looked at "hundreds of houses" in a two-week search. One problem was that she had no idea of the cost of living in Britain. The ICI grant was for £800 per year. When she asked if a family could live on this amount, a don's wife replied that "clergymen have very often to live on less."[1]

The quest for housing produced the first cultural shocks. Arriving from Germany, they looked for a house with central heating. Lotte had grown up with it in Berlin and also had it in Breslau. One prime reason for her parents' move from a fashionable older building in Friedrichstrasse to a newer flat in Pariserstrasse had been that the former was still warmed by classical, elegantly tiled corner stoves. In Oxford, neither of these comforts was available. Central heating was unavailable, and even deemed unhealthy.

A more central consideration was the style and quality of the house that they should select. Lotte, fearful for an uncertain future, inclined towards a smaller, less expensive house. Franz thought in terms of a house of the standard they had known in Breslau. To settle for less, he reasoned, would make them feel degraded and unhappy, and might lead to misunderstandings on the part of new colleagues and neighbours. "We must take a house we really like," he said.[2]

In the end, they decided on a house that Lotte had seen at the beginning, one of two then available in Belbroughton Road in north Oxford. They arranged to rent it, with an option to buy after one year. The ICI grant was for two years. Their future was precarious, but so too was the future of Europe itself in August 1933. They moved in on 11 September.

Another priority was to get the children settled in school. The two girls, aged seven and five, were soon enrolled in the junior school of the Oxford High School for Girls, two streets away. Neither child had any English. Lotte, whose Berlin diary had carefully recorded her daughters' acquisition of language as infants, described in her BBC interview their first experiences in an English-medium school. For Kathrin, who had learned basic reading and writing in Breslau, the transition went smoothly. She began to talk in class from the first day. "She just got a list of words to learn every day. And if she didn't know a word she asked."

For Dorothea, the experience was more traumatic. Arriving in England without previous schooling of any kind, she faced her first separation from the family in a situation where everyone else chattered in an unknown language and no one understood hers. By Lotte's account, "She spent the first day crying," and Lotte was asked to sit in to help her daughter get settled:

> So for six weeks nearly I went to kindergarten every day . . . . She never opened her mouth, just watched. And after six weeks when she started to join in she knew exactly what the other children knew.

In Dor's recollection, the role of comfort and support was often filled by her paternal grandmother, Amama Simon.

When the new language did come, it came in whole sentences. Nevertheless, it had been a difficult time. The persona that emerged in her adult years was different from that of her sister, quieter, more reserved, reticent. While Kathrin would retain her German and build on both languages as she matured, her younger sister locked away the German side in the recesses of the family circle and built a new identity modelled on the classmates around her.

On 28 September, Lotte found time to write to a friend in Berlin, Lili Friedländer, who had evidently been ill. The letter is revealing for its description of Lotte's first fortnight in Oxford. She apologizes for missing "Peter's birthday," but there had simply been too much to do. "In our place things are going very well," she reports. Her parents want to visit them, and the children are in school:

> Our house is very nice (*hübsch*), but fairly large, and English houses are more difficult to manage (*bewirtschaften*) than ours [in Germany]. There is therefore considerably more work. But this does not trouble me at all. I am glad if I have a lot to do . . . We have many visitors. Many of our acquaintances are coming to London, and then make a side trip [to Oxford] . . . Franz is in Zurich at the moment for a discussion about an invitation to Constantinople! He also has several bids to America. I await quietly to see where we may land next.

Even before the house was up and running, the first Oxford visitors had called to welcome the new arrivals. The first was the Lady Mary Murray, the wife of the classicist Gilbert Murray, soon followed by the wife of Franz's new colleague, Alfred Egerton, the university's Reader in Thermodynamics. These and others introduced Lotte to the social language of a new academic circle. As she recalled much later for the BBC, she found strict conventions that prescribed exactly what to do in various situations. Invitations and replies were couched in formulae that everyone understood. The formulae extended to children's parties, and children too knew what to do on these social occasions. By comparison, the social milieu in Weimar Germany had been one of fewer social conventions and larger areas of uncertainty. Once these English conventions were learned, she said, "I found it made social life very easy."

There were other early impressions that did not appear in the BBC tape. In her later years she sometimes spoke of the code of manners she found at the time of the family's arrival in Oxford. One did not issue invitations by phone in those years; the telephone was considered too intrusive. Instead, one used the post, and replies came back in writing, sometimes the same day. One also made frequent use of visiting cards.

This part of north Oxford in the 1930s had elements of paradox. Many of its inhabitants were living lives of the highest intellectuality, but the area was just two generations removed from its Victorian roots, the period when fellows of colleges were first permitted to marry and have families. In Belbroughton Road, the rural fields and riverside meadow, complete with farm animals, ran up to the last house on the street. To the amazement of the newcomers from Germany, milk was still delivered by horse and cart, unpasteurized, and ladled from large milk cans into the householder's container. Orders for the greengrocer, the butcher, and the fishmonger were taken by tradesmen with order books, and delivered by bicycle. Laundry was collected and returned in vast wicker baskets. Rubbish was collected by horse and cart.

For the newcomers, the greatest problem was the absence of central heating. The housing market in Oxford offered no alternatives. Though most houses in Belbroughton Road were less than 10 years old, almost none had any central installation for heating. Number 10 had been built in the late 1920s with eight open fireplaces and chimneys, four on each floor. Some of these fireplaces had gas fires, the others used coal.

Winter weather brought extreme, unaccustomed hardship. Lotte's parents, Omama and Opapa, came to Oxford in the first year for an extended visit from November to February, but returned to their warm apartment in Berlin. Franz, the expert on thermodynamics who would later campaign against the wastefulness of British open fireplaces, began the revolution in his own house. At an early date, the fireplace in the entrance hall was replaced by a coal-fired round stove, which radiated most of its heat into the upper and lower hallways instead of directly up the chimney. Another so-called cosy stove, rectangular, with opening front doors, replaced the fireplace in the living room.

A major problem of adjustment was the absence of domestic help, a situation that Franz had reported from California but Lotte had never experienced. On the BBC tape, Lotte recalls how she learned to cook:

> I still remember how proud I was when I made the first scrambled eggs. I looked up two cookery books. And it really was, to my great surprise, it was scrambled egg.[3]

Her experiment with eggs represented an advance over that of the preceding generation. By family legend, when boiled eggs arrived too firm at the breakfast table of the newly married Opapa and Omama, she returned them to the cook for *more* boiling to soften them. In time, Lotte became a superb cook, capable of dinners of gourmet quality.

When the philosopher Ernst Cassirer and his wife Toni came to Oxford from Hamburg in the same term, they were invited to the Simon house for dinner. For the occasion Lotte prepared a special meal of pheasant and sauerkraut, the latter unavailable in Oxford but brought from London. As recounted by Lotte, the menu proved too grand for Toni Cassirer, who protested that this was not the style in which refugees should live. Lotte disagreed: "If one has children one has to live as if everything were normal, because otherwise you don't give them a sense of security."[4] Despite their initial disagreement on this occasion, they developed a close, lasting friendship. But for Lotte as for Franz, the issue was an important one: the welfare and comfort of the generations to follow became an unstated but deeply held operating principle, even if on occasion this required them to be shielded from reality.

The Cassirers, who lived in rented furnished rooms on Banbury Road, did not stay long in Oxford. In 1935 Ernst accepted a personal professorship at a small Swedish University, Göteborgs Högskola, where he could lecture in German instead of in English. In 1941 Ernst and Toni would depart from Sweden for the United States, leaving behind both of their sons with their families—one in Oxford, the other in Sweden. In an autobiographical memoir, Toni Cassirer explains at some length their disillusionment with Oxford: initial misunderstandings with the head of their college (All Souls); insecurities about their future; lack of recognition of Ernst's continental reputation; disadvantages in lecturing in English; the absence of their own belongings and household; the reserved English temperament; the college system, with its scant regard for formal lectures; and the "unbelievable *Primitivität*" of ancient college buildings.[5] Overall, the Oxford milieu was too much at odds with Toni's values and Viennese upbringing.

One further step helped Franz to integrate into his new surroundings. As a researcher at the Clarendon, he had no official connection with a college, a vital part of university life. This was remedied when his friend from Berkeley days, the classicist Cyril Bailey, at this point senior Fellow of Balliol, introduced him to the college. In 1934 Franz became a member of Balliol senior common room, and in 1935 he was admitted by decree to an Oxford MA. The common room in the same year saw other interesting additions. Another Mendelssohn, the distinguished jurist Albrecht Mendelssohn Bartholdy from Hamburg, became a senior research fellow. The Eastman Visiting Professor for 1934–5 was Arthur Compton, the Nobel prize-winning physicist from Chicago. A postdoctoral student, R.V. Jones, became Skynner Scholar in Astronomy. Within a decade, the careers of Compton, Jones, and Simon would intersect in important ways.

Franz's colleague Alfred Egerton, the university's Reader in Thermodynamics, was also at Balliol, and when Egerton moved to a chair at Imperial College in London in 1936, Franz, still supported on the ICI stipend, was appointed his successor as Reader. Though his search for a regular chair elsewhere continued, his roots in Oxford were growing stronger.

Growth at the Clarendon

On the research side, the transfer from Breslau to Oxford went smoothly. Two low-temperature liquefiers had been purchased for the Clarendon before the Nazi seizure of power, and the second of these had been installed by January 1933. Kurt Mendelssohn, who had helped to install it, returned from Germany to Oxford by April to become one of the first ICI grant recipients. Because his politics were leftist, he was considered to be in greater danger of arrest than the others. By October, the physics group from Germany had increased to five—Mendelssohn, Simon, Heinrich Kuhn (who had been an assistant to James Franck in Göttingen), Nicholas Kurti, and Fritz London (an assistant to Erwin Schrödinger in Berlin). London's younger brother, Heinz London, would follow in 1934 after his Breslau thesis was completed.

Kurti, who had been Franz's doctoral student in Berlin and his assistant in Breslau, arrived on 16 September. Celebrating his 90th birthday in 1998, he recalled that arrival:

> I arrived in Oxford by train at 10 p.m. one Saturday evening; it was dark and raining. I took a taxi to Mendelssohn's flat in Headington. Next morning was bright and sunny, and Mendelssohn took me to the laboratory on his motor bicycle; there was little traffic, and as we crossed Magdalen bridge I thought 'This is fairyland – why should I ever leave it?' And I did not.[6]

For Lindemann, another coup was close at hand. During his first foray to Germany in April 1933, he had called on Erwin Schrödinger, successor in 1927 to Max Planck's Chair in Theoretical Physics in Berlin University. Lindemann mentioned a pending offer to Fritz London. When Schrödinger heard that London was still undecided, he said, "Offer it to me. If he does not go, I'll take the position."[7] But London did accept, and Schrödinger, as an Austrian citizen and non-Jewish, did not fall within the ICI criteria. The astonished Lindemann did not give up the idea. Back in Oxford, he persuaded Magdalen College to establish a supernumerary fellowship for Schrödinger, which was supplemented by extra funding from ICI.

On 3 October, Lindemann wrote to Schrödinger in Zurich to announce the news: "Cher ami, I am glad to tell you that you were today officially elected a fellow of Magdalen." He goes on to discuss the "formalities" that will follow and adds a paragraph of reassurance about the customs arrangements, referring to the recent cases:

> You need not worry in the slightest about the permit. Simon, Kuhn, Mendelssohn and London are all here and have had no difficulties whatever. I am also glad to be able to tell you, there should be no trouble about bringing your belongings to England. Simon has brought in all his property without any difficulty in demands for duty on the assurance that they were for his personal use and that he did not propose to make any commercial use of them. This allowed him even to bring in scientific instruments which in the ordinary course are definitely dutiable so that I am sure there would be no trouble whatever about your things.[8]

The Schrödingers arrived in Oxford on 4 November. At the welcoming ceremony at Magdalen on 9 November, news came that the 1933 Nobel Prize in Physics had been awarded jointly to Professor Schrödinger of Berlin and Professor Paul Dirac of Cambridge. Early in 1934, the Schrödingers settled into a house at 24 Northmoor Road, just around the corner from the Simons.

By year's end, Professor Lindemann could look back upon 1933 as a year of solid and even spectacular accomplishments, the best since his election in 1919 to the Chair in Experimental Philosophy. The original Clarendon Laboratory had been completed in 1872, and had been used up to 1919 primarily for teaching undergraduates. Lindemann's predecessor, Robert Clifton, who held the chair for 50 years, had been a conscientious teacher but looked on research with disdain, and even suspicion. In 1919, the old Clarendon was still lit with gas lamps, and had no mains electricity. The workshop tools were primitive, and the annual budget, mainly for routine items, was £2,000. A separate laboratory for studies in electricity and magnetism had been created in 1910 under the direction of a second chair in Physics. For teaching purposes, the two laboratories were separate and distinct by university statutes.

From this unpromising point of departure, Lindemann had set out to raise physics in Oxford to a level rivalling any university in Britain, and the Clarendon Laboratory to parity with the Cavendish in Cambridge. The 1920s saw modernization of the laboratory's basic utilities and equipment. Lindemann's widening social and industrial connections, which he cultivated carefully, led to donations of equipment and research money. His correspondence shows an immense talent for persuasion. His early doctoral students, carefully selected, had spectacular careers in science. Lindemann had given priority to low-temperature research before 1933, as shown by his purchases of liquefiers from Berlin and Breslau, but acquiring a full internationally known team in this field, complete with leading-edge apparatus and supported mainly by ICI funding, was a huge, almost unbelievable, triumph. It had been due not to chance but to Lindemann's continuing connections with Nernst and other German scientists.

He was unable to resist crowing a little over this progress. The issue of *Nature* for 11 February 1933 contained a two-page article celebrating the formal opening in Cambridge of the Royal Society Mond Laboratory, adjacent to the Cavendish, on 3 February. Built primarily for Peter Kapitza, the Russian scientist who had worked in Cambridge since 1921, the new laboratory was generously equipped for low-temperature and magnetic research. At opening day, the liquid hydrogen plant, compressors, and gas storage systems were all in place, but the helium liquefier, based on a new design by Kapitza, was not yet ready though expected "within the next few months."[9]

By a strange conjunction, however, the same issue of *Nature* also carried a short research note *preceding* the Mond article, written by Lindemann and Keeley of the Clarendon Laboratory. It described a simple new helium liquefier that had been installed in Oxford in early January. Built in Breslau at a modest cost of 400 marks, or about £30, it had been accompanied to England by Kurt Mendelssohn, who guided its installation. This liquefier, the note explains, had made it possible "to obtain, without hitch or trouble, liquid helium within one week of the arrival of the apparatus in Oxford." It is noteworthy that Lindemann's first helium liquefier became operational *before* Hitler's seizure of power, three months before the first of the refugee scientists—Mendelssohn himself—would settle in Oxford, and 15 months before Kapitza would produce the first liquid helium at the Mond Laboratory on 21 April 1934.[10]

The new arrivals from Germany and the students they attracted led to severe overcrowding in the old Clarendon Laboratory, and this was soon recognized by the university. A special fundraising issue of the Oxford Society's magazine of February 1937 gives equal recognition to the needs of the humanities and the natural sciences. On the humanities side, the top priority was the Bodleian Library's need for more storage space and the projected design for the New Bodleian building. For the sciences, the top priority was new laboratories for physics, physical chemistry, and geology. Construction for the New Bodleian began in 1937 and for the new Clarendon Laboratory in 1938. The latter was completed late in 1939, but the outbreak of war forced suspension of normal academic research while most scientists worked on war projects.

For the low-temperature group, basic research after 1933 had to proceed by the best means available. Experiments for which equipment was lacking in Oxford had to be carried out in other laboratories. By January 1934 Franz wrote to ICI to ask permission to continue high-pressure experiments at Dr Michels's laboratory in Amsterdam. He estimated absences from Oxford for 1–2 weeks every quarter. Within four days, ICI approved the request.

In 1935 Franz and Nicholas Kurti made the first of a series of visits to Professor Aimé Cotton's electromagnetic laboratory at Bellevue near Paris. By this date the Clarendon team was using the method of magnetic cooling first suggested by Pieter Debye in Zurich and William Giauque at Berkeley independently in 1926–7, a method that would give access to temperatures closer to absolute zero. In Lotte's miscellaneous papers is a useful handwritten list of Franz's and her own absences from home from 1922 to July 1940. The list records seven research visits to Paris between 1935 and 1938, usually in April and December. These absences for laboratory research were typically for a month at a time. The visits to Amsterdam were less frequent.

The exodus continues

Franz and Lotte wrote to each other almost daily during these absences, as they did during the Berkeley visit in 1932. This time, however, the settings were familiar to both of them, and the focus was on people and events rather than places. On two occasions Lotte is the absent partner, with or without the children. These letters give intermittent portraits of daily life and adjustment to the new environment. Letters or postcards appear even when the main message is "nothing new," or "nothing new from Berlin." They can be supplemented by surviving letters from colleagues, friends, and relatives. In combination, the letters from these years show events evolving on several fronts at once.

The core of their lives centred upon immediate family. School reports and encounters with teachers were major news, and a source of pride for the parents. The children's social or cultural activities ranked close behind. The "children" now included Franz's 13-year-old nephew, Thomas Frank, who accompanied the family to Oxford in 1933 and continued his schooling there until 1936. After Thomas, 10-year-old Curt Lipmann, Omama's nephew and Lotte's cousin, would be sent by his parents from Germany in 1938 for schooling in Oxford.

In 1933, the family still had close relatives in Berlin: Thomas Frank's parents and his brother Stephan; Franz's younger sister Ebeth and her husband; Franz's mother; and Lotte's parents. The longer these remained in Germany, the more they became hostages to an increasingly repressive regime.

The Henschels were the first to leave. Elisabeth Henschel-Simon, known to most as Ebeth, had studied art history under Paul Frankl at the University of Halle, and had married a fellow student and artist, Albert Henschel. After her doctorate, she worked as a curator in the Berlin state castles and gardens. Following her dismissal in 1933, the Henschels decided to try Palestine. Just why they opted for Palestine is not clear from the documents. Ebeth was not a Zionist, and Albert was from a Christian family. They had no specific occupational objective. But Palestine was in the air at this point. They had no dependants, and they were not too old to venture to something new. They found a tenant for their house in Tannenbergallee, bought a second-hand Ford car, packed it with their belongings, and set out from Berlin on 5 October 1933.

The plan was to drive through France and Spain to Gibraltar, where they would board a ship for Haifa. They were leaving the comfortable world that they knew for something totally different. Ebeth produced a detailed report of their odyssey, which she typed in four copies and arranged for each copy to be circulated in sequence to five or six friends and relatives. Among Ebeth's letters to Franz is included a copy of this report that ended with him at Delbroughton Road. With it is a second report, circulated in the same way, on their first weeks in Palestine. The two reports together comprise about 15,000 words of closely typed text.

Once on the road, they headed south across Germany, with a few brief stops for visits, towards Konstanz, where they stayed with Albert's parents, taking care to reach the Swiss frontier before their exit permit expired on 15 October. At the border, every piece of baggage was unpacked and examined. Even unexposed photographic paper was checked in a darkroom for hidden currency. Once cleared, they proceeded to Basle to visit old friends. Over the next four weeks, they alternated between farewell visits to friends or relatives and a tour of artistic monuments. The itinerary included Paris, Amsterdam, Bruges, London, Oxford, and many smaller towns along their route that were famed for churches or altarpieces.

For Albert, they also visited the wartime battlefields, including the exact spot near Arras where, 18 years earlier, he had been taken prisoner and incarcerated for the duration of the war. In this region they passed "immense and unspeakably sad military cemeteries . . . every couple of kilometers." Ebeth found the new houses adjacent to these cemeteries "especially chilling." They had been rebuilt exactly as before the war, with half-destroyed buildings deliberately left standing here and there as war memorials, when modern, friendlier housing could easily have replaced the areas destroyed. "The whole land," she noted, "is inconsolable."

The drive through northern France and Belgium was delayed by recurrent car problems. They attributed this partly to the 30,000 kilometres showing on the odometer. There were stops for repairs in Arras, Bruges (twice), and Lille, each costing time and money, but after the visit to England they still found time for an extra week in Paris to visit more friends and arrange their sea passage before heading south on 15 November.

The next day, in high spirits, with good weather, and on straight roads, they were approaching Mâcon when the car suddenly shuddered and veered sharply to the left. Albert, trying hard to correct, lost control. The car rolled, and with a loud, splintering crash came to a stop on its roof, wheels in the air. After they climbed out of the open doors, they gradually realized that they had been fortunate. Although both had severe bruises, no bones were broken. The accident had resulted from a failed rear wheel rim and a blown tyre.

The car had serious damage but was repairable. This, however, would take some time and would have to be done in Lyons, about 60 kilometres away. While waiting and recuperating, Ebeth and Albert spent a few days in Avignon. Repairs went slowly, and were not completed until 3 December. Even before this date, they both realized that "our beautiful travel plan" to drive through Spain to Gibraltar would have to be changed. Since the ship they were boarding would also call at Marseilles, their sea voyage could begin from there.

It was too late for Spain, but by switching they could have a few days longer in the south of France. They found a hotel in Arles, and from there visited half a dozen nearby towns. As time ran out, they felt an urge to see the Riviera, and drove across to St Raphael to spend their last night in France. For Ebeth the area fully lived up to its fabled reputation—"flowers blooming in the garden, the blue sea, and the warm air." Her description resembles Lotte's reaction to Capri in 1921. In the morning, they drove back to Marseilles via the coast road:

> And this drive, partly directly along the sea and then at the end over a high, barren mountain by a very difficult, serpentine road, was scenically the most beautiful of the entire journey to this point. At the end we were a few kilometers from Marseille high up on the mountain, and the town lay by the blue sea almost directly underneath us. It was a beautiful ending for Europe and a compensation for missing out on Spain.

In Ebeth's account one can sense her reaching out to build memories of the sweetness of Europe, memories to sustain them in exile.

The second report, dated 2 February 1934, summarizes their early impressions of Palestine. On arrival in Haifa, they watched the unloading of cargo from the *Excambion* into lighters, which in turn relayed it 500 metres to shore. Later, in Jaffa, they would see other lighters offload their freight in even more primitive fashion to Arab labourers standing in shoulder-deep water at the shore. After viewing Haifa, they set off by car through Nazareth—"an uninteresting town"—to Tiberias on the Sea of Galilee. After a brief visit they then drove south towards Jerusalem, where they spent two weeks, with an excursion to Bethlehem for Christmas Eve services, and another to the Dead Sea. The long stony descent towards the Dead Sea, with its distant view of equally high mountains across the valley in Transjordan, left them with the impression of a moonscape. Jerusalem, however, proved chillingly cold in January, and they moved on to Tel Aviv and its warmer Mediterranean climate.

Their disillusionment with Palestine began from the first day. In Haifa, they found houses in the Arab sector to be badly built, neglected, with filth and dirt everywhere. The new Jewish quarter at Carmel was only partially developed, saddled with high, uncontrolled land costs and unplanned building activity, with some small houses positioned precariously on the steeper slopes without gardens. In the other cities, their disillusionment continued. Jerusalem faced an acute water shortage, and its housing lacked central heating despite cold winters. Tel Aviv, just 15 years old, was marked by unplanned streets lacking proper drainage for the winter rains, houses badly built by unskilled workers, and expensive rents. Clearly Ebeth and Albert had envisioned the new Palestine from afar as a projection of the best in recent European urbanism. They were unprepared for the hasty, improvised conditions of a frontier society. Their high expectations before arrival were crushed by reality.

More serious than these structural deficiencies was the cultural gap. Their enlightenment on this point began before leaving the *Excambion*, when they were victimized, like many other "unwary newcomers," by a local Jew posing as a customs official. What they discovered through further experience was a societal ethic far different from the concepts of dealing in good faith (*Treu und Glauben*), reliability, order, and cleanliness that they had known in Germany. Instead, they found a relaxed Levantine culture of bargaining, inefficiency, and acceptance of things as they are. Some of the deficiencies they blamed directly on officials of the Mandatory authority: a delivery time of five or six days for local post, the same as for letters sent from Europe; a year's wait for a telephone; an inadequate water supply in Jerusalem. They suggested that the colonial power had only a marginal interest in developing Palestine, and that colonial officials viewed the Jews as nothing but "natives with occasionally a lighter complexion."

For Ebeth the most painful part of this experience was coming to know her fellow Jews. Next to the "giant-sized" gulf between employers and workers, the report notes, there are serious differences stemming from the settlers' earlier homelands. "The big division understandably is between *Ostjuden* and *Westjuden* (mainly the recent German immigrants)." While the latter bemoan the absence of their cherished values, the former remind these Germans that they were the first to arrive and that they had done the early, hard, pioneering work. "Moreover, the Eastern Jews come from such a different civilization and culture that they feel what they have produced is already paradise." Ebeth's values were more German than she knew.

The report also describes economic conditions and the language milieu in Palestine. On the economic side, possibilities for intellectual and professional work appeared very limited.

As for language, while German and Yiddish were still widely understood, they noted strong pressures to promote Hebrew, and even to replace German personal names with Hebrew ones. Although the Mandate recognized three official languages—English, Arabic, Hebrew—they noted instances of "atrocious chauvinism" around the language issue. New firms from Germany with international names and reputations faced demonstrations and smashed windows if they resisted pressures to post names and signage in Hebrew only. At a personal level, their own car was bumped and damaged three times in two weeks because of its German plates.

These events had a sequel. Two years later, when Franz was approached concerning the Chair in Physics at the University of Jerusalem, his sister's report on Palestine became a background factor in his response. In a letter to Lotte from Paris on 21 March 1936, he wrote of his willingness to learn Hebrew and to lecture in the language eventually, but he rejected meeting any preset language deadline, or giving language learning priority over the build-up of physics. His central concern was to build a first-class, generously supported institute where one would have "a free hand in essential matters," unhindered by "small nationalistic interests." In the end, the Jerusalem prospect did not go ahead.

Concluding their second report, the Henschels attempt to strike a fair balance for those who may follow. "Who should immigrate, and who should not?" Immigration is risky, they suggest, for people over 30. Those who know Hebrew have a powerful advantage, but "one who does not bring an unflinching Zionist idealism will find it difficult to settle in, and even many Zionists have gone back disillusioned." Certain occupations were needed more than others, but intellectuals and members of the "free professions"—doctors, lawyers—had in general no possibilities. The final sentences balance pain and hope:

> This report is obviously subjective, yet we find our interpretation confirmed by many other people. It has often been difficult for us to write so little that is gratifying about the land to which we came with so many hopes. Yet in spite of this one must make it clear that very many people have found a new homeland here and consider themselves very fortunate.

When they sent this second report to friends abroad, they had no intention of staying in Palestine beyond the coming summer. Yet stay they did, no doubt because they found no better alternative.

By September 1934, their mood was softening. They were finding modest prospects for intellectual work. After a summer visit to their families in Germany, Ebeth immersed herself in studying archaeology and prehistory. She had heard of future openings at the new Rockefeller Museum of Archaeology, where she was first offered a junior post in August 1935. For several seasons both she and Albert were involved in nearby excavations, most notably in Aqaba. In a milieu that was predominantly male, British, and bureaucratic, she faced a triple barrier as a female, a married woman, and a German–Jewish immigrant. For several years she endured low pay and precarious year-to-year appointments, but eventually was given more responsibility and more absorbing work as assistant keeper. When war came in 1939, it closed off any further move until 1945.

The departures of the Simons and Henschels in 1933 led to further changes in family housing arrangements. Some changes had already occurred. The vast Simon house in Landhausstrasse, which was shared with the Henschels before Ernst Simon's death in 1925, and then with the Franks as well, was too expensive for the family's reduced means. It was sold in 1930. As a family friend remarked, "All those Simon millions down the drain."

The Franks returned to a flat in the Kaiserallee, in the same building as Ludwig Frank's medical practice. The Henschels and the widowed Anna Simon moved to a small modern house in Tannenbergallee. In 1932, the Franks moved to a house with larger rooms in Kaubstrasse, and Anna Simon moved again to live with them. Lotte's parents had also downsized from their massive flat in Pariserstrasse to one in Carmerstrasse in 1931.

In the summer of 1934, there were partial reunions in Berlin. Ebeth and Albert made a summer visit from Palestine, to see their relatives and renew professional contacts in publishing. Thomas Frank returned from Oxford for the school holidays. Lotte returned to Berlin to visit her parents.

Lotte travelled by air on 20 July, her first venture in a plane, and this sharpened her emotional responses. In daily letters and postcards to Franz and the family, she describes these two weeks in detail. A two-hour crawl across London was followed by a quiet four-and-a-half hour flight from Croydon to Berlin. Over Germany the weather was clear. "We flew directly over Sans Souci, the Glienicke bridges, Babelsberg, Sakrow, Peacock (*Pfauen*) Island – like a dream . . . . Berlin appears unchanged."

She had much to do in Berlin. She found her father fine and her mother "tolerably" well, but she signed power of attorney forms concerning their care for each of them. She had two evenings with the Simon relatives in Kaubstrasse, and joined the Franks and Henschels again for a Sunday outing to Ferch. She looked up her closest childhood friends, delaying her return to Oxford by a few days to do so. She mentions meetings or telephone conversations with more than a dozen others, some with relatives already in England. With two or three invitations every day, she writes to Franz, "the days are very full."

She went one afternoon to the Wilmersdorf cemetery to visit the grave of Lili Friedländer, accompanied by Lili's sister. A photograph of Lili was found in the photographic collection in Oxford, a formal portrait of an attractive woman in her thirties, marked with the year of her death, 1933. The graveside visit with Lili's sister may explain how the portrait, and Lotte's original letter to Lili—postmarked in Oxford on 29 September 1933—were found in Oxford among Lotte's papers.

A year after her departure, Lotte's emotional ties to her native city were still strong. She could still be captivated by its July sunshine, its orderly tree-lined streets and parks, and its unchanging appearance. "Here it is as if I had never gone away – like a dream! Very quiet on the streets, people elegant and chic." What she wisely did not write—or may not have known—was that her arrival on 20 July had come just 3 weeks after the infamous purge of the Night of the Long Knives—30 June 1934—when the Nazi party turned against itself in a massacre of its own leadership. It is understandable that Berliners were quiet. Ebeth put the summer in clearer perspective in September, in a letter written after her return to Palestine: "Berlin was wonderful, but that is no ground for wanting to go back."

With the summer visitors gone, life went on for those in Berlin. Ludwig Frank continued with his private medical practice and laboratory in Kaiserallee 205, established there since 1919. With a consulting room and a laboratory, an assistant physician, and a dispenser, the practice flourished. Ludwig specialized in internal medicine, while Mimi, who had studied both medicine and economics, looked after the financial side. Their younger son Stephan, now 12 years old, continued at the Fichte Gymnasium. Lotte's parents, living in comfortable retirement in Carmerstrasse, saw no immediate cause for concern.

Life continued against a backdrop of increasing and officially encouraged antisemitism. For the anti-Jewish boycott of April 1933, a wartime comrade of Ludwig's removed the family to a safe retreat outside the city until the danger passed. In Stephan's school, the teachers behaved with integrity, but in the streets there were unpleasant incidents. What he remembers to this day is one occasion when the outcome was unexpected. He was being chased on his way home by a group of bigger boys shouting "Get that Jew!" When he turned to face them, the leader unexpectedly called a halt, saying "Not this one." He had recognized Stephan as the little boy who, with his older brother, had allowed him to share their toboggan one snowy winter four years earlier.

In the medical profession, doctors were affected by anti-Jewish measures in different ways. Those in state or academic employment were dismissed in 1933 along with other officials, subject to the Hindenburg exemptions. Professional associations quickly came to be dominated by the Nazi-controlled association of health insurance physicians, which spearheaded the pressures for change. In the absence of a legal basis for ousting Jewish doctors in private practice, these faced a continuing pattern of "capricious as well as systematic oppression,"[11] as a study by Michael Kater concludes. Measures included denial of panel employment, refusal of insurance payments for services, dismissal from medical associations, trumped up infractions, false accusations from patients, and outright defamation from hate groups. When patient loyalty to Jewish doctors remained strong in spite of this, state employees were forbidden to consult them by a law of October 1936. As Kater also notes, anti-Jewish measures were supported by many younger, unemployed non-Jewish doctors and by disgruntled older colleagues. Both groups saw them as a means of personal career advancement in dismal economic conditions.[12]

The Nuremberg racial laws of September 1935, which ended the Hindenburg exemptions and extended racial restrictions to those of half or quarter Jewish ancestry, further undermined the position of Jewish doctors, but still did not close the door to private medical practice. This option remained open even for dismissed panel physicians. Full Jews, however, were deprived of any rights of German citizenship by the Nuremberg laws, becoming mere residents in their country of birth. Among other restrictions, they could not employ non-Jewish female staff. When patient loyalty continued despite increasing obstacles, a systematic registration was ordered in April 1937 of all Jews active in the medical profession. A year later a second decree ordered the delicensing of all Jewish doctors, with loss of all privileges of professional status, from 30 September 1938.

Before this endpoint was reached in 1938, however, Ludwig Frank's career in Germany was derailed in another way. One bizarre paradox of the totalitarian regime was that—even as it was preparing to sterilize and euthanize deformed, unwanted, and unproductive individuals—it invested heavily in public health programmes for the privileged German *Volk*. The central theme of a public campaign against cancer was that "cancer is curable (*heilbar*) if detected early."[13] In typical bombastic style, the campaign proclaimed that nothing is impossible for true believers. Ludwig is believed by the family to have dismissed these claims as nonsense, and his sentiments were reported to the authorities. To question party orthodoxy on such matters had cost other doctors their practices or their lives.

In July 1936 a letter addressed to Ludwig arrived, and was opened as usual by Mimi Frank. It summoned Ludwig to appear before the minister, Leonardo Conti, the infamous Prussian Commissioner of Health. At that moment Ludwig was away in Britain, visiting

Franz and Lotte. His visit coincided with a medical convention in Oxford, and he was clearly looking at prospects for emigration from Germany. Mimi advised him to remain in Oxford.

Britain's initial policy towards refugee physicians had been to require study at a British university and medical re-examination before being allowed to practise in Britain. During the first 2 years of Nazi rule, about 200 doctors had been allowed in, but by 1936 entry was more restricted. As a condition of asylum, refugee doctors were being asked to sign a commitment never to apply to practise in Britain. When Ludwig applied to remain in Britain, he discovered this unexpected barrier.

At this point Professor Lindemann intervened. In a letter to the Home Office on 8 July, he outlined a research proposal whereby some of the new radioactive isotopes currently being produced at the Clarendon Laboratory might be used in biomedical research for both diagnostic and therapeutic procedures. The project would be developed from Dr Frank's wide experience in Germany on the effects of therapies using naturally recurring radiation in the waters of some European spas. Lindemann strongly endorsed the proposal and Ludwig's own qualifications and character, promising close cooperation with the Clarendon. The project would be facilitated, he noted, if Dr Frank were on the British medical register, and Lindemann asked that the standard prior commitment never to apply be waived in this case. On 20 July the Home Office did so, allowing Dr Frank to remain in Britain without signing the usual undertaking. Lindemann's research proposal was not only a support for Ludwig Frank's entry without medical restrictions, it also staked a claim for the Clarendon laboratory to take a leading role in developing isotopes for nuclear medicine.[14]

Following Mimi's advice, Ludwig did not return to Germany. Mimi and Stephan left Berlin in November, travelling by plane to Amsterdam, and then by boat from the Hook of Holland to Harwich. After a visit to the Simons in Belbroughton Road, they settled into a flat in Holland Park Avenue in London. In the meantime, Ludwig had registered at the University of Edinburgh, the only place that offered refugee physicians the possibility of completing British medical examinations in 1 year. After a year of hard study, mainly in London but with periodic trips to Edinburgh, Ludwig passed all sections on the first try.

On 3 November 1937, Lindemann intervened again, urging the Home Office to permit Dr Frank to set up a practice in London:

> London seems to be the only place where there is a prospect of his being able to earn enough to keep himself and his family and to obtain a suitable clientele so as to be able to carry out research on the new methods in which I am so much interested . . . .
>
> Dr. Frank is a man of considerable eminence in his profession and I am quite confident his presence in London would prove an asset to medical research in this country.

On 12 November, the Home Office granted this permission. In January 1938 Ludwig's new practice opened in London. Located in Devonshire Place, adjacent to Harley Street, it would grow into a distinguished consultancy in internal medicine. As previously in Germany, the new office suite included a consulting room, a laboratory with its own technician, and X-ray and electrocardiographic equipment brought from Berlin. Ludwig's reputation soon gained him many prominent patients, among them Lindemann himself. However, the radioactivity project failed to move further in the early months. In December 1937 Lotte wrote to Franz that Lindemann had not yet responded to Ludwig's invitation to see the new practice rooms. Lindemann did eventually go to see them, with Franz, on 1 February 1938, but, in a world falling into chaos and war, future research was losing ground to more urgent issues. By late 1937, with Ludwig qualified to practise, the Frank family moved to larger quarters in Langland Gardens in Hampstead. Thomas Frank left Oxford to join the family in London, and the Franks were united under one roof again for the first time since 1933.

After 1936, Lotte's parents were the only close relatives still in Germany. Her father saw no immediate reason to leave. Both parents had experienced a winter without central heating in Britain in 1933–4. His two older sisters, Pauline and Rosa, were still in Berlin. In March 1937 Lotte became worried when her father was reported to be "not entirely well." She wondered if she should not visit Berlin again. Franz, replying from Paris, was vigorously opposed, arguing that it was too risky and not strictly necessary. Since her parents were still capable of travelling to meet her outside Germany, it was unreasonable of them to put Lotte at such a risk.

Through 1938 the political situation worsened. In January, Lotte, on a ski holiday in Switzerland with the children, relayed home reports from the *Neue Zürcher Zeitung*: on 24 January—the blocked mark at 13 to the pound—she comments "What will that be later?"; on 27 January, increasing restrictions on Jews in the German economy, and for those emigrating, a lowered ceiling of 30,000 marks, or 50,000 for a couple —"Now will really come the great catastrophe that one has long been expecting." In March, after Germany annexed Austria, the flow of Jewish refugees became a flood. In May, the British authorities, unable to cope, reinstated visa requirements for holders of German or Austrian passports. Visas between several European countries had been in abeyance by diplomatic agreements since the 1920s.

The Münchhausens' visa application to Britain was lost in the flood. It remained lost until Lotte, after weeks of anxious waiting, heard of the delay and appealed for help to the Master of Balliol, A.D. Lindsay. With his intervention by telephone to the Foreign Office, the visa was agreed to at once and picked up by Lotte in London the same day. By this point exits from Germany had become more difficult. The regime permitted its border guards to seize all currency beyond travel requirements, which they did, leaving the elderly couple only enough cash for a fearful, frantic taxi ride from the Dutch frontier to their ship. They arrived in Oxford without money, in the autumn of 1938. The exact date of their arrival is not known, but evidence places it during the gloomy weeks that followed the Munich settlement. Writing to his colleague V.M. Goldschmidt in Norway of a situation that "now appears truly hopeless," Franz finds one small bright spot: "We have our close relatives all out, some of them only in the last few days. But still heaps of friends and acquaintances left behind." The reference is clearly to Lotte's parents.[15]

The letters of 1937 and 1938 have many references to emigration plans of other family members and friends. All too often, the story is one of inaction, indecision, or increasing administrative barriers. Lotte reports that Omama's niece, Grete Aberle, and her family are still making arrangements to leave. "He always thought of Palestine. Grete has no will of her own, but puts herself completely under him." Otto Müller, a nephew of Omama, in Antwerp telephones that the emigration to Britain of Carl-Heinz Lipmann, another nephew of Omama, will go ahead. Later on, his sister Agnes is coming to visit England. With siblings already there, she will probably be allowed to stay. Otto Müller, trying for the United States, requires a new legal basis in order to keep his position in the line. The Koppels, another Lipmann connection, want at some point to join their son Ernst in the United States. "I have always wondered," Lotte adds, "why they have never done so."

In a letter to Ebeth in Palestine on 3 August 1938, Franz reports hearing from "our cousin Elisabeth" and her music-teacher husband. They too are anxious to emigrate, but Franz is not optimistic:

> From the testimonials that she sends, her husband is a competent teacher and historian of music, and to judge from her letter he appears to be *sympatisch*. But what one can do for him is unclear.

This cousin is Elisabeth Mendelssohn, Anna Simon's niece, who married Erich Werner. By some means or other they reached New York, where he would later produce a landmark scholarly study of the life and music of Elisabeth's cousin, Felix Mendelssohn.

Many friends still in Germany were also on the move. Franz reports a talk with Nernst's daughter Angela and her husband, Albert Hahn, who have to leave because the husband is Jewish. The Eisners, Lotte writes, are trying for the United States. Presumably she means her cousin Paula Eisner (née Münchhausen) and her family. The Hirschbergs—not otherwise identified—have sent a telegram from Shanghai. "They have a terrible time behind them," she adds. This mention of Shanghai in the letters is a reminder of a door that remained open while most others were closing.

Many of Franz's academic colleagues from the first exodus in 1933 were at this point still seeking permanent posts. The grants from ICI and the Academic Assistance Council were intended to be temporary, for two or three years only. Some recipients got extensions, while others received new university posts financed in part by grant money. Of the Clarendon group, Heinz London moved to the University of Bristol in 1936; Fritz London to the Poincaré Institute in Paris in the same year, where he saw Franz and Kurti during their research visits to the Bellevue magnetic laboratory. Fritz London's Paris appointment did not work out well, owing to his diffidence in speaking French: "He envies me," Franz writes to Lotte, "that I can make myself understandable." By 1938 London had an offer from Duke University, which Franz thought he should accept, but here too London hesitated, at first accepting only a visiting term there, until the approaching war put the matter beyond personal choice in the summer of 1939.

A few of these academic moves were more noteworthy. A letter from Franz on 5 December 1937 reports that the Ruhemanns would be returning from Kharkov. A letter from Lotte dated 8 December reports that "Crowther was here today and has brought five copies of Ruhemann's book for storage." The matter struck close to the family, because Ruhemann's wife and Kurt Mendelssohn's wife were sisters, while a third sister had married the scientific journalist James Crowther. Lotte had immediate questions. Where could they live? With the Simons at Belbroughton Road? With Kurt Mendelssohn's family? Franz replied in a later letter that the housing question was not straightforward, since the Ruhemanns' marriage was also troubled.

Martin Ruhemann gives his own account of their Kharkov experience in his unpublished memoir "Half a Life." The Stalin purges created a climate of hostility to foreigners, and Ruhemann accepted his dismissal in the summer of 1937 with regret but without surprise. "So we started to detach ourselves from the Kharkov environment, which was facilitated by the fact that most of our friends and colleagues felt obliged to detach themselves from us as undesirable foreigners."[16] Though given notice in the summer, they had to wait until December for exit visas. When these came, they made a safe exit by train.

The Schrödinger case is more widely known, more complex, and it touched the Simons more closely. It was carefully researched by Schrödinger's biographer Walter Moore, whose sources included an interview with Lady Simon in 1985. Erwin felt underused and underappreciated in Oxford, and soon looked for a more prestigious position. In 1936 he was offered the important Tait Chair in Edinburgh, but refused it in order to accept a chair in Graz, coupled with an honorary chair in Vienna, in his native Austria. The supreme unwisdom of this choice became apparent in March 1938, when Austria's annexation to Germany left him in deep trouble with the new Nazi masters. Naively, he compounded it. Speaking out as "an old Austrian who loves his homeland," he composed a fawning letter to make peace with the Nazi regime. In this "repentant confession," published in a Graz newspaper on 30 March, he declared that he had "misjudged up to the last the true will and the true destiny of my country. I make this confession willingly and joyfully . . . . and I hope thereby to serve my homeland."[17] Once published, the letter circulated widely among his scientific colleagues.

Not surprisingly, his attempt failed, and on 26 August he was dismissed from Graz by the Austrian Ministry of Education on the grounds of political unreliability. By now aware of their peril, he and his wife slipped out of Austria on 14 September for Rome, taking only light luggage and abandoning all else. Once in Rome, Schrödinger prepared three letters: to Lindemann in Oxford, to Richard Bär in Zurich, and to Prime Minister de Valera of Ireland. To avoid Italian censors, they were written and posted from the Papal Academy in the Vatican. Eamon de Valera, then in Geneva as current president of the League of Nations, had heard in April of Schrödinger's difficulties in Austria and—like Lindemann in 1933—had perceived an opportunity to obtain a scientist of international repute for a planned Institute of Advanced Study in Dublin. The Schrödingers had been informed of de Valera's interest through tortuous oral channels long before they left Graz. In June, Anny Schrödinger signalled a provisional acceptance back to de Valera at a meeting with Richard Bär in Constanz, but Erwin seems to have wanted, somewhat naively, to test the waters in Oxford before visiting Dublin.

Schrödinger's letter to Oxford enraged Lindemann. He declined to help. It fell to Franz to send a quick reply on 20 September:

> Lieber Herr Schrödinger,
> Early today I had a letter from Bär which brought the news about you, and then your letter to Lindemann. He thinks it best to ask Gordon [President of Magdalen] to inform the Embassy in Rome so that you obtain a visa. The French will then pass you without anything further. I am writing this only hastily, since Lindemann apparently will not answer right away. He is now of course very busy, as you may imagine and – besides – frightfully depressed . . . .
> So hopefully we shall see you soon, all else by word of mouth.
>
> Mit herzlichen Grüssen

To understand what happened next, one should bear in mind that the Munich crisis was just entering its final phase.

On September 28—the eve of Chamberlain's flight to dishonour in Munich—a telegram arrived at 10 Belbroughton Road. Dispatched from Paddington at 5.43 p.m., it said "Leaving 605 for Oxford – Schroedingers." Whether Erwin and Anny may have arrived at the house before or after their telegram—which is marked as delivered at 8.14 p.m.—is not known. Moore mentions neither Franz's letter of 20 September nor the telegram of the 28th, and probably he had seen neither.[18] His account of the meeting is based on his interview with Lotte, a first-hand witness and participant:

> Then suddenly, about the first of October [sic], they appeared. They came to the Simons and it was an uncomfortable evening for everyone. Franz Simon told Erwin, 'Well, we have tried very hard to tell people here that this letter was written under duress. We don't know the conditions, we said, probably somebody with a gun stood behind him and said "you sign this".' First he said 'What letter?' and then he became quite excited and said, 'What I have written, I have written. Nobody forced me to do anything. This is supposed to be a land of freedom and what I do is nobody's concern.' The Simons told him with more emotion than logic that he had made the situation more difficult for all the other refugees. They said that if war broke out tomorrow, he would be in serious difficulties. Erwin said he was grateful that they had made him aware of the matter, but it concerned only himself. Anny supported Erwin strongly – whatever he did was right. The next day he saw Lindemann who told him the same thing as Simon.[19]

Following the first meeting and the rebuff from Lindemann, both Franz and Erwin described the encounter in separate letters to Max Born in Edinburgh, who managed to stay on friendly terms with both parties. On 7 October, Simon comments to Born that Schrödinger "does not have it in his heart to confess that what he has done is wrong . . . he behaves like a spoiled boy . . . . These happenings have disturbed us greatly." The Schrödingers remained in Oxford for about two months, staying with the Whiteheads around the corner at 22 Charlbury Road as "paying guests." But the doors in Oxford remained closed. On 20 October Franz comments further to Born: "I met Schrödinger once in college. He is especially mad at Lindemann who is not willing to see him again. Yet Lindemann is certainly the Englishman who has done the most for him."

From November onward, Schrödinger committed himself to de Valera's project. He visited Dublin for a few days to discuss details. But the project was delayed in the Irish Parliament when opposition parties attacked it as an unnecessary vanity for a poor country. To fill the interval awaiting its passage, Schrödinger accepted an offer to teach in Belgium at the University of Ghent for the balance of the 1938–9 academic year. Erwin and his family stayed in Belgium until September 1939, by which point the war had begun. As Austrians, they were enemy aliens in Britain or France, requiring a special arrangement to reach Ireland. Lindemann, requested to help, intervened once again, and a 24-hour transit visa enabled them to reach Dublin on 6 October. There, Erwin would pass six wartime years in relative tranquillity and unfettered academic activity, while the war raged around the Irish coastlines and in the skies overhead.

An eye on Germany

Franz also attempted to keep in touch with scientific developments in Germany. Among his papers is a letter of June 1934 from his former colleague Suhrmann in Breslau. In it he summarizes the year just past. He is glad to know of Franz's continuing activity from a colleague and from publications seen in *Nature*. Since Easter, Suhrmann has had to teach the students in the university as well as those in the Hochschule, together with a heavy administrative load. Apparently the amalgamation is still on hold. To solve the space problem, the new construction at the institute will "hopefully" begin soon. Herr Straus (the Jewish organic chemist also targeted by the Breslau students) "has been transferred (*versetzt*) to Berlin, and his successor will arrive soon."

In the Suhrmann household, everything is going well, the baby now 9 months old and beginning to crawl, their oldest has been at school since Easter, and the land around the house has ideal play areas for children. He sends two photographs of the house and garden, a conventional modern, stuccoed, semidetached house, with a square, barren-looking back garden overlooking a field with a few trees. The photograph was taken in winter, Suhrmann explains. The general tone is friendly: "In the hope that all is well also at your place ... with cordial greetings from our house to yours."

Suhrmann's letter to Franz shows that these two were not in regular contact. Franz's main continuing contacts were with Walther Nernst and Max von Laue. Both were anti-Nazis, but their situation was different. After 1933 Nernst had only emeritus status in Berlin and spent most of his time at Zibelle, his estate in the countryside southeast of Berlin, where he hunted almost daily. He stayed in regular contact with Lindemann, in a series of letters extending from 1913 to 1939, including during the First World War. After 1933 some letters include greetings to the Simons, and others mention letters sent to Franz directly. The letters to Lindemann portray the life of an isolated academic, opposed to the regime but too prominent to be attacked directly.

By 1935 two of Nernst's grandsons, Charles and Peter Cahn, were at St Edward's School in Oxford and were frequent guests at the Simon household for Sunday lunch. In a letter to Lindemann of 26 January 1936, Nernst expressed his pleasure at seeing "the effects which their study in England had on our grandchildren."[20] The high point of Nernst's connection with Oxford came in June 1937, when he received an honorary doctorate from the university. Marking the occasion, several pictures in the Simon family albums show three Nernst generations and other guests in the garden at Belbroughton Road.

The visit had been long in preparation. In the autumn of 1935, Lindemann had invited Nernst to Oxford to give a lecture, but Nernst replied cautiously that it "takes time" and requires ministerial consent.[21]

Something was then planned for the spring of 1936. Among the Simon papers in the house are some printed invitation cards for a rather elaborate event:

> Mr. & Mrs. F. Simon
> At Home
> in Balliol College Hall
> on Tuesday, May 12, 1936, 4–6 P.M.
> The Vice-Chancellor will be present
> Academic Dress need not be worn
> To meet Professor & Mrs. Nernst   R.S.V.P.
>   10 Belbroughton Road
>   Oxford

Since the Master of Balliol, A.D. Lindsay, was also Vice-Chancellor of the university at the time, the proper protocols had to be specific on the invitation.

There seems no evidence that the Nernsts were in Oxford for this occasion, or even that the event took place as planned. Kurt Mendelssohn's biography of Nernst mentions only one visit to Oxford, that of 1937, and Nernst's grandson, Charles Cahn, confirmed this by interview. But news of the event spread afar. Two days *before* its scheduled date, Ebeth wrote to Lotte from Jerusalem:

> We had a great laugh about your description of your At Homes . . . . But 130 people is a fair number. We have also attended one that large here, but we are not yet – thank God – in a category to have to hold one ourselves.

Even if the Nernsts were absent, the Simons had something of their own to celebrate, for Franz was about to succeed Egerton as Reader in Thermodynamics. The Nernst visit and honorary degree would take place, with German ministerial consent, a year later.[22]

The other continuing contact with German science was through Max von Laue, who remained a central figure in Berlin physics from 1919 to 1945. Laue was a consistent anti-Nazi who disguised his feelings only to the extent absolutely necessary. His high professional and social position gave him a degree of freedom denied to others. In several roles, he fought with partial success to preserve the institutions of academic science from complete subordination to the Nazi party. As a frequent traveller, von Laue met Franz on occasional visits to Oxford or Paris. The photograph album contains a record of at least one visit to Belbroughton Road, probably in late May 1934, when von Laue stayed with the Schrödingers just around the corner in Northmoor Road.

Laue was also a trusted correspondent, from whom Franz received information on developments in Germany and internationally. On 21 December 1937, Franz reports to Lotte from Paris:

> Letter from Laue. In the *Reichsanstalt* something must be done. Seven people have given notice, and 'from the closing of the Fifty-Year Festival one hears remarkable (*merkwürdige*) things, so remarkable that one doesn't believe one's ears.'

Ridicule was a favourite instrument for Laue, and Lotte recalled in her BBC interview his practice of forwarding to Oxford copies of *Das Schwarze Korps*, the weekly newspaper of the SS, when that organ printed absurdities about science.

The *Reichsanstalt*, or National Physico-Technical Institute, had been headed with distinction by Nernst until his retirement in 1933. Thereafter it was run by Johannes Stark, an ardent Nazi and critic of Einstein, whose goal was to develop a purified "German" physics cleansed of all Jewish influences, including the theory of relativity. In a *Schwarze Korps* article of 25 July 1937, entitled "White Jews in Science," Stark denounced Werner Heisenberg as "just one example among many" of those who "collectively are representatives of Jewishness in German intellectual life. They must disappear, just like the Jews themselves." Heisenberg, stung by the attack, appealed for protection to Heinrich Himmler as head of the SS, with results that will be examined more fully in Chapter 6.

The darkening political scene

The letters between Franz and Lotte in the later 1930s show an increasing emphasis on the political. Those of 1934 and 1935 contain almost nothing on politics. On 7 December 1936, Franz notes from Paris that Prime Minister Baldwin cannot remain much longer, and "then Churchill might succeed him, which for us for various reasons can only be welcome." He was doubtless aware of Lindemann's personal friendship with Churchill, a connection going back to 1921.

On 10 December, Franz mentions the abdication crisis, which had been studiously avoided to that point in the British press: "Are the English people very excited about the matter, or only the newspapers?" Lotte, 2 weeks later, refers to a *Times* leader on "Germany's Choice" (a role in Europe vs war-like autarchy) and a social note that Mrs Simpson is on very friendly terms with German Ambassador von Ribbentrop.

In 1937 the political comments originate chiefly from Lotte. In April she mentions twice the Franco campaign against Bilbao. The Spanish Civil War is at a critical stage. She also thinks that the Schuschnigg–Mussolini talks do not bode well for the future of Austria, as events would soon prove. On 29 November, she reports the arrival by post of issues of *Das Schwarze Korps* from an anonymous sender, containing a "marvellous" piece called "The Other Side of the Medal."

Three days later, she reports at length to Franz on a lecture she has attended by a certain Paton, entitled "Fashions in Philosophy." Its theme was the "readjustments" in philosophy following its separation from the natural sciences. Philosophy must give up its position "behind a wall of metaphysics" far away from the real world, "otherwise catastrophe is unavoidable." A special lack of interest in political philosophy has led to a dangerous influence had of "pseudophilosophers," such as Gobineau, Houston Stewart Chamberlain, or Hitler. "In spite of differences in ideology, the Russian situation is identical with the German one . . . . I must tell you more of this in detail. Finally someone has said sharp and clear what you have always tried to explain to people." Two days later, he thanks Lotte for the report on Paton but does not comment further. Herbert James Paton, a Kantian scholar, was the newly appointed White's Professor of Moral Philosophy at the university.

In January 1938, the family went on a ski-ing holiday in Switzerland. Lotte and the children stayed on after Franz's return to Oxford. In spite of the international situation, this holiday at Montana in the Valais was a good one, with daily reports to Franz of "heavenly weather" or "beautiful sunset" and long talks with "Onkel Carl," Opapa's brother, who joined them for several days. These few idyllic days stand out sharply against her daily reports of difficulties for friends and relatives still in Germany. Beneath her felicity lies a feeling that this may be the last such experience for some time.

Franz, from Oxford, was also watching Germany. On 29 January he writes:

> In Germany something unusual appears to be going on. First the Streicher affair, then the sudden countermand (*Absage*) of the Reichstag. Here there are rumors of another June 30, but who murders whom is still not clear. I shall hear news today.

On February 3 he notes further rumblings in the press about internal disunity. Franz watched closely for cracks in the Nazi regime, but some rumours turned out to be unfounded. Something, indeed, was up in Germany. On 4 February an administrative shake-up put the army and the foreign service more firmly under the control of the Nazi party. Five weeks later, on 12 March, an emboldened leadership invaded and annexed Austria, without resistance.

Late in March, Franz returned with Kurti to Paris, and daily letters were resumed. The work at Bellevue was intense, but abbreviated to about 10 days. The research connection appeared to be winding down. Professor Cotton had been inactive, and the other senior French colleague, Lainé, who coauthored six articles with them on the Bellevue experiments, was about to leave for a provincial university. Franz was extremely tired from overnight experiments, and "fed up." Though Cotton invited them to come again, one can sense a loss of momentum in the Paris connection. Besides, there was a prospect of better facilities in Oxford at the new Clarendon Laboratory, scheduled for completion in 1939.

On the political front, the deterioration continued through 1938. The German leadership, flush with success from the Austrian annexation, concentrated its attention on Czechoslovakia, building an escalating crisis, backed by military mobilization and Sudeten extremist incidents, to its ugly climax in September. Lotte wrote from Oxford on 31 March of widespread "discontent over Chamberlain's limp stance (*schlappe Haltung*)." In reply, Franz's view of France was just as pessimistic:

> Politically, it looks very bad here. Many think that fascism may break out here this year. Blum is much too weak. Further, people on the right continue to act very unacceptably. For them, outright fascism at any price is preferable to socialism.

Even before the Munich crisis, both Lotte and Franz saw serious political weakness in both countries.

In Munich, in September, the leaders of four great powers—Britain, France, Germany, Italy—reached agreement on the dismemberment of Czechoslovakia, which was unrepresented at the conference. The territories in dispute were divided among Germany, Poland, and Hungary, but principally Germany. On 30 September, Prime Minister Chamberlain flew home to a thunderous welcome from the London crowds gathered as though in celebration of a great victory. His brief announcement in front of 10 Downing Street concluded: "I believe it is 'peace for our time.' Go home and get a nice quiet sleep." It seems typical of democratic leaders to believe the best of dictators long after these have repeatedly and irrefutably shown themselves to be monsters.

The agreement did, indeed, defuse the immediate crisis and gain a little time. For Czechoslovakia, stripped of its military fortifications in Sudetenland, the respite was less than six months, until Bohemia and Moravia were overrun in March 1939. France and Britain gained 11 months, to 1 September 1939. In the interval, Hitler, celebrating a huge double victory, stepped up his campaign against Poland.

The celebrations after Munich did not impress the refugees. They suffered from the curse of Cassandra, knowing the truth about Germany but seeing their warnings go unheeded. Deeply depressed, Franz wrote to Max Born in Edinburgh a letter reproduced in part by Nancy Arms:

> I have been intending to write to you for several days, but I have not been able to work myself up to it; everything is so hopeless. It is still worse for us since we know the Nazis and have foreseen everything so clearly, and since it could all have been so easily avoided. I believe now that England will sink rapidly or that, when more people open their eyes, a war will still come which will be much worse than that now avoided and in which the chances of winning are slim.
>
> The number of people who sympathize with Churchill is undoubtedly increasing and especially here in Oxford where one finds few people who approve of the present government.[23]

There were no absences of Franz or Lotte in this post-Munich period, and no letters passed between them. Yet one file retrieved from a remote storage area of the house is both eloquent and enigmatic—a plain legal-size folder containing complete copies of *The Times* for 29 September 1938 and *The Times Weekly Edition* for 6 October. The issue of 29 September is folded into two parts, to leave it open at the diplomatic documents of the crisis (pages 17–18) and at Chamberlain's speech to the House of Commons on 28 September, just prior to his departure for Munich (pages 6–7). The weekly issue on 6 October has broader coverage, with pictures, of the "anxious days" of the month-long crisis: queueing for gas masks; digging trenches in the parks; removing stained-glass windows; homecoming celebrations for all four leaders returning to their capitals; thanksgiving ceremonies in the churches on 2 October; and Hitler's triumphant entry into Sudetenland on 3 October. Both papers are in pristine condition.

The enigmatic element is the file folder itself. On the cover, someone—neither Lotte nor Franz—has written "München 1938" in a fine, clear, careful hand. Is it Opapa's? Probably not. Too many minor variations, and he would probably not have read sources in English. Others passed through the house in this turbulent year, but the file yields no other mark or clue.

Six weeks after Munich, the German state unleashed its most thuggish elements into a nationwide orgy of violence against its remaining Jewish shops and businesses—the so-called *Kristallnacht* of November 9/10. On the following day, 11 November, Franz wrote a furious 10-word letter to the German Physical Society: "Herewith I announce my resignation from the *Deutsche Physikalische Gesellschaft*." His post-Munich depression turned to unrelenting anger.

By resigning from the German Physical Society in November, he sidestepped an oncoming scandal. The president of the society, Peter Debye, sent a terse letter to the membership dated 9 December 1938, declaring that "the membership of German Reich Jews in the sense of the Nuremburg Laws in the German Physical Society can no longer be upheld. In agreement with the Board of Trustees I therefore summon all members who fall within this provision to inform me of their withdrawal from the Society."[24] Foreign members, including refugee scientists, were not directly affected by the order, but many resigned in protest and in solidarity with those expelled. This abject submission to the regime cost the society about 10 per cent of its 1938 members.

After Munich, the flow of refugees from Germany, Austria, and now Czechoslovakia became a flood. The remedy for personal depression was to help ease the desperate situation of those around them in any way possible. One case close to the family is that of Claire Ash, a friend of Lotte's of long standing. Dor and I had been privileged to hear the essentials from Claire herself. Escaping from Germany, she had headed in 1939 for South America, but permission to land was denied her at several ports. The ship then headed back towards Germany, stopping briefly in Southampton. On hearing of this, Franz and Lotte drove to Southampton, where Franz was able to arrange for Claire's release by posting a bond for her as a visitor. "Quite literally," she said, " they saved my life."

In the Simon Papers two letters supplement Claire's account. Franz and Lotte, newly naturalized themselves, sought help from a British colleague in Southampton to act as a witness for them when they met the ship. This back-up plan failed because this colleague was out of town at the critical time. A second letter then followed to tell him of the outcome. Claire had sailed for South America on the *Cap Arcona*, the largest, fastest, and most luxurious vessel of the Hamburg–South America Line. She had obtained a visa from the Paraguayan consul in Paris. On arrival in South America, she and some 25 other *émigrés*, victimized in the same way, were not permitted to land. When the ship arrived in Southampton on her return voyage to Germany on 27 March 1939, Lotte and Franz were at the pier with the necessary guaranty:

> We did not succeed in getting our friend to England but we managed that she was allowed to land in France. In meantime we got the permit for her and now we are expecting her in the near future. The whole story was very exciting and too long to tell in a letter.[25]

In due course, Claire disembarked in France, crossed the Channel and stayed on in the Simon house until she could arrange to join a brother in the United States.

In the writing desk was a small cache of recent letters from Claire. Lotte kept them and clearly valued them. Claire followed a regular pattern, writing sometimes at Christmas, always before Lotte's birthday in June. The latest of the series was dated 7 June 1991. In it Claire congratulates Lotte on her ninety-fourth birthday, and hopes for her good health. "Who would have thought that we would still be congratulating ourselves on this day when we were getting acquainted 89 years ago at Fräulein Hanke's English lessons!"

A farewell to peace

Within the maelstrom, scientific life continued. On 1 January 1939, Franz was sent a handwritten invitation to a conference in Strasbourg. The Institute of Physics of the university was planning an international *Réunion d'étude*, with invited participants from several countries, to review recent research on magnetism, including sessions on paramagnetism and on the effects of low temperatures, to be held at the university in late May. It would be subsidized by the International Institute of Intellectual Cooperation of the League of Nations. Franz accepted, proposing to send a paper by mid-March. The work at Bellevue would be a central feature of the session on low temperatures.

When the deadline arrived, however, the refugee crisis was still in full flood. He had to write to the conference director to ask for more time:

> Dear Professor Weiss,
> I am very sorry not to be able to keep my promise to send my report by the 15th of March. This is due to the fact that my whole free time is taken in trying to help the refugees who are pouring in to England in ever increasing numbers. We have now about 250 in Oxford! There is so much urgent and important work to be done for them, especially for the children who have come without their parents, that you will understand that I had to postpone all other things.
> I shall try, however, to send the report as soon as possible although I cannot promise a date.

When Weiss enquired again on 23 April, Franz reported that writing was in progress and delivery would be "within 15 days at the latest." In the end the paper was delivered, translated into French, printed, and distributed just in time.

The conference was primarily a meeting for experts. Of the 50 or so participants mentioned in the press coverage, about half were from France, nine from the Netherlands, five from Germany, four from Britain, two from the United States, and a few from other countries. The low-temperature session was held on the third day, with papers by Simon from Oxford and Casimir from Leiden. Despite its troubled birth, Franz's paper was a significant contribution, and even the popular press picked up his suggestion that the new magnetic methods could improve efficiency of cooling procedures by "up to 1000 times."[26]

After four days, the conference ended on May 25 with a six-course banquet, complete with wines and coffee, and featuring a menu that could have been devised only in France. With the dessert came speeches, toasts, and rounds of thanks. The Rector of the university then closed the conference, raising his glass in a final toast "to the honour, and preservation, of free scientific thought" as it had just been exemplified at the conference itself. It was a noble thought, grotesquely at odds with a world collapsing around him. Franz put all of the Strasbourg conference correspondence into a separate file folder and kept it at home. Like the newspapers from the Munich crisis, the file stands out today as another memento of a world living in illusion after war had become inevitable.

5

ANY CAPABLE PHYSICIST 1939–1941

> ... The lines on which we are now working are such as would be likely to suggest themselves to any capable physicist.
>
> Maud Report, Part I, text in Gowing, 1964, 395

The war that had become inevitable began on 1 September 1939, when Germany invaded Poland. On 3 September, Britain and France declared war on Germany. At 9 p.m. that evening, the *Athenia*, a passenger liner heading out from Glasgow for Montreal, was torpedoed and sunk by a German submarine northwest of Ireland with a loss of 112 lives. For the second time, Franz and others of his generation were entering a major war. This time, however, there were two major differences from 1914. In a war that depended heavily on science and technology, Franz was in a position of knowledge and power at the peak of his career. Second, he now had major family responsibilities that would be tested in new ways under wartime conditions. As in 1933, the safety of the near family would be paramount.

On the domestic front, the first consequence of the declaration of war was a total night blackout in all the towns. Yards and yards of black fabric were purchased to make blackout curtains for every window in the house. Each window was also taped with brown paper tape to reduce the danger from shattered glass. Householders in Oxford built air raid shelters, usually in their gardens. Because of Lotte's aged parents, the one for the Simon household was constructed of brick and concrete above ground, in the angle between the living room and the original dining room, directly adjoining the house. The children carried their gas masks to school. The atmosphere was defensive, preparing against attack from the air and from submarines. When Winston Churchill visited French military defences in August 1939, he noted the impregnability of the Maginot Line along the entire Franco-German frontier and the defensive mentality that accompanied it, but he also noted France's vulnerability to attack at either end, through Switzerland or Belgium.

A German "secret weapon"?

On the scientific front, interesting developments were unfolding. Those relating to air defence had been in progress for several years, but others were new, and even speculative. In the Cherwell Papers, there is a document that seemingly stands alone. It is typed, unsigned, though someone—probably an archivist—has labelled it "Atomic Energy" and added a tentative date: 1939 superimposed over 1938. In four short paragraphs, the memorandum attacks a report in "one of the Sunday papers" about an "immense amount of energy which might be released from uranium by the recently discovered chain processes" and the consequent possibility of "some terrible, new, secret explosive capable of wiping out London."

The document then states the "true position": that "only a minor constituent of uranium is effective," and its extraction "will be a matter of many years"; that any resulting explosion would be only a "mild detonation" similar to existing explosives; that such experiments would have to be done on a large scale and therefore could not be kept secret; and that Berlin in any case "controls only a comparatively small amount of uranium." The Nazi threat of a secret weapon, therefore,

> is clearly without foundation. Dark hints will no doubt be dropped and terrifying whispers will be assiduously circulated, but it is to be hoped that nobody will be taken in by them.

Much, much later, the key to the mystery turned up in an unexpected source: a similar document is printed in full in Churchill's history of the war. Closer comparison showed the two documents to be identical, except for one sentence added and another rearranged in the Churchill version.[1] Churchill explains that "after a conversation" with Lindemann, he sent this letter to the Air Minister to stiffen cabinet resolve in backing Poland. Churchill's text is dated 5 August 1939. The memorandum in the Cherwell Papers is undoubtedly a draft prepared by Lindemann for Churchill. The minor changes are literary rather than scientific, apparently Churchill's own revisions. While the document had a political purpose, it doubtless represents Lindemann's own views and those of other British scientists at that point.

In September, seven weeks after Churchill's letter to the ministry, Franz received a letter from a refugee scientists' organization which had been asked by the British government to enquire into "what is known in German circles about the weapon threatened by Hitler." His reply summarizes in layman's terms developments in nuclear physics since the famed experiments of Hahn and Strassmann with uranium in Berlin late in 1938: "In all probability Hitler had hinted at the uranium bomb." A large enough lump of uranium, Franz explains, could be made to explode, but for this purpose

> only the rare isotope that constitutes 1 per cent comes into the question. I believe myself that, even with the use of very large equipment, nothing could be ready under one or two years.
>
> However it is not at all certain whether the thing would go off.... If everything worked well, then such a bomb, or a few, could decide the war.
>
> The Germans were working on this; much has been done in America, France, and here as well. Thus the thing is not limited to the Germans. I might mention once more that every physicist who follows modern nuclear physics can give this information and that the Germans cannot know more about it than others.[2]

In most respects, Franz's letter is similar to the Lindemann memorandum, but his shorter timescale and its possible relevance to the war are striking differences.

British initial reactions to the uranium question proceeded slowly, cautiously, reactively. When Joliot-Curie and his Paris colleagues raised the possibility of a chain reaction in *Nature* in April 1939, several scientists were moved to action. Professor George Thomson of Imperial College in London immediately raised the question of uranium supply, which was referred to Sir Henry Tizard's Committee on the Scientific Survey of Air Defence. The main supplies of uranium in Europe were owned by the Union Minière in Belgium, which had stocks of uranium residues and some refined oxide as a byproduct of its radium mining in the Congo. Tizard, initially favourable to a purchase, decided after consultations and a meeting with the Union Minière president that expenditures to buy available uranium stocks could not be justified. He rated the odds against any successful military application at 100,000:1. The British Admiralty declared that if "foreign nations"[3] were not buying uranium, there was no need for Britain to do so. In May 1939 Professor Arthur Tyndall of the University of Bristol wrote a cogent brief for the Chemical Defence Committee on the possibilities of an atomic bomb, but this committee and other government bodies were advised to leave the subject to Tizard's committee so as to avoid duplication.

Despite almost universal scepticism, some research was initiated at this point. George Thomson in London and Mark Oliphant in Birmingham began experiments on uranium in their respective laboratories, but this work was slowed when both became involved in war projects. When the matter was raised in the War Cabinet, a new enquiry went out through the Department of Scientific and Industrial Research (DSIR) to Professor James Chadwick, Britain's best known nuclear physicist, of the University of Liverpool. Chadwick's reply came in two stages. In the first, in October 1939, he remained sceptical about a workable bomb but promised to look further. By December, he had read the recent literature and modified his position: "I think one can say that this [fission] explosion is almost certain to occur if one had enough uranium. The estimates of the amount necessary vary from about 1 ton to 30 or 40 according to the data adopted in the calculations."[4] He stressed, however, the lack of any experimental data and his inability to answer the central question of a practicable application.

Lord Hankey, a non-scientist in charge of this file for the War Cabinet, found this scientific vacuum strangely comforting: "I gather," he wrote to another minister, "that we may sleep fairly comfortably in our beds."[5] His comment reminds us of a young American who at this moment was completing an honours thesis at Harvard. It would be published, famously, in 1940, under the title *Why England Slept*, by John F. Kennedy.

After several months of experiments with uranium oxide, George Thomson was ready to abandon the project as unworkable by February 1940. His team had tried both fast neutrons and slow neutrons with either paraffin or water as moderators, but none of their efforts achieved a chain reaction. The alternative of using heavy water as a moderator, as in the French experiments, was not available in Britain. A report on these experiments was prepared and submitted to the Air Ministry in April 1940.

Before this report was received, however, the situation changed dramatically: in a three-page self-typed memorandum submitted to Mark Oliphant and passed by him to Tizard and on to Thomson, two theoretical physicists in Birmingham, both refugees ousted from Hamburg University in 1933 by the anti-Jewish legislation, made a few calculations that would change the world. Sometime in February or March of 1940, as Peierls remembered it, Otto Frisch and Rudolf Peierls approached the uranium problem from a different angle: what would happen if the experimental material were pure uranium 235, and what critical mass of this isotope would be needed for a chain reaction? Their mathematical calculations argued persuasively that the critical mass for the pure isotope would be measured not in tons but in far smaller units: "one might think in terms of about 1 kilogram as a suitable size for the bomb."[6]

When the Frisch–Peierls memorandum reached Thomson, action began in several directions. Under Tizard's Air Ministry Committee, a subcommittee on uranium was formed, which met twice in April 1940. Enquiries were sent to Washington about ongoing work in the United States, and to intelligence about activities in Germany. The Ministry of Economic Warfare was asked to secure the uranium oxide stocks in Belgium. They moved too slowly, however, and most of it was seized by the German army when it invaded Belgium on 10 May. Early in April, a visit from a French intelligence officer revealed earlier German attempts to buy Norway's entire stock of heavy water—the sole European supply—and to order 2,000 litres more from future production. The French government forestalled this by arranging removal of the existing stock to Paris before the Germans invaded Norway on 9 April.

Franz's involvement in uranium research began modestly. His own letter of September 1939 on Hitler's threatened weapon may have led him to think about the issue in experimental terms. In the same period he was also developing another idea with possible wartime applications closer to his own field: to extend the flying range of aircraft by using liquid hydrogen as fuel. The first experiments concerning uranium had very modest beginnings. In the spring of 1940 Franz appeared at the laboratory with an ordinary kitchen sieve from the Simon household. He had hammered the wire mesh to reduce the holes to pinpoint size, and he used these to see under what conditions a mixture of carbon dioxide and water vapour molecules could be separated by repeated passages through the holes. Without official support, these experiments continued through the summer.

For security reasons, the British government was reluctant to employ any refugees, whether aliens or those recently naturalized, on wartime research. The uranium subcommittee, which soon became camouflaged as the picturesquely named Maud Committee,[7] was at first composed only of British-born scientists. When it met to consider the Frisch–Peierls paper, the two authors were not allowed to attend, or to explain what they had written. When Peierls questioned the wisdom of this, Thomson agreed that they could be consulted further, but not added to the committee. On being consulted, Peierls urged that the leading experimentalist to explore isotope separation would be Franz Simon, whom he had known in Berlin in 1929, in Odessa in 1930, and in England regularly since 1933. The question of Simon's involvement was raised at the June meeting of the Maud Committee, and again at the July meeting, but on each occasion misgivings were expressed and no action was taken.

An insight into Franz's state of mind during the summer of 1940 may be found in letters he wrote to Max Born. The first and longest of these letters, dated 4 June, gives his view as an individual scientist, an outsider, on the uranium issue:

> It will interest you to hear – if you have not heard already – that the uranium bomb seems to become a practical proposition. It has been proved experimentally in America that U^{235} is responsible for the fission phenomena. Frisch and Peierls have calculated that a mass of a few kilograms of U^{235} would blow up and that the times would be quick enough to ensure a real "detonation". Of course one cannot predict this with absolute certainty, but they believe that the probability is about 50%. Oliphant is interested in it and lends it his support. The chief point is to separate the isotopes (U_{235} is only present in normal U to about 1%). This is possible with very much money, the order of 1 million £! But it would be worth while as it could decide the war....
>
> The point now is to separate as quickly as possible a few milligrams and to develop the methods for separating kilograms. I have tried hard since the beginning of war to persuade Lindemann to let me go on with this matter on a biggish scale in the Clarendon, as I always believed that something should be prepared for the U bomb in case it should work and as moreover the separation of isotopes will prove valuable for many other purposes also. I had no success, however, as L. concentrated all efforts on the Admiralty work now carried out in our Lab. I shall now try again after the U bomb seems so much nearer realisation and after having the support of the Birmingham people.[8]

In five remaining paragraphs he discusses the difficulties of finding public support; the "lack of vision . . . of the class ruling this country"; a "growing animosity against the refugees which is worked up by the cheap newspapers"; the impact of the war on the family and children; and the forced cessation of "our experimental work" when the outbreak of war interrupted the move from the Old to the New Clarendon. He feels frustrated and underutilized. "From time to time I have a little work for the Ministry of Aircraft Production – as it is called now; but we all long to use our whole force in this struggle for this country! As it looks now, however, this seems to become increasingly more difficult and improbable." This letter is scientifically extraordinary, but Franz's post-Munich pessimism still overhangs it like a dark cloud.

The prospect of German invasion

On the war front, the face of Western Europe was being altered beyond recognition. On 10 May 1940, the *Sitzkrieg* became a *Blitzkrieg*. The Netherlands capitulated in four days, Belgium in 18 days. The collapse of the Low Countries endangered a British expeditionary force of some 250,000, the bulk of whom were successfully evacuated, though with serious losses of men and equipment, from the beaches of Dunkirk during a few days of calm weather in early June. German armoured units then struck southward rapidly on 5 June, forcing the French Marshal Pétain to ask for an armistice on 16 June. Germany's lightning war had overrun and occupied Denmark, Norway, the Netherlands, Belgium, and most of France in just 10 weeks. With these operations completed and the British army in severe disarray, Germany began to plan for Operation Sea Lion, the invasion of England.

The prospect of imminent German invasion spelled extra dangers for Jewish refugees. If examples were needed, Austria, Czechoslovakia, and Norway had provided ample evidence. For Franz and Lotte, the safety of their daughters, aged 14 and 12, again became their supreme concern, as it had been in 1933. Two options opened up. After France fell, Oxford friends came to the house offering to adopt the two children and send them to a boarding school under a different identity. Then a second possibility appeared: they could join a group of University of Oxford faculty children being sent to Canada through contacts with faculty at the University of Toronto. The source for both options was Kenneth Bell, an Imperial historian and senior Fellow of Balliol who had taught in Toronto and had retained ties with Toronto colleagues. The Oxford–Toronto group, some 15 children including Bell's own son and daughter, was one of several privately organized child evacuation arrangements between British and North American universities.

On very short notice, the Oxford–Toronto group of children sailed from Liverpool late in June, unaccompanied by any adult. After they had gone, Franz wrote an anguished letter on 27 June to three friends and colleagues in the United States, Percy Bridgman and John Van Vleck at Harvard and Otto Stern at Pittsburgh, asking them if they would be willing to help the children should this be needed. The joint letter first explains the essentials of the Oxford–Toronto scheme and then looks beyond:

> Dear Bridgman, Stern and Van Vleck,
> I am writing in a great hurry as we have not much time for thinking now, and, moreover, we do not know how long the mails will go regularly . . . . As the situation is, it seems rather certain that we shall be separated from the children for a considerable period and of course one has to keep in mind the possibility that

we do not survive the things to come, one way or another. We think it therefore necessary to entrust some friends with some kind of responsibility as we are not now in a position to advise and help our children. May I ask you three to take this responsibility together or singly, as you think it right, or to appoint another person whom you think suitable. I know we are asking a lot – we ourselves have acted in this capacity in the last years in several cases. If you believe you cannot accept the responsibility, please accept it temporarily at least until we can make other arrangements.

I think you will realize without many words what it means for us to send our children into an uncertain future. It would be a very great relief for us to know that friends, whose general outlook to life we know to be very similar to ours, would take a personal interest in their welfare.

Yours, F.E. Simon.[9]

At the end, a postscript lists seven other friends who might assist in an emergency: "Lawrence and Giauque in Berkeley, Urey and Michaelis in New York, Swain in Stanford, Brickwedde in Washington, London in Durham." Reading this text several decades later, one can sense the parallel in Franz's mind with the Kindertransports of 1939, the sad trains from Prague or Vienna that brought several thousand unaccompanied refugee children to London after tearful partings from parents whose chances of survival were remote.

Nancy Arms's transcription of this letter omits one paragraph that can be discussed more openly today. Franz writes in part:

> There is one particular point which is worrying us. We have heard that there is rather widespread Antisemitism in Toronto. Thus there is the possibility that the children might come to a family which is not too happy to get "such" children or the other possibility that they might resent their "German" origin.[10]

Should any such mismatch or other difficulty arise that they could not resolve in Toronto, "I told them to get into touch with one of you." It is interesting that Franz relies on his physicist colleagues ahead of his own relatives at a time of need. Toronto at this date had extensive segregation between Jews and non-Jews in hotels, restaurants, beaches, and private clubs, but its wider reputation for antisemitism seems to have stemmed from an infamous six-hour riot at the Christie Pits in August 1933, which erupted during a softball series between a Jewish team and one known as the Swastikas and their respective supporters.

The anxiety eased a little a few days later, when Lotte agreed to join the children in Toronto. Although she was extremely reluctant to leave Oxford, Franz insisted strongly, and in the end prevailed. Her permit to leave arrived on the day that the children landed in Canada, 2 July 1940. On 8 July Franz wrote to Born again to tell him of these departures. By this date the children had arrived in Toronto and Lotte was at sea on the Polish liner *Batory*. The new and acute problem in Oxford was internment of male refugees and the dire situation of their dependants left without support: "On account of the external and especially of this internal situation we have decided to send our children away . . . . I think we have done the right thing."[11]

Born replies two days later: "I think it most reasonable that you have sent your children away and that your wife has followed them. The new regulation that we had to report to the police has shaken me rather much, and if I had an opportunity to send my family away I would do it."[12] The rest of the letter has its quota of gloom: several relatives interned; his students reduced from eight to three, with two interned, two removed from the Edinburgh "restricted area" to Birmingham, and one returned to India.

Franz had a special reason to be apprehensive for the safety of his family. After the war, German records confirmed that both he and Kurti were on the Nazi Black List for Great Britain, and targeted for retribution.[13] The most likely explanation for this listing comes from R.V. Jones, a Clarendon Laboratory colleague with a dangerous penchant for practical jokes. When he delivered the Cherwell–Simon lecture for 1965, Jones briefly recalled a bizarre incident of November 1935:

> Let me also acknowledge my debt to Sir Francis Simon.... His name was among those of British citizens to be rounded up by the Nazis if "Operation Sea-Lion" had succeeded; and I must now confess that this may have been due not only to his own eminence but also to a hoax that I extemporized on an agent of the German Secret Service whom I found wandering around the old Clarendon Laboratory in 1935, and whom I persuaded (in an effort to distract his attention from some air defence equipment) that there was a great anti-Nazi organization in Britain with Simon at its head. I hope that Simon forgave me – especially since I stand indebted to him for support at important stages in my own career.[14]

In a more elaborate version of the same incident in his own later memoir *Most Secret War*, Jones adds corroborative details: that his Oxford visitor, Hans Thost, was a correspondent for the Nazi daily *Völkische Beobachter* and was expelled from Britain 3 weeks later; that in 1937 a German emigré newspaper uncovered the story of Thost's report to Germany on a "great anti-Nazi organization" headed by Simon in Oxford; that Simon and Kurti were baffled to encounter this hoax during their 1937 work at Bellevue in Paris; and finally that Thost himself had published in 1939 an account of his five years in England up to 1935 as a Nazi "reporter." On closer study, however, his expulsion in November 1935 is traceable to non-journalistic activities of a somewhat different kind. In addition to propaganda and organizational work for the Nazi party and regime, these included surveillance of German citizens and Jewish refugees in Britain, with adverse reports sent back to Germany through Alfred Rosenberg, his editor and inner-circle Nazi boss.[15]

When the German offensive rolled over the Low Countries and France in May 1940, the situation of refugees in Britain worsened sharply. Technically, all German or Austrian citizens, including refugees, had become enemy aliens at the declaration of war, but for several months of the "twilight war" most of them were subject only to police checks. Churchill, who became Prime Minister on the first day of the German attack, ordered a rigorous policy of general internment. All males, from schoolboys to those in their sixties, whether Jewish refugees or other German nationals, were herded into hastily improvised camps. In the universities, the sweep removed lecturers and scholars from their college rooms, and scientists from their laboratories. Only a few women were interned, but thousands were left destitute by the removal of their husbands and sons. By early July, the round-up was complete.

In the general panic of May and June, public support for internment reached a peak. An undated clipping from the local paper reports on a debate in the House of Lords on the alien situation in Oxford. The town police were concerned over 204 male and 277 female enemy aliens living in Oxford, and were suggesting tighter restrictions such as those adopted for Cambridge and the coastal counties. The university registrar, Douglas Veale, speaking for the Vice-Chancellor, backed the police position:

> Aliens are a potential menace and we feel that they should be interned. They can be sorted out after internment if necessary.... These are abnormal times and demand special regulations. Steps have already been taken to guard vulnerable points in the university. They are being patrolled to prevent possible sabotage.

The popular press went further. On 28 May a female columnist in the *Daily Mirror* called for an extension of Churchill's policy to women as well:

> There are 30,000 women enemy aliens at liberty in this country – 30,000 points of danger to our soldiers at the Front and to our lives at home . . . . To expect German refugees fleeing from Hitler to cease loving Germany is like asking a critic of Mr. Chamberlain to hate England. It cannot be done. I DEMAND . . . THAT THESE ALIEN ENEMIES BE ALL INTERNED.

In such a climate, even recently naturalized refugees could feel insecure.

In the transit camps, male internees were separated by age. Those between 40 and 70 were sent to the Isle of Man, where some would experience for a second time the same confinement they had faced during the First World War. Of those under 40, thousands were removed to camps in Canada or Australia. At the beginning of July, three ships were loaded in Liverpool with internees and prisoners of war for Canada. One of them, the *Arandora Star*, was torpedoed and sunk off Ireland on 2 July with a loss of over 800 lives. Though Franz and his immediate family had been naturalized early in 1939, those interned or shipped abroad included cousins, personal friends, and academic colleagues. Helping them wherever possible became a priority concern.

As the panic subsided, public sentiment mounted against the irrationality and waste of human resources involved in blanket internment. By late summer, the government recognized a need for concessions, particularly for the Jewish refugees, whom Franz considered "Hitler's oldest enemies."[16] With thousands of files to be handled on a case-by-case basis, releases became a matter of weeks or months, and even longer for those recently arrived in Britain. Through the summer, those who knew the German academic scene were frequently called upon by refugee assistance organizations to help vouch for individual cases. One by one, those previously established in academic life reappeared, often unannounced. Churchill himself was "distressed" about the violation of British traditions, but without regret. In December he urged the Home Secretary to ease conditions for detainees: "The public danger justifies the action taken, but that danger is now receding."[17]

Through all the turmoil and disruption, Franz's scientific mind remained active. On 7 May, three days before the Germans struck at France, he wrote to Lindemann enclosing a clipping from *The Times*, a report from its New York correspondent on atomic energy. The clipping mentions recent experiments on uranium 235 at Columbia University, in Minnesota, at the General Electric laboratories, and—more ominously—reported work in Germany ordered by the German government. Franz comments:

> The remark about the activities in Germany seems quite plausible to me. You know that I have often emphasised that in my opinion the Germans will spend any amount of money even if they would not give a higher probability than 10% to the successful working of the uranium bomb. I would not be astonished at all if they were already engaged in erecting a big factory for the separation of U^{235}.
>
> Should not something be done in this country also? I know that experiments are carried out in Oliphant's laboratory, but I do not know on what scale. In any case such experiments should not be restricted to a single laboratory; more people find more things. Should we not try to get something done in the Clarendon? The High Tension Lab. is the ideal place! I personally would be only too pleased to undertake research in this direction.[18]

By this date Lindemann was in a position of influence in charge of Churchill's statistical unit, a unit formed by Churchill while at the Admiralty in 1939 and transferred with him when he became Prime Minister.

On 14 June, with France collapsing and Paris evacuated, Franz wrote again to Lindemann to inform him about current work at the Clarendon. First he encloses copies of recent correspondence with David Pye, Director of Research for the Ministry of Aircraft Production, about the liquid hydrogen project for aircraft, and then adds another paragraph:

> At the beginning of the week I paid a visit to Birmingham and saw Oliphant and Peierls. I wonder what you think about the new situation of the Uranium question. I am trying at present together with Kurti and Arms to construct a separating column for the U-isotopes using the gaseous UF_6 (as far as this is possible without a mechanic). We also will try to make experiments on the best conditions for this method, about which not much is known at present.[19]

A few days later, Lindemann received another letter, handwritten by James Chadwick and dated 20 June, stressing the absolute urgency of keeping France's heavy water out of German hands, and if necessary "to put it down the drain." Chadwick concludes: "I hope you don't mind my worrying you on this matter but there is a very real danger that the uranium process can be made to work."[20] As it happened, the heavy water had already left France. On 17 June it had been quietly evacuated from the Bordeaux region, along with Halban, Kowarski, and other French scientists, on the *Broompark*, a battered, nondescript British coaster, in one of the most dramatic rescues of the war.

Transatlantic lifeline

When Lotte sailed for Canada on 4 July 1940, she and Franz resumed their old habit of writing to each other almost daily. From then until March 1944, their letters paint a vivid chronicle of their daily lives. The relevant letters and cables back and forth were preserved in five durable, used manila envelopes. One of these envelopes, dated October 1939, is marked International Institute of Intellectual Co-operation of the League of Nations (which sponsored the Strasbourg conference). Two others are postmarked from the 1950s, and two show no dates. The smallest is marked "cables." A second contains mainly administrative or financial material (medical, dental, foreign exchange control, etc.), but it also yields some press clippings and warm responses of Franz's colleagues in the United States to his appeal regarding the children's evacuation to Canada. The remaining three envelopes, tightly packed and sealed with tape, contain Franz's and Lotte's letters to each other. It is very likely that they have been sealed for a very long time.

There are some differences from their earlier letters. The selective month-long portraits of life in the 1930s turn into continuous dialogue, a long-running feature film. What emerges is a complex chronicle of family separation and wartime stresses, sometimes daunting, sometimes comical—Franz's sense of humour often saves him from despair—in a household where Lotte's guiding touch has suddenly disappeared. Within this kaleidoscope of daily activity, a thin thread of discreet information on work hints at Franz's links to the war.

For Franz there is a single series of letters numbered up to number 277 of 19 March 1944; for Lotte the numbering is broken and restarted several times. The letters also refer to a comparable flow of postcards, but these have dispersed. Some letters went missing in transit, and delivery dates became irregular. As Atlantic sinkings increased, both began to send second copies of important information, and Lotte began to type her letters to avoid writing the same letter twice. Each one recorded and reported those received and those still missing.

From August onwards, there is a third party to their dialogue, unseen but always in mind. On 25 August Franz writes:

> Knowing that all the last letters have been opened by the Censor, I can't express myself freely. As, however, he knows already from your letter that you love me, he may also know that I love you (I hope it is the same Censor who is reading this)! And he may also know that I would like very much indeed to have you in my arms again.

Lotte replies on 5 September: "Thank you so much for your letter 20. I know what you feel like about the Censor but I hope he won't mind!"

Unlike earlier series, these letters are invariably in English. To avoid suspicion and possible delay, foreign words are avoided, and many proper names are brushed up a little. Thus Ludwig Frank is always Louis, and the Cassirer son, who lives nearby, is simply Cass. Mrs Wohl becomes Mrs Well and Mrs Cahn becomes Mrs Barge. Sometimes one must guess a little. Mrs Carpenter is probably Mrs Zimmermann, who helped in the house. When Franz reports on 3 October that Mr Jacobs-Valley has returned, he means that the 60-year-old Professor Jacobsthal has been released after 12 weeks of internment.

When Lotte reached Toronto on 16 July, after a ten-day voyage to Halifax and train from there, she found the children already placed in a host family. By a strange coincidence, the diaspora that had scattered her extended family over five continents now worked to their advantage. The children from Oxford were among the first war guests to reach Canada, and their arrival was featured in Toronto newspapers. One photograph in the *Toronto Star* showed a small group including the Simon children with the university's President Cody, each identified by name. The picture and story were seen by one of Lotte's more distant cousins, Fritz Jaffé, now living with his wife in Toronto while other family members were spread from England to Shanghai to South Africa. The Jaffés, in difficult circumstances, could do nothing themselves, but Gerda Jaffé was doing housework for a research assistant to Dr Bruno Mendel of the Banting Institute. When the Mendels heard of the connection to the Jaffés, Mrs Mendel telephoned Mrs Cody and offered to take the Simon children into their home, an offer quickly approved by the committee.

After Lotte met the Mendels and was offered hospitality in turn, she wrote to Franz in glowing terms the day after her arrival in Toronto. The three Mendel children were close in age to the Simon daughters, the children were happy together, and the two families had remarkably similar backgrounds:

> It seemed to me as if fate had something to do with it, we could not have dreamed of a place better suited for the children. Very much the same atmosphere as our own, the same background in every way. M[endel] is the son of a collaborator and great friend of Morgenroth, his father was a medical man of Essen. She is his cousin, daughter of Fischbein and M, they lived in Wannsee . . . . Both very sympathetic and very eager to help.

The medical background was immediately familiar to Lotte from her work for Professor Morgenroth after 1914. Bruno Mendel, after active military service and medical studies, had also worked in a Berlin teaching hospital before developing a private medical practice and his own laboratory for research, where he worked in close association with the renowned biochemist Otto Warburg. In 1933 Mendel left Germany with his family to work in the Netherlands. In 1936 he moved to Canada, where he joined the Banting Institute in

Toronto and did important work on cell physiology. The link with the Simons established in July 1940 would grow into an enduring friendship between the two families, which continued after the war when the Mendel parents returned to the Netherlands.

In the same first letter from Toronto, Lotte mentions a promising idea from Professor Mendel. "After what he found out since yesterday, there seems to be a growing feeling here, that long-range scientific work which under the present circumstances cannot be undertaken in England should be transferred here." Excited by any possibility for family reunion, Lotte swung into action. Briefed by Dr Mendel, she arranged to see Eli Burton, the director of the McLennan Laboratory and head of the Physics Department, and also Banting. "As he [Burton] and Sir Frederick Banting are the two Toronto members of the main council in Ottawa, it seemed the best to talk to both of them." At this date Franz was still balancing his two projects with possible wartime applications, liquid hydrogen as aviation fuel and his more recent experiments on isotope separation.

Burton received Lotte without delay and showed interest in the low-temperature work, but he emphasized that it should be arranged through government-to-government channels via the Canadian High Commissioner, and originating with a British request. He gave her a brief written memorandum suggesting how this might be done. "About the other possibility," Lotte writes, "he was less interested, there is no cyclotron in Canada . . ." Banting, already briefed by Burton, came to the Mendels' house on Bedford Road to meet Lotte on 19 July, but "he was very noncommittal." The subject was clearly far from his own field.

From a British standpoint, the times were far from propitious for discussing long-range research. Even as Lotte wrote, the Battle of Britain had begun, and when Franz received these letters in August the life-and-death struggle for air supremacy over Britain was escalating towards its peak in September. Franz reacted cautiously to Lotte's efforts. He was concerned to avoid misunderstandings over well-intentioned short-term efforts to help by North American colleagues, but he felt the Burton connection in Toronto worth consideration. To this point in mid-August, neither of his current projects had official support.

In two successive letters to Lotte, Franz summarizes his position. In an undated letter (probably 14 August), he writes:

> The day before yesterday I had a letter from Thomson asking me whether I would like to cooperate in their scheme. He would like to see me and I am waiting now for a date in London this week. Yesterday Oliphant came and we had a very interesting discussion about this matter, but also of a more general character. I assume that I shall see Thomson by the end of the week and I then will probably have to go for some days to Liverpool. – Just now a letter from Lindemann that he could see me in London about the other thing. I don't think it impossible to follow up both things at the same time as in the long run they both have to be transferred to the same place.

In his next letter, dated 16 August, he reports on these meetings:

> Yesterday I went with the 8.40 to London. First I met Thomson and we had a rather longish discussion about the possibilities for my co-operation in their scheme. At first it will be more on theoretical lines, but later probably also experimental. After a few weeks we will see how things develop. I don't bring any special qualities for this particular job, but I certainly can help them and it can become quite interesting.
>
> Then I saw L[indemann] in his office, he is now in a very high position. I talked for an hour with him about the whole question and interested him again in it. We discussed also the possible transfer of the experiments and I have just sent a short resumé for his brother Charles, who is now attached to the Embassy in Washington. I mentioned there also Burton's willingness. We must wait now. L sends his regards.

The doors hitherto closed were now beginning to open. In early September Franz prepared a report for Thomson and went to London again to discuss it with him. On 12 September he reports to Lotte:

> As you will have seen from my cards, I was in London on Tuesday to meet Thomson . . . . I am now definitely entrusted with an important part of this business and Arms and Kurti are to co-operate with me. We are beginning with the experimental work here very soon, and we are trying to find now mechanics. As we don't need many in the beginning, I think we can manage until we have found some. Well, I am glad we have got so far and we are already very busy.

On 18 September, he reports an all-day meeting the previous day in London, "quite busy . . . Blackett and Halban send their regards." The meeting on 17 September was Simon's first appearance at the Uranium Technical Subcommittee. He made detailed notes on the topics discussed: methods of isotope separation, chemistry of uranium, its compounds, graphite.

With official doors opening, others had to be closed. On 4 September he wrote to Born again, mentioning briefly among other news that "It seems that I am accepted now for co-operation in a hush-hush business, the one I mentioned in my letter of June 4th."[21] Franz, hitherto an outsider, had become an insider, unable to discuss any aspect of his own classified work. Their wide correspondence on other topics could continue, though it would be impeded at times by Franz's increasing work load.

The Maud Committee at work

The formal contract with the Oxford group arrived only on 22 October. It brought salary support for Kurti and Arms and, as Franz reports a few days later, "I have now my own funds, a great simplification . . . . [and] I got additional petrol coupons for 500 miles per month, so that I can do all the necessary travelling in the car." As a naturalized British citizen, Franz could own a car; aliens could not. By working through contracts with four universities (Birmingham, Liverpool, Oxford, Cambridge), the Ministry of Aircraft Production could gloss over the alien question. The Ministry itself employed only persons of British parentage, but of 31 scientists listed by Margaret Gowing on the four Maud Committee teams, 13—or 42 per cent—had alien or refugee backgrounds.

With increasing involvement in the Maud project, Franz's "other project" faded into the shadows. On 31 August he tells Lotte of a hurried conversation with Lindemann: "his people" did not want to forward Franz's résumé to Washington without a "complete agreement with that side," and instead of the Burton option in Toronto he "prefers the thing to be done in O[xford]." He observes, "It seems to me that my chances are better with the work I am beginning now, although I don't give up to press the other also."

On September 14 he reports again:

> Long discussion with L[indemann] about my present work and also about the other one which I planned together with him. I don't think he can be really of much help. All the ways which he proposes in order to accelerate it, are bound to offend other people, who also are not without importance.

From this point on, the project on liquid hydrogen as fuel seems to have been shelved. With many colleagues away, the university term also had to be organized. On 17 October he writes: "Today the first lecture, crowded like never before, certainly something like 80 people."

From early September, the Maud work grew in intensity. Franz's letters now refer routinely to "a busy day" or "extremely busy week." New experiments are started. A mechanic is hired. There is more travel to Birmingham (in the car with Kurti and Arms), to Liverpool, or to see Thomson in London. Thomson visits Oxford, along with a representative of the Ministry of Aircraft Production. More complex apparatus is planned, priced, and discussed with the firm that would manufacture it. The reports to Lotte on work are generally positive. On 10 October:

> My work is getting on very well, Th[omson] seemed to be much impressed. We shall begin to construct a model very soon. The final experiments are, however, on a very big scale, and cannot be carried out here.

Around these brief work reports, Franz paints in colourful detail elements of the human comedy around him. The air raid shelter is now comfortably furnished, stocked with various drinks and Opapa's wireless set, though not yet needed. The garden is yielding potatoes, other vegetables, a "really wonderful" crop of apples, and some pears. The base family consists of Lotte's parents, Franz, and "the boy" (Lotte's young cousin Curt Lipmann). Others include a succession of German-speaking household helpers—Omama and Opapa as yet have no English—and usually one or two shorter term visitors. On 23 September a cousin of Lotte's, never named and unidentifiable to us, arrives "with two trunks" for an indefinite stay. Franz openly disliked her, and after two weeks, when Omama invited Franz to address her using the familiar "Du" (in the letter he writes "tu"), "I refused politely but energetically."

On the morning of 2 October, Mimi Frank rang up from London. Their house in Langland Gardens had been bombed and was uninhabitable. Fortunately, the family had escaped injuries. From that evening, the Simon house would have to accommodate four extra people, Ludwig and Mimi, their son Stephen, and Amama (Anna Simon). "How good that you three have left," Franz notes, "how else could I put up all these people?" On 6 October, he begins: "To-day it was like in olden times, 11 people for lunch!" Every bed in the house was occupied, and extra camp cots were pressed into service. The two grandmothers developed a weekly rotation for the cooking, with each trying to outdo the other. The extra people brought increased noise, confusion, arguments, and misunderstandings. Franz tried to maintain a level of detachment from the daily fray, but the *ménage* had become a menagerie, a word he uses more than once.

He sought ways to alleviate the daily tensions. The late evenings, in the quiet of the night, were a time to unburden his frustrations in letters to Lotte. They exchanged notes on reading— Balzac novels, Heine, a new book by the young Kennedy, "son of the American ambassador." His social life with refugee friends and colleagues continues. The men are reappearing from internment, with tales of worse hardships than his own. He dines frequently in Balliol, and sometimes as a guest in other colleges, reporting stimulating conversations with new and interesting people. "These evenings in college are hardly changed at all . . . the whole atmosphere still the same, it really is astonishing." On one occasion he misses the name of his neighbour at the table: "I tried in vain to find out in the discussion what he was . . . it did not fit to any profession that I could imagine. Later . . . it turned out that it was the Crown Prince of Norway, a rare profession indeed!" The Prince was a Balliol man, and they became friends. At a subsequent dinner they had "a very interesting talk" on Norwegian surplus water power. "He seemed to be quite familiar with technical questions which concerned Norway. Of course, he knew our friend [V.M. Goldschmidt]; in these small countries the Crown Prince knows every scientist personally."

A more grave theme from these letters is the occasional comment on the ups or downs of the war. On 18 August, Franz notes that "everybody here is very much impressed by the achievements of the RAF," which have boosted morale and made a German invasion unlikely. "The whole atmosphere is one of great confidence and very much better than in the weeks after the French collapse." After visits to London he comments on the visible damage there, at first dismissing the press reports as overblown, giving "a quite wrong impression" (12 September). On 18 September, he notes "a little more damage, but chiefly to the glass." On later visits, he describes more serious damage, but Londoners are carrying on with their daily lives nonetheless.

With many refugee friends awaiting transatlantic passages, the war at sea is a frequent preoccupation. On 23 September Franz's letter begins: "This morning the terrible news about the torpedoed liner was in the papers. It seems that Olden died on a lifeboat . . . ." Lotte, visiting New York and staying at this point with the Ureys, wrote to Franz on the same day: "I just saw in the paper that the Oldens were on the torpedoed boat. It is all so terrible that I cannot say anything about it . . . . I am so upset about the Oldens that I can hardly think of anything else." For five days longer, however, until a cable proved otherwise, she was in anguish that her close friend Grete Wohl and her son, whose arrival in New York was expected imminently, might have been on the same ship.

Rudolf Olden, a journalist and author, had been until 1933 political editor of the *Berliner Tageblatt*. In exile he became secretary of a new branch of the international PEN clubs, founded in London for German authors in exile. In 1940 he was interned along with other refugees living in Oxford. On grounds of poor health, he was released to go to America, but he and his wife died in the sinking. Their deaths left their two-year-old daughter Mary, known as Kutzi, the youngest of the Oxford–Toronto evacuee group, an orphan.

Two days after the first report, the lost ship was identified in the press as the *City of Benares*. She had sailed from Liverpool for Canada in convoy on 13 September, carrying 197 passengers, of whom 90 were government-sponsored child evacuees. Four days out, in a force ten gale, she was struck by a torpedo at 22:05 on 17 September, but the sinking was not announced until 23 September. Of the 90 child evacuees, 77 were lost, and this led the British government to end its child evacuation programme.

Franz's letters also contain occasional glimpses of life in Nazi-dominated Europe. On 12 December he reports the imprisonment by the Vichy government of his elderly Paris colleague, Paul Langevin, of the Collège de France. On 20 December he notes the shutdown by the Germans of Leiden University and the Delft Institute of Technology: "They are pigs!" In the same letter he mentions a second issue of *La France Libre*, "which also contains a few words from me."

On 19 November, Franz writes a letter to Lindemann on behalf of the committee: "There is an urgent need for about three kilograms of the fluoride in order to obtain the necessary preliminary data for our work." A bureaucratic deadlock has developed, he explains. The ministry cannot give the order until they know the cost. ICI cannot set a price until they know the size of the larger order that may follow. The scientists cannot determine this until they have experimented with the initial quantity. "You expressed a willingness to help in this matter, so I am writing to ask if you will use your influence with Lord Melchett . . . so that we may get the substance as soon as possible and break this deadlock . . . . Our greatest need is to get the substance in a reasonable quantity, and to get it quickly."[22] Lindemann intervened successfully, and a contract was signed in December, but the three kilograms of uranium hexafluoride required new ICI plant investment and the price was £5,000. Up to this point, the research contracts with the four universities for the project had cost the government a modest total of £8,500.

In the laboratory, work continued. On 19 December, after a five-day gap, Franz writes: "I have to finish a report by the end of the week and was extremely busy. Today I dictated the last pages and tomorrow it is hectographed." Three days later, he writes that the report is finished but there are problems with the hectographing apparatus—"I was touring the shops yesterday to find spare parts"—and it cannot be done elsewhere because of its "secret nature." With these delays, Christmas Day found Franz working all day in the laboratory with Kurti to finish reproducing copies of the report; he had already celebrated the holiday on Christmas Eve, with presents for the younger folk and a goose for dinner. Owing to disruptions from air raids, the postal service was unpredictable. On 27 December he describes delivery of the report:

> I was today in Cambridge as I wanted to hand my report directly to Thomson, who spends his holidays there. We left very early, when it was still rather dark and arrived there just before 11 o'clock – I was with Kurti. We left in time to get back before the blackout, as driving in the dark is rather unpleasant. We had lunch with the Halbans who asked to be remembered to you.

With the report out of the way, the end of December was a time to reflect on a year marked by deadly conflict, family separation, and dedicated research.

The year 1941 would prove to be momentous at several levels: for Franz himself, for the Maud Committee, for Britain, for the global widening of the war, and—as later events would show—for the postwar international system. The transatlantic letters continue. Those from Franz's side are his carbon copies, and thus may include a few that were delayed or never delivered. His urge to write to Lotte seems at times to increase as the pressures and tensions grow. The letters are a release, an outlet for his frustrations, but they are also a mirror of his complex world from week to week. Though their references to "the work" are often cryptic, they can be matched almost day by day and meeting by meeting with the detailed postwar histories.

From January to July the defining events were the meetings of the Maud Committee. At a meeting on 8 January, the report that Franz had laboured with Kurti over Christmas Day to deliver to Thomson was discussed, along with reports from the other university groups. Among these was a report from the Halban team that uranium oxide, using heavy water as moderator, would be "95 per cent certain"[23] to produce a self-sustaining chain reaction if the two ingredients were available in sufficient quantities. Through those meetings, each team involved could obtain a view of other parts of the project. There is no evidence at this stage of compartmentalization, and this was a factor in the Maud project's rapid and conspicuous success. On 11 January, Franz reports to Lotte as follows:

> Our meeting was very successful; L[indemann] was there also and Lord Melchett. Their presence had good catalytic effects and I think it was a good idea to bring the latter into the whole business. This renewed, by the way, my contact with ICI and I am going to see them next week – I am quite glad about this by-product. As a consequence of the discussions at the meeting I will be now consulted also by other departments and I will be quite busy travelling about . . .

The presence of the ICI executives at the January meeting meant that the committee was now beginning to consider problems of large-scale industrialization of processes hitherto attempted only in research laboratories. From an early stage, ICI became interested in Halban's Cambridge work on slow neutron research— "the boiler"—because of its potential for power generation. On 16 January, Franz reports another, more specialized meeting: "I was yesterday the whole day with the ICI people, with whom I am at very good terms. We had discussions the whole time, interrupted only by a good lunch in the Junior Carlton Club." In postwar sources,[24] the purpose of this meeting is clearly set out: the requirements for an isotope separation plant—ground area, power supply, manpower, and money.

The ties with ICI became closer. On 16 February, Franz writes to Lotte:

> On Friday I got a telegram from Lord Melchett that he would spend the weekend with his son who is up at Christ Church and that he would like to see me. Thus we spent the whole afternoon together in the Lab. He is very reasonable, sees the point quickly and is not afraid of taking responsibilities; in addition, and this is important, he is accustomed to think in different orders of magnitude. He will be a great help in the whole thing and I'll see him probably at regular intervals from now on. L[indemann] was also present for a short time.

On 10 March he writes of a two-day working visit to the farm of R.E. Slade, the research controller of ICI, at the end of another visit to Birmingham and Cambridge:

> Then to Slade. He lives on a farm . . . . Before dinner we went shooting wood pigeons, who do a lot of harm . . . . The house very nice and well furnished. We were alone, his wife was away, and we had a good talk, beginning with general topics and ending at our subject. Next morning a walk and then Perrin, his assistant and Halban arrived and we worked the whole day, interrupted by one walk. In the evening H. left and I stayed with Perrin the night. Again talks into the late night. It was a very nice time, with important results – they will help in our work with all energy and I am glad that I had the idea of taking them into this thing. Incidentally, he also asked me whether I could advise them in some of their work which has nothing to do with this. – One incident is worth mentioning. We were sitting and having our whisky when the maid – an old nanny – came in telling that they had just rung through from London that his London flat had been completely destroyed. He only said "That's that" and afterwards once "A pity, the nice furniture" and then it was not mentioned any more! Imagine how a Frenchman for instance would have reacted!

From this point things began to move more rapidly. ICI set up an internal committee to coordinate the firm's nuclear efforts, which met on 20 March. On 27 March, Franz tells Lotte of changes in the Maud Committee structure. He is now on both the Technical Committee and the smaller Policy Committee, originally limited to British-born scientists. The next meeting of the Technical Committee was on 9 April, and Franz reports to Lotte two days later:

> We had a good meeting in London and the question of the transfer was discussed in greater detail for the first time. No decision has yet been taken, but it is more probable than not. In any case, the time is ripe for taking the decision which we can expect in the near future. – The day before the meeting we met the industry people . . .

The question of the transfer, that is, of shifting portions of the work to North America, was of special importance for them personally, for it held out hope for a family reunion.

Twenty years later, Ronald Clark gives details of this meeting,[25] which was attended by 10 project scientists, Dr Slade and Michael Perrin of ICI, and an invited American observer from the National Defense Research Committee, Kenneth Bainbridge. The work on nuclear fission, for both "the bomb" and "the boiler," was reviewed in its theoretical and practical aspects. Among others, Simon and Arms described the Clarendon experimental work on isotope separation. Most of those present now believed that "Britain, with the help of Canada," could produce a nuclear bomb before the end of the war, estimated minimally at another three years. The majority also felt that Britain had the technology to develop nuclear power but lacked the electricity needed to produce enough heavy water or graphite to use as a moderator. This gap could be remedied in North America.

Between the April and July meetings of the committees the interest of ICI in nuclear energy developed rapidly, and Franz was actively involved. On 19 April, he reports:

> I think I have not yet written that our meeting with Melchett was very satisfactory, L[indemann] was also present. We have another meeting on the coming Thursday, but nearly every day before will also have its meetings here. And now soon my lectures begin again – Thursday week; not to think of the May lecture!

Six days later, on 25 April:

> Yesterday I had discussions with Melchett and some of the "big shots" and we agreed about the plans for co-operation; I get some of their people now to help us. It is quite difficult to find accommodation for them, particularly as term begins.

Over the next few weeks the pace only intensifies, with a never ending round of meetings, visitors, ordinary lecturing, and a looming, widely advertised public lecture on low-temperature research that he had agreed long before to deliver in late May. On 18 May his letter begins: "I have spent the whole day in the Lab – although it is Sunday and wonderful weather – but I had to prepare things for a meeting tomorrow in London."

For the content of this meeting we have to look elsewhere. Clark reports that the Maud Policy Committee, at its meeting of 19 May, considered a "final detailed proposal" for Simon's isotope separation plant. After discussion, the committee agreed that the plan had "more than a reasonable chance of success," and that a pilot plant should be built by ICI and Metropolitan–Vickers to advance the project.[26] The decision was conveyed to ICI on 20 May, with promise of a contract to follow.

The same day, but without a hint of the main reason, Franz's letter to Lotte announced: "The next 7 days will probably be the busiest of my life . . ." For 2–3 weeks the pressures of the coming public lecture—still unprepared—had been rising towards the top of an unusually busy schedule. His letters mention headaches, inability to concentrate, sleeplessness, and bad nerves as the deadline approached. As late as two days before the lecture, not a line of text had been written. Then, by a supreme effort, in isolation from the laboratory, he drafted a text and delivered it to a capacity audience. "I was definitely not at my best," he wrote on 30 May, "but the people seemed to be pleased – I am sure when it is printed, it will be good." With this lecture behind him, the tensions diminished.

Even through the crisis weeks, the letters continue. When there is a gap, the essentials are filled in later. The letters are not an extra burden but a remedy, a lifeline to a pleasanter world currently beyond their reach. Letters from the family were eagerly awaited, and provoked anxious cables from Franz when they were delayed. Both dreamt of the day of family reunion. On 19 March Franz notes that a few evacuee wives were returning, as Lotte had wanted to do as soon as return passage was permitted:

Our cases are, however, different in one very essential point: it is not at all impossible that my duties will necessitate my going to Canada or the States and it would be awful if you then would be here and I there! We have to wait still for some time before we can make a decision.

The uncertainties around the future of the nuclear projects became encapsulated in the code word "holidays," which signified some future reunion in North America in conjunction with developments in the work. This dream would persist over several months in Kafkaesque fashion, surfacing a dozen times, always in sight, always postponed, always dependent on decisions by "those higher up," until almost the end of the year.

These stressful weeks also saw one important gratification. On 19 March Franz arrived at the laboratory to receive warm congratulations: he had been nominated for election to the Royal Society. It is not yet official, he explains to Lotte, but "in the last 100 years" no nominee of the committee has ever failed to be elected. The announcement of his selection evoked a wave of letters of congratulation, which are duly reported to Lotte as they arrived. Those from Born and Polanyi, he adds on 24 March, are "very interesting"; he promises to send details later but seemingly fails to follow through. Two months later, on 29 May in London, the new fellows signed the Charter Book of the Royal Society and were admitted to fellowship at the hand of the president, Sir Henry Dale.

The letters of congratulation survive in a separate file, containing 48 letters, notes, and cables. Many are routine and brief, but others add comments from varying perspectives. One colleague writes: "I am particularly pleased that the honour comes at this time. It emphasizes once again the international aspects of physics, and is, I believe, a hopeful augury for a better future." Another writes: "May I congratulate you on this gesture of recognition of yourself as a British physicist & of the work you have done amongst us since you came to this country."

The comments that Franz had reported as "very interesting" stressed the significance of his award for other refugee scientists. Michael Polanyi wrote: "It is also a satisfaction to have now two of us refugees, yourself and Born, accepted in this fashion." Born wrote a longer letter, containing news of family and current work, but begins: "Cordial congratulations to the F.R.S. I felt always depressed that I had it and not you, and I am extremely glad that your brilliant work has been officially appreciated. It will help the whole community of refugee scientists." In the Charter Book their signatures appear on adjacent pages, Born (elected 1939) on page 103, Simon on page 104.

The rewards for what we do often come unobtrusively. There is a hint of this in a letter that Franz received from his Harvard friend and colleague Percy Bridgman just before the Royal Society announcement. Franz relayed its essentials to Lotte on 5 March, including its final sentences:

I see R.H. Fowler [British scientific representative in Canada] from time to time and he tells me that at last you are fully occupied with military problems. I envy you the opportunity to be engaged in such work. I have not yet been called on, except in a very minor way, to assist with any of the defence problems, and I find it hard to keep going in the old routine when so many more important things are crying to be done.

In just over 12 months, Franz had moved from being a frustrated bystander in an unfinished laboratory to become a critical link in a vast and unknown project of discovery. Had he had time to reflect amid the pressures of these months, he might have realized that to be engaged to the limits of human endurance in a cause to which one is totally committed is to experience a peak of human felicity.

In June, Thomson prepared a draft of the Maud Report for discussion at the committee's meeting on 2 July. Those attending included Dr Slade of ICI and another American observer, Charles Lauritsen of Cal Tech. For 2 days the members examined this draft carefully, giving special attention to clarity in all technical details. The fate of the final report would be in the hands of non–specialists and even non–scientists. In redrafting, the report was split into two independent reports concerned, respectively, with fast and slow neutrons—the "bomb" and the "boiler"—each one with appendices. On 6 July Franz reports to Lotte that he was "very busy yesterday . . . I had to make some changes on my report." On 24 July, he writes that "our final report left only yesterday and it seems very optimistic to expect a reply before the end of August." Among other, larger consequences, their long-delayed hopes for a "holiday" in North America were bound up with the governmental response.

The Maud Report and its reception

The Maud Committee's report, separated into two parts, and long restricted from publication, was printed in full in Gowing's official history in 1964. The first addresses directly the questions posed to the committee in 1940, and its three recommendations are clearcut: "A uranium bomb is practicable and likely to lead to decisive results in the war." Further, the work should "be continued on the highest priority and on the increasing scale necessary to obtain the weapon in the shortest possible time." Finally, "the present collaboration with America should be continued and extended especially in the region of experimental work."[27]

The non–technical part of the first report, just five pages long, begins with a "General Statement" in layman's terms. To underline the urgency of the question, it refers to Germany's known efforts to secure supplies of heavy water, and emphasizes that "the lines on which we are now working are such as would be likely to suggest themselves to any capable physicist."[28]

The rest of this part, together with five appendices, is a careful review of relevant work to date and problems still to be solved. The 10-page Simon Report on the design and requirements for an isotope separation plant, apparently a version of the one delivered to Thomson in December, appears as Appendix IV. It has plans for a plant to produce one kilogram per day of U^{235}, including design diagrams, site requirements (20 acres), construction costs (£5 million.) and operating costs (about £1 million per year). Appendix I, prepared by ICI, estimates that "with highest priority" for the project, "the first bomb could be produced by the end of 1943." Optimistically, the report itself incorporates this date.

The second part is briefer and less precise. While recognizing the future importance of uranium as a source of power, it does not foresee this development as achievable within the duration of the war. As such, it lies outside the Maud Committee's mandate. It also notes, however, more distant military possibilities of "certain of the substances which would be produced in a machine of the Halban type." This refers to the "transuranic" elements 93 and 94, which members of the Cambridge team in January 1941 had begun to refer to, in planetary sequence, as neptunium and plutonium.[29] This report recommends continuation of the Cambridge research on these elements, but also recognizes that larger scale development involving heavy water cannot be done in Britain because of the large quantities that would be needed. It suggests that Halban and Kowarski therefore "should be allowed to continue their work in the U.S.," where new production of heavy water is expected.

The second report is followed by two appendices, both from ICI, which emphasize the importance of nuclear power development for the world at large, the importance for Britain and the Commonwealth of remaining at the forefront of this development, and the close scientific relationship between the military and peacetime projects. Both documents broadly support an idea presented formally to the Maud Policy Committee on 24 April: that ICI would be willing to underwrite the past and continuing research of Halban and Kowarski and the subsequent development of nuclear power on a commercial basis, even if this work had to be moved to North America. For the committee, this was a matter for the government to decide.

Once delivered on 29 July, the Maud reports had to navigate official channels. Since they originated from a modest subcommittee in the Ministry of Aircraft Production, these channels were complex. The reports were referred first to the ministry's controller of research, who consulted Tizard before sending them to his own minister, who in turn forwarded them, with the ministry's own views, to Lord Hankey on 27 August, with a request that a panel of the Scientific Advisory Committee(SAC) consider them and make recommendations. At every stage, much scepticism remained. The SAC panel held seven meetings during September, heard expert witnesses—including Lord Cherwell and Tizard—and reported its conclusions on 25 September.

The SAC Report tried to balance urgency and caution. For the bomb, the panel considered it "a project of the very highest importance." It urged pressing on as rapidly as possible with the early research and design stages, including design and construction of a pilot isotope separation plant—which was already under contract—but called for "careful and independent review" of results from this stage, plus additional research on radiation effects of the bomb before construction of a full-scale separation plant. On time requirements, which different Maud scientists had estimated at 2–5 years, they rejected the lower estimate as "too short."[30]

The panel's deliberations on location of the full-scale separation plant were more useful. Some opposed building it in Britain, stressing the vulnerability of such a plant to air attack. Others, including Cherwell, were reluctant to place Britain's survival and postwar position in the hands of the United States, still a neutral country in the widening conflict. Still others, including Chadwick, were doubtful about Canada as an alternative location, on account of that country's low level of industrialization. In its report, the panel noted "strong technical arguments" for locating "one pilot plant and the full-scale separation plant" in Canada, using components manufactured in the United States.

The panel's concluding summary gave only a scant six lines to the "power project": "We regard this as a long term project." In obvious rebuttal to the ICI bid, it "should not be allowed to fall into the hands of private interests" but should remain under the "close control" of the government and be developed in "close collaboration" with the Canadian and American governments. The transuranic "substances" discussed in the second Maud Report are not mentioned at all.

The Prime Minister did not wait for the scientific panel's report. Late in August, before the panel had even met, Churchill received a minute from Lindemann urging action on the Maud reports. On 30 August, Churchill sent it on to the Chiefs of Staff with a famous memorandum of his own:

Although personally I am quite content with the existing explosives, I feel we must not stand in the path of improvement, and I therefore think that action should be taken in the sense proposed by Lord Cherwell, and that the Cabinet Minister responsible should be Sir John Anderson.

I shall be glad to know what the Chiefs of Staff Committee think.[31]

On 3 September, the chiefs of staff met and agreed that the project ought to proceed, with the highest priority, in secrecy, and in Britain rather than abroad. The most incisive argument for proceeding was in Lindemann's minute: "It would be unforgivable if we let the Germans develop a process ahead of us by means of which they could defeat us in war or reverse the verdict after they had been defeated."[32]

The outcome of Churchill's decision and the panel's report was a series of meetings in mid-October to establish a new organization that would move the project from a research focus to an industrial one. Sir John Anderson, the minister in charge and a member of the War Cabinet, had earlier taken a doctorate in chemistry in Leipzig. Although he remained sceptical of the Maud time-frame, he understood that nuclear development was too important to ignore. The new structure was set up as a new division of the Department of Scientific and Industrial Research (DSIR), complete with a directorate, an advisory council for policy questions, and a technical committee. Although ICI had been rebuffed in its offers to develop nuclear power on a commercial basis, the top managers of the new directorate, Wallace Akers and Michael Perrin, were both seconded from ICI. The new organization was given a plausible but suitably disguised name, Tube Alloys. Until 1945, it operated in strictest secrecy, unknown even to Anderson's colleagues in the War Cabinet.

The new organization had been imposed from the top down, without any consultation with the Maud scientists. The technical committee included Halban, Peierls, Simon, and Dr Slade from ICI, but other Maud committee members were dropped. Some were aggrieved at the new industrial model. Mark Oliphant, a charter member of the uranium group in 1939, and recently returned from a visit to the United States, openly protested against the new structure: "This organisation is tantamount to that which exists in the United States where the whole thing is in the hands of non-nuclear physicists and is therefore being badly mismanaged."[33]

A further consequence of the changed structure was that Franz became a full-time employee of DSIR. In a letter to Lord McGowan dated 9 December, he explains that in this capacity he is "no longer free" to continue the connection with ICI through the research grant. He thanks the firm for the past eight years of research support for himself and Nicholas Kurti, and especially "for the free hand you have always given us with regard to the nature of our research."[34] In a letter to Lotte on 6 December, he explains that his termination of the research grant followed a visit from Lord Melchett and "a long talk about all kinds of things . . . I had insisted on [terminating the grant] in order to be absolutely independent. It is true they never have tried to influence me in the least, but I thought it better for many reasons." Quite possibly his acute political senses warned him of shoals ahead. His university readership remained unaffected by the changes.

On 9 November, Franz had written to Lotte in a good frame of mind of a first meeting of the new committee, and a recent visit to England by Harold Urey:

Our discussions were very satisfactory, as well those with Urey, as those concerning the new arrangements. I will be extremely busy during the next weeks to get everything started, but then I'll have my holidays at last. The time is not yet quite decided upon . . .

On 3 October, George Thomson had delivered copies of his Maud Committee's reports to the heads of the American government's new defence research body, Vannevar Bush and James Conant. Within a week, two Columbia University uranium experts, Urey and George Pegram, were selected for an official visit to England to investigate the findings of the Maud teams at first hand. Franz knew both men, and Urey was a close personal friend with similar research interests. Lotte also had stayed with the Ureys during her visit to New York in September 1940.

Franz was informed of the impending visit on 10 October. The ministry arranged a tour of the project sites, which now included factories as well as research laboratories. Arriving on 27 October, the two Americans first visited the Tube Alloys directorate in London, then the Chadwick group at Liverpool University, and next a printing firm in Bradford doing experimental work on fine membranes for gaseous separation. The middle phase of the visit is described in Franz's letter to Lotte of 4 November:

> Well, Thursday [October 30] Urey and Pegram arrived, and we have been working until Sunday. Urey stayed with us – he slept in your room and I moved to my study –, Pegram stayed in College and Pyers [Peierls], who was also present, with Kurti. On Thursday night I gave them a dinner in Balliol in a separate room and after dinner all my collaborators joined us for coffee and port – it was a very agreeable evening. Friday lunch at the Mitre; in the evening I had the two for supper at home and afterwards again some people came to work. On Saturday we lunched at the Golden Cross (where I had never been before, an old-fashioned, nice place). In the afternoon a long discussion with Cherwell. We then had supper in the Indian Restaurant, rather dirty, but not bad; we then went home and talked until late in the night. Sunday morning we had a walk first – Marston Ferry, Univ. Park – and then did a bit of sightseeing. Then P. left for London and U. came to have lunch with us. We had tea with Sidgwick [professor of chemistry] whom I had not seen for a very long time and then I took U. to have dinner in Hall. Very interesting discussions with the Master and some others. Then at home some talk including looking at Low's cartoons, which U. did not know and which he liked extremely, particularly the Blimps which in his opinion have so much bearing on the conditions in his place. On Monday we had to visit the chemists in Birmingham and so we drove there. We started rather early as U. wanted to see a bit of Stratford and arrived at 12. After having finished U. left for London and I stayed the night at the Pyers as I did not want to drive in the black-out.

The visitors then went to see the Halban team in Cambridge. A highlight of the visit there was when Urey, discoverer of heavy water and familiar with it in micrograms in his own laboratory, was handed a metal canister containing a gallon by Kowarski.[35]

Urey and Pegram returned to the United States enormously impressed by what they had seen. On 1 December, Urey sent Bush a preliminary report on the British projects.

In the meantime the Tube Alloys Committee, at its 11 December meeting, decided on a reciprocal visit of a British team to the United States. On 12 December Franz writes:

> We have decided at our meeting that three of us should pay a visit to the States – I will be one of them – and that we should arrive there at the middle or end of January. I think we'll stay for about four weeks and it is not impossible that I'll have to spend a week at the end of visit in Canada also. All these things are of course not absolutely certain but the probability is now very high. I write already now as you'll have to make some preparations: I assume that my "headquarters" will be in New York and I think you should stay there with me during that time, and possibly a part of the time the children should also come. Thus it would be good if you would take steps to get the visa – ask Urey to help you.

After several postponements, the long delayed "holidays" for Franz—days of family reunion—were now in sight. When the news arrived, Lotte cabled back on 10 January:

"Delighted with letter 189." As plans evolved, four British scientists would travel at this stage—Akers, Halban, Peierls, Simon—though not simultaneously, and by different routes.

The year winding down at this point had seen spectacular developments for the British project: the Maud reports in July, the decision to proceed in September, followed by a new structure to manage the project in its new phase.

The year 1941 had a few other noteworthy developments. A prominent feature in Franz's letters is his increased contact with Lindemann. The latter's visits to Oxford were frequent. In normal times, he lived and worked there. On 12 April, following the April Maud meeting, Franz reports: Lindemann "turned up suddenly and thus I went to Christ Church and spent three hours with him. He is now very much interested and we discussed the points also which I mentioned yesterday [concerning transfers of work to North America]; he seems to be reasonable." On 14 June: "The most exciting news this week was of course that L. had been created a Baron. I saw his Lordship to-day for a moment and congratulated."

Further meetings followed: "Had a rather long talk with L. yesterday" (22 June); "Had a long talk with Lord Cherwell - this is the name our friend has accepted" (6 July); ". . . we had a rather extended talk [with Peierls also present]. . ." (14 July); "In London . . . saw Lord Cherwell. I had a talk of one hour with him, I believe a successful one" (30 July); "Spent the greater part of yesterday afternoon with L. who, incidentally, always asks how you and the children are and who wants to be remembered to you. He told me our propositions have been approved and that we'll soon hear more"(24 August); "Had a long talk with Ch. yesterday. Next week will be very interesting for us" (11 October). These "long" or "longish" talks continue to the end of the year, and often they can be keyed in to important meetings. In the midst of all else, their mutual mentor, Nernst, died at his country estate in Silesia, Zibelle, on 18 November, as Franz reported in a letter to Lotte four days later. The obituary of Nernst as a Fellow of the Royal Society would be jointly authored by Cherwell and Simon.

An area of wider concern for the Maud scientists was the German question. Were the Germans already leading in the race? After the July meeting of the Committee, Thomson wrote to Lindemann (whose title was not yet chosen) on 4 July:

> Several members of the Committee raised the question of getting better information as to what is going on in Germany, and I wonder if you could help us in this respect . . . .
>
> I feel the question is an important one, as if definite evidence of German activity on the isotope lines as opposed to heavy water are forthcoming, it makes the whole thing much more urgent and would be a strong argument in pressing for large scale work.

Thomson mentions "certain facts" about Norwegian heavy water supplies and some "rumour and speculation" about isotope separation. He appends a further page of "rumours," notes in point form set down by the physiologist A.V. Hill and attributed to one Veblen von Neumann. These notes report that the Kaiser Wilhelm Gesellschaft's nuclear physics laboratory in Berlin is actively pursuing uranium research, for purposes of both power and explosives; and that the physicists working there might be merely doing research to avoid military service, or

> It may be that there is really something in it . . . . There was no doubt in V.N.'s mind that the nuclear physics is going on at hot speed at Dahlem: and this is not likely to be due to a disinterested love of Science by the Nazis.[36]

Franz comments less often on the wider war in his 1941 letters, probably because of his own work pressures. On 13 May the news of Rudolf Hess's flight to Scotland is noted in passing with a single word: "astounding." The German invasion of Russia, an affair of a different order, evoked a longer, more prescient comment on 22 June: "Just now the news of the invasion of Russia! It is astonishing after all. It must be pretty bad in Germany if Hitler is forced to such a step. What about Japan now, and how will the reaction be in USA?" When the Japanese did attack Pearl Harbor without warning on Sunday 7 December, Franz commented only briefly on 12 December: "Needless to say that the Japanese business and its consequences overshadowed all other events." Among immediate consequences was his committee's decision on 11 December to go ahead promptly with the visit to North America.

Unlike 1940, when Franz's comments on bomb damage in London were frequent, the 1941 letters seldom mention the domestic scene even when he travels to towns like Liverpool that had the heaviest raids. There are rare exceptions. On 20 June, as the final meeting of the Maud Committee approached, he writes of a day in London to confer with Thomson, have lunch with ICI colleagues, and then attend a Royal Society meeting as a newly elected F.R.S. He describes the scene to Lotte:

> Dirac gave the lecture, on a new quantum electrodynamics. It was a remarkable atmosphere: he talked for 1½ hours on this highly abstract problem and you could see through the windows the balloons of the barrage; on the streets the newspapers with the reports on Russia, the "ladies" walking along Piccadilly like always, life going on in the old way, although a house does not exist any more at certain intervals and very many of the glass windows are replaced. It must be difficult for you to realise it; after what I have seen these last months, I have the impression that the bombing of towns cannot be of decisive influence in this country.

Throughout 1941, the two-way flow of news about family, friends, and visitors continued unabated. One event, however, is revealing. On 7 January word had come that Omama's younger brother, Carl Lipmann, severely disabled by a stroke, was leaving Germany with his wife Ellie by the only exit still open at this late hour, across Russia by the Trans-Siberian Railway. Franz disapproved:

> A wire from the boy's parents. They are on their way to her brother. I think it is foolish that these two old, ill people embark on such a journey; first the long days on the train and then the sea passage. And what is the whole good for?

Seven weeks later, on the occasion of a visit from "the boy's" sister Hannah, Franz refers to the journey again:

> I am convinced that her parents should not have gone, they must become a burden to the children, it is unavoidable and what was their life there and what can it become in the new surroundings? Such considerations may sound cruel, but they are therefore not less essential.

The next day, 1 March, letters from Ellie arrive with good news. They have reached Kobe, and the worst of the journey is now behind them. On 19 March, 10 weeks after leaving Hamburg, a telegram announced their arrival "in good health" in Montevideo.

Franz's disapproval of the journey is revealing. The three Lipmann daughters in Leeds were struggling economically. Franz had accepted financial responsibility for young Curt, and Hannah too requested his help occasionally. A cruel economic calculus suggested that the family welfare could only be worsened by the move, and that the disabled elderly parents could only fare worse by moving than by staying in Germany.

On this point Franz would be proved mistaken. Although he had the lowest possible opinion of the Nazi regime, he could not yet conceive of the plan that would be unveiled discreetly before a select group of high Nazi officials in conference at Wannsee on 20 January 1942. The minutes of this meeting were recorded and sanitized by Adolf Eichmann, and the resulting documents show that the plan—authorized by Hermann Goering in July 1941—aimed at a "final solution of the European Jewish question" regarding some 11 million Jews in 30 European countries, including 175,000 still in Germany or Austria and 330,000 in England. Against such a background, the Lipmann parents' exodus must be accounted a success, for although Carl died in Uruguay in April 1944, Ellie would later rejoin the family in England and live in London until 1969.

Beginnings: United States

To assess the American situation at this point, it is useful to go back three years. Interest in nuclear fission in the United States began properly on 26 January 1939, when Niels Bohr, on a visit to the United States, reported on the Hahn–Strassmann uranium experiments of 1938 at a conference in Washington of the American Physical Society. His report touched off a wave of experiments on the new phenomenon and led to almost a 100 publications on the subject in 1939 alone. In the midst of the excitement, three Hungarian refugee physicists, Leo Szilard, Eugene Wigner, and Edward Teller, worried about the possible misuse of the new discoveries by Nazi Germany. Over the summer they developed a plan to alert the American administration to the potential energy from uranium by having a letter signed by Einstein and conveyed to President Roosevelt. After delays, the letter was presented on 11 October directly to the President, who acted immediately to set up an Advisory Committee on Uranium chaired by Lyman Briggs, director of the Bureau of Standards.

The Briggs Committee met on 21 October 1939. The three Hungarian scientists, invited guests, explained the quest for a chain reaction amid considerable scepticism from committee members. Nevertheless, the committee recommended further investigation and eventually obtained $6,000 from the military for experimental materials. Research continued in the major universities, but the committee did not meet again until 27 April 1940. This meeting also favoured small, incremental steps. By this time discontent was mounting among scientists at the slow, reactive pace of the committee. When German forces swept over Belgium and France in May, it was time for more decisive action.

In June 1940 Roosevelt established a National Defense Research Committee, headed by Vannevar Bush, to mobilize scientific resources for defence purposes over a broad front. The Uranium Committee was henceforth to report directly to Bush. By early autumn, Bush had reorganized the committee to remove the military representatives, add several leading nuclear scientists—Jesse Beams, Ross Gunn, George Pegram, Merle Tuve, Harold Urey—and exclude foreign-born scientists on security grounds. Research results on topics of military significance, including nuclear uranium fission, published openly until the spring of 1940, were henceforth to be screened by a committee of the National Research Council.

Under the new structures, research continued in the universities for another year, with primary emphasis on developing a chain reaction by slow fission for power development, but also with a growing interest in isotope separation. The Uranium Committee had increased money available for grants, but did little towards coordination or policy development. The neglected part of the American effort was work on fission by fast neutrons, with its potential for explosives. The country was still at peace.

By early 1941, many scientists were discontented and frustrated by slow progress and the lack of any clear programme. Their frustration grew while evidence filtered in of the work of Britain's Maud Committee. This evidence multiplied in 1941 through several channels: from contacts with Britain's scientific representatives in North America; from American scientists attending Maud Committee meetings as observers in April and July; from Mark Oliphant's visit to the United States in August and September, which included talks with Ernest Lawrence at Berkeley; and from the draft Maud Report itself in July. Through all of this, all minutes of Maud Committee meetings and reports to the committee had been sent to Washington, where for security reasons they were safely locked up in Dr Briggs's safe.

Vannevar Bush, as head of the National Defense Research Committee, was aware of these shortcomings and prepared to act more positively. In April 1941 he called on the National Academy of Sciences for a complete review of the uranium programme. In June he secured by executive order a new structure, the Office of Scientific Research and Development (OSRD), to operate within the executive office of the President under a more direct defence mandate. The NDRC would continue on an advisory basis within OSRD, and the Uranium Committee would reappear in disguise as the S-1 section of OSRD. The new structure placed Bush, and his deputy James Conant, who replaced him as head of NDRC, in a position of considerable power, with direct access to President Roosevelt when needed.

The National Academy committee reported on 17 May 1941. Its recommendations stressed full funding support for developing a chain reaction in natural uranium, and continuing work, but with *reduced* emphasis, on isotope separation. It also called for more interchange of research information, and suggested a visit from Halban to hear more of his team's work in Cambridge. It made no recommendation about explosives. To remedy the gaps, Bush asked the Academy that this first report be reviewed from an engineering standpoint. The Academy panel, enlarged by adding two engineers, reported a second time on 11 July, largely endorsing the first report. It stressed again the priority for a chain reaction in natural uranium, but because any increase in concentration of U^{235} would be important, work on isotope separation should not be reduced. Once again, the potential for explosives was not discussed.

Bush was prepared to try again. On 10 July he had heard directly of the Maud Committee conclusions from Charles Lauritsen, the American observer who had attended the crucial Maud meeting of 1 July. Through the summer more British evidence arrived, and on 3 October James Conant received a copy of the final Maud Report from George Thomson personally. On 9 October, Bush moved on two fronts: he called for a third report from the Academy of Sciences, and met with President Roosevelt and Vice-President Wallace at the White House. This time his questions for the Academy panel were very explicit: the feasibility, size, cost, and destructive effect of a bomb based on U^{235}.

In the meeting at the White House, Bush outlined to the President and Vice-President Wallace the data and conclusions of the Maud Report. He emerged from this meeting with a clear mandate from the President to expedite the research and the production planning phases of the uranium project in every possible way. The plant *construction* phase would require further authorization from the President, probably under some new structure. Given the high cost of the construction phase, "it would be best if the necessary construction were done jointly in Canada."[37] Bush was directed to draft a letter to open a top-level discussion on the matter with Britain. Finally, Roosevelt emphasized secrecy, reserving discussion on policy issues to himself, Wallace, Bush, Conant, Secretary of War Henry Stimson, and Army Chief of Staff George Marshall. Two days after this meeting, Roosevelt wrote to Churchill, proposing an exchange of views "in order that any extended efforts [e.g. plant construction] may be coordinated or even jointly conducted."[38]

The third report from the Academy of Sciences, requested on 9 October, came in on 6 November. It was clear and unequivocal, giving Bush everything he needed. Uranium–235 could produce a bomb of "superlatively destructive power." The critical mass needed might be between 2 and 100 kilograms. The separation of isotopes was feasible. With best efforts, the time required might be three or four years, and the overall cost might range between 80 million and 130 million dollars.[39]

On 16 December, in the shadow of Pearl Harbor, Vice-President Wallace held a small, top-level meeting to review the Academy of Sciences report, the management structure being developed by Bush, and the project in general. There was agreement to expedite fundamental research and engineering planning, including construction of pilot plants. On other points, the group concurred with Bush that, when full-scale construction began, the Army should assume direction of the project, and also that, in relations with Britain, general policy questions should be left to Roosevelt. This meeting also looked again at the earlier proposal for a joint Anglo-American production plant in Canada, as envisaged by Roosevelt and Bush at their key meeting on 9 October. With the United States now fully at war, the situation had changed radically.

For the United States, more than for Britain, the closing weeks of 1941 marked a sharp turning point. The official history of the American atomic project labels this point as the "end of the beginning."[40] From August until December, with the Maud Report pointing the way for both countries, prospects for close collaboration on a common project looked attractive for both sides. Yet Britain held back. President Roosevelt's letter of 11 October was not answered by Churchill until December, and then only in generalities. One problem was that Britain, with its indecision over locations, had postponed hard decisions concerning the "extended" or construction phase, whereas Roosevelt had authorized Bush on 9 October to make plans for this phase as an outgrowth from continuing research. The idea of a joint project jointly managed, discussed first in the Maud Committee and welcomed in August by the American leadership, was allowed to wither away to an agreement in November to continue free information exchange while each country pursued its own project and priorities. The November agreement gave the British exactly what they wanted at the time. A joint project with an ally still nominally neutral, and not focused on weaponry, appeared too risky for Britain's embattled leaders. In the light of what would come later, however, Margaret Gowing's official history of the British project identifies British hesitation during these critical months as a major missed opportunity.[41]

Beginnings: Germany

In Germany, the application of nuclear energy to weaponry began in the spring of 1939, initially taking two separate paths. Two Göttingen physicists, Georg Joos and Wilhelm Hanle, pointed out the economic and military potential of uranium in a letter to the Education Ministry. This letter was passed on to Abraham Esau, head of the physics division of the Reich Research Council, who organized a "uranium club" of specialists that met on 29 April to consider the question. On 24 April a similar letter was addressed to the Army Ordnance Office by Professor Paul Harteck of Hamburg University and his associate Wilhelm Groth. Harteck, who had studied with Rutherford at Cambridge in 1932 and later coauthored with Mark Oliphant a paper on his work there, had been since 1937 a consultant for the German army on chemical explosives.

When war broke out in September 1939, Harteck was invited by Army Ordnance to a meeting of nuclear experts in Berlin. It was held on 16 September at the initiative of Kurt Diebner, the head of the ordnance nuclear physics group, and attended by about a dozen leading experimentalists, including Otto Hahn. The topic was how to harness the energy potential of nuclear fission. Despite some scepticism, the group agreed to meet a second time 10 days later, with added theoretical representation by inviting Werner Heisenberg. The invitation to Heisenberg met with some resistance. In 1939 he still had enemies among the Deutsche Physik faction, but his stature as a leading theoretician, in a group composed mainly of experimentalists, gave him a central role in the project.

On 5 October, the Army Ordnance Office requisitioned the Kaiser Wilhelm Society's Institute for Physics for war work. In realization of a lifelong dream of Max Planck, it had recently been completed, with Rockefeller Foundation money, opening in January 1938. Its first director, the Dutch physicist Peter Debye, was told to become a German citizen or resign. In the end, he secured a leave of absence with pay and a guest professorship at Cornell University.[42] For the time being, Army Ordnance was firmly in control of the uranium programme, and Esau's group at the Reich Research Council was ordered not to proceed further in this area. Diebner became the new administrative head of the institute, with Heisenberg and Hahn as senior scientific advisers on fission studies. The rumours passed on in Thomson's letter to Cherwell of July 1941 concerning takeover of the institute by the Army were indeed largely correct.

Other parts of the German programme were less obvious to the external observer. As in Britain and the United States, research was carried on in universities and other institutions. Heisenberg worked on chain reactions at his institute in Leipzig. Harteck pursued isotope separation in Hamburg, and Walther Bothe, a leading nuclear physicist of long standing, directed research on nuclear properties in his laboratory at the Kaiser Wilhelm Institute for Medical Research in Heidelberg.

Bothe's career to this point is noteworthy. An early pupil and friend of Max Planck, with a distinguished research career in Berlin during the 1920s, he was appointed professor and Director of the Institute of Physics at Heidelberg University in 1932 on the retirement of Philipp Lenard. After the Nazi seizure of power, Lenard and his colleague Johannes Stark were influential enough to force Bothe's ousting from the university as an exponent of the suspect "Jewish" physics. Max Planck, as head of the Kaiser Wilhelm Gesellschaft, intervened successfully in 1934 to find Bothe an alternative post as director of the Physics Department at the society's nearby Institute for Medical Research. Freed from teaching and academic infighting, and better funded than before, Bothe profited from the change, isolating himself and his team from the ideological battle in the universities to pursue fundamental research.

Heisenberg and Harteck, from a later scientific generation, were more actively interested in applying the new discoveries. Heisenberg began immediately to prepare two reports on energy possibilities from uranium fission, which he delivered to Army Ordnance on 6 December 1939, and 29 February 1940. They concerned both energy from natural uranium through a controlled chain reaction and nuclear explosives from enriched uranium. Distributed in the restricted Army Ordnance series, they were welcomed by colleagues as a blueprint for the road ahead.[43]

As formulated by Heisenberg, the choices ahead took the form of a dilemma. If ordinary uranium were used, it would require either heavy water or pure carbon—both scarce and expensive—as a moderator. If natural uranium could be enriched sufficiently, a reactor might run on ordinary water. By either route, the end product could be a nuclear explosive. Thus from an early stage the power project and the quest for explosives were recognized by German scientists as complementary, and this was an argument for continuing Army support. When carbon was ruled out in favour of heavy water in January 1941, the choice became a narrower one between more heavy water or more enrichment of the uranium fuel.

Among the enthusiasts for Heisenberg's blueprint was Paul Harteck, who in 1934 had succeeded to the Hamburg Chair in Physical Chemistry left vacant by the resignation in 1933 of Franz's friend Otto Stern. Harteck and Wilhelm Groth began work on uranium isotope separation in their Hamburg laboratory. Their chosen method was by the thermal "separation tube" developed in 1938 by two Munich physical chemists, Klaus Clusius and Gerhard Dickel.[44] In a circuit joining parallel heated and cooled vertical tubes, the lighter molecules of a gas mixture would predictably rise on the hot side and sink on the cooler side, allowing lighter isotopes to be progressively concentrated and collected. After successes with other elements, the Clusius–Dickel tube was considered superior from the standpoint of initial cost, energy needs, and space requirements to separation by gaseous diffusion through a porous barrier, a method developed by Gustav Hertz in the early 1930s. Harteck and Groth had successfully separated isotopes of mercury vapour by the Clusius–Dickel method in 1939.

Harteck, already well connected with Army Ordnance, kept several wartime projects moving simultaneously in his laboratory. For uranium isotope separation, he designed a tube apparatus, 6 metres tall, which required 65 kilograms of scarce nickel. After substantial delays, Army Ordnance intervened to obtain the material for him and construction began in October 1940. He obtained a uranium sample to make the hexafluoride, and in May 1940 he was allocated 200 kilograms. of uranium, a significant proportion of the total stock available at that point. In February 1941 he reported to Army Ordnance that prospects for separation looked very favourable, but from this point onwards his reports become more muted. With the hexafluoride, the nickel tubes developed serious, uncorrectable corrosion problems, and, worst of all, no enrichment could be shown at any temperature range. By September 1941 the Harteck team had given up on the Clusius–Dickel method.

With growing disillusionment over the Clusius–Dickel tube, attention turned to other plans for isotope separation. These included centrifuges, electromagnetic techniques, and an ingenious proposal by Erich Bagge for an "isotope sluice," based on revolving shutters that would separate molecules travelling at different speeds along a molecular beam. As 1941 drew to a close, a centrifuge for the Hamburg group was under construction but not yet tested, Diebner had authorized a contract for a prototype of Bagge's isotope sluice, and at least four scientists in different locations were experimenting with electromagnetic approaches, though Germany still had no operative cyclotron. The problem was that some methods would not work with uranium or its compounds, and others could separate only minute laboratory samples. By December 1941, Harteck had to admit that prospects for industrial-scale separation of U^{235} were nowhere in sight. In this context, Heisenberg's alternative option of a heavy water reactor using natural uranium, with later prospects for fissile material from transuranic elements, looked more feasible.

Something, however, is missing here. One form of separation not being attempted in Germany at this date is Gustav Hertz's method of gaseous diffusion, which Simon had been experimenting with in Oxford since the spring of 1940. The contrast with Britain stands out more clearly if we look again at his first meeting with the Maud Committee on 17 September 1940. Simon's notes from that occasion show that the two separation methods were discussed and compared one after the other at the start of the technical meeting: "Discuss methods of isotope separation. All agree that Cl[usius] D[ickel] does not seem too hopeful for big scale, but might do for small, await Frisch's experiments (have impression that no result before 1–2 months can be expected)." In the next paragraph: "Diffusion, Hertz is discussed." The discussion moves to ideas for suitable filter materials, and Simon adds: "I give results for rolled gauze wires . . . . Give data for possible improvements . . . . Show sketches, how I believe procedure can be carried out technically . . . . Seems to find no objections."[45] Though solutions might be far in the future, a key issue had been identified and addressed.

But what had happened to Hertz, who was still in Germany? Gustav Hertz, born in 1887, a nephew of the discoverer of Hertzian waves, completed his doctorate in Berlin in 1911 and shared the Nobel Prize in Physics with James Franck in 1925. He became Director in 1928 of the Physics Institute of the Berlin Technische Hochschule, where he separated isotopes of neon by a diffusion cascade. After the Nazi seizure of power, he was at first protected from dismissal as a war veteran, but in 1934, after President Hindenburg's death, he refused to take a new loyalty oath to the Führer—Hitler's new title for the combined office of Chancellor and President of the Reich. In November of that year he was investigated by the Nazi Teachers League but was defended by none other than Johannes Stark, the Deutsche Physik partisan who three years later would denounce Heisenberg as a "White Jew." There is nothing "Jewish" about Hertz's science or conduct, Stark informed the League, and it would be senseless to ban him from teaching on the ground that one grandfather had been Jewish. Despite this backing, he was stripped of the right to teach and to examine students.

Hertz then resigned his university posts and in 1935 joined the Siemens firm as director of an industrial research laboratory. As a one-quarter Jew by the Nuremberg categories, he continued living and working in Berlin until 1945. At war's end, he would move with other German scientists to the Soviet Union, where he would head a large research laboratory and direct work—successfully—on uranium isotope separation by gaseous diffusion.

In the first week of December 1941, the top physicist in Army Ordnance, General Erich Schumann, announced a thorough review of the nuclear power project, with Army Ordnance scientists preparing their own assessment for senior military planners. That report, completed by February 1942, would emphasize once more the importance of nuclear energy for Germany's economic future and military potential, and it would press urgently for its further development, on an industrial scale, with the highest priority possible. The same report, however, had to acknowledge that uranium isotope separation was not yet achievable, and that a working reactor was a prerequisite to discovering the potential of element 94. In short, the ultimate goal was too important to abandon but not immediately achievable. Army planners, with a horizon focused strictly on research relevant to the war, were forced by increasing constraints to review all their options. Through three winter months, the nuclear scientists would have to defend their claims to continuing support from an army facing increasing burdens.

The background for this review was a slowing down of the German offensive on the Eastern Front and increasing Soviet counterattacks. The German war economy faced growing shortages of strategic materials and manpower. In the universities, physics students were being called up for military service upon completion of their doctorates. Early in January 1942, Harteck and von Weizsäcker, key figures in the nuclear programme, were themselves called up for service on the Eastern Front. Through Heisenberg's influence, their deferred status was reinstated, but only with difficulty.

On 11 December, four days after Pearl Harbor, Germany declared war on the United States. The war was taking on vastly different dimensions. Just as in Britain and in the United States, the closing weeks of 1941 would mark a turning point in Germany for its nuclear programme.

Beginnings: USSR and Japan

Other countries, most notably the Soviet Union and Japan, had their own capable physicists. Prerevolutionary Russia's scientific tradition, already noteworthy, was pushed strongly after the 1917 Revolution. By the early 1930s, the USSR had several institutes of nuclear physics. By 1940, the Radium Institute in Moscow had a working cyclotron, and Joffe's institute at Leningrad was building the largest cyclotron in Europe, to be completed in 1941. Soviet physicists, like others worldwide, had been excited by the discovery of nuclear fission, and in 1939 they plunged into experiments exploring the new discoveries.

It was William Laurence's front-page story on atomic energy in *The New York Times* on 5 May 1940 that transformed the subject from exciting laboratory experiments to an issue of state policy. His article had highlighted the vast potential of uranium 235 as an explosive, and the 1939 takeover for war work of the Kaiser Wilhelm Institute in Berlin.[46] The *Times* article was forwarded to a veteran mineralogist and Academician, Vladimir Vernadskii, by his son, then teaching at Yale. The same story at one remove—it was sent to *The Times* of London by that paper's New York correspondent—had prompted Simon's letter to Lindemann on 7 May to urge that something should be begun in the Clarendon. Laurence later wrote that, after his article appeared, "nothing happened for two months,"[47] but he was mistaken. His message had been heard—and acted upon—in Britain and in the Soviet Union.

Vernadskii, excited by the story, swung into action. On 25 June, a "troika" was formed in the Academy under Vernadskii's chairmanship to study the issue. They wrote memoranda to the Academy, and on 12 July a letter to Deputy Premier Bulganin to alert the government to the importance of fission. The government approved formation of a Commission on the Uranium Problem, and its membership—nine Academicians and four professors—was announced on 30 July. Through the autumn, winter, and spring the new commission developed a research agenda: understanding the chain reaction, selecting methods for isotope separation, completion of the cyclotron projects, locating necessary raw materials. Uranium supplies in the USSR were extremely limited, even for laboratory-level needs.

Through 1940–1, work went on actively but "without any great urgency."[48] The USSR was at peace—after brief campaigns against Poland in 1939 and Finland in 1939–40—and the military aspect of nuclear energy hardly surfaced at this stage. Yet scientists who followed foreign scientific journals were aware that articles on fission suddenly disappeared in 1940, and drew conclusions accordingly. For the 77-year-old Vernadskii, his excitement over fission was fuelled by a grander, longer view. In a letter thanking his son for sending the clipping on 5 July 1940, he wrote: "I think now that the possibilities that are being opened up for the future here are greater than the application of steam in the XVIIIth century and of electricity in the XIXth."[49]

On 22 June 1941, Hitler unleashed against the USSR a blitzkrieg like the one that swept over the Low Countries and France a year earlier. The Soviet forces, although alert and apprehensive after the German conquests of 1940, were unprepared, and fell back in disarray. Within days of the attack, scientific research was reorganized under a science "plenipotentiary" and redirected to urgent military problems. The Uranium Commission ceased its work, and nuclear scientists were reassigned to more pressing problems or to military service. Thus, at the end of 1941, Soviet nuclear development had entered what one Western analyst has called "the dormant period."[50] Even for a project in dormancy, however, its proponents can dream. One who did was Georgii Flerov, a young researcher from Igor Kurchatov's nuclear group at the Leningrad Physicotechnical Institute. Together with a colleague in Moscow, the pair had won distinction in 1940 as the first to discover and measure spontaneous fission in uranium. As an air force lieutenant, Flerov was allowed a one-week pass in December 1941 to attend a physics seminar in Kazan, and to make a strong, well-grounded plea to Joffe and others for renewal of uranium research. With German forces close to Moscow and the government evacuated, the moment was not propitious, but Flerov would continue his campaign in 1942, and carry it up to the level of Stalin himself.

In Japan, as in Germany, the military authorities acted promptly from the beginning.[51] General Yasuda, the Imperial Army's Director of Aviation Research, had followed the literature on fission of 1938–9, and in April 1940 he commissioned a member of his staff to investigate military possibilities. By October, Lieutenant Colonel Tatsusaburo Suzuki reported back that Japan had access to sufficient uranium in Korea and Burma, and that a nuclear bomb was therefore possible. A leading physicist, Yoshio Nishina, who had trained in Copenhagen under Niels Bohr, was chosen to direct research towards a bomb for the Army. At his Tokyo laboratory, located in the Riken scientific research complex, he had built a small cyclotron and in 1940 was building a larger 60-inch model, with the aid of an assistant who had trained at Berkeley with Ernest Lawrence. In December 1940 work began on measuring nuclear constants. In April 1941—three months ahead of the British Maud Report—an order from the Army Air Force authorized research towards development of a nuclear bomb.

Nishina's team at the Riken was not the only Japanese group investigating uranium. On 23 May 1941, Tokutaro Hagiwara, a physicist at Kyoto University, gave a lecture to the Second Arsenal of the Imperial Navy titled "On the Super-Explosive Atom, U^{235}," which was printed two months later in the unit's internal journal. This lecture reviewed existing fission literature and emphasized the urgent need to achieve isotope separation. The key message in this printed version was the possibility of a super-explosion: "If in some way it becomes possible to manufacture a fairly large amount of uranium 235 and mix it with suitably concentrated hydrogen on an appropriate scale, the uranium 235 is expected to have a high probability of causing a super explosion."[52] From 1942 onwards, the Japanese Navy would support nuclear research at Kyoto University, but this phase and further work by the Army-supported Nishina team in Tokyo belong properly in a later chapter (see chapter 7).

6

INDUSTRIAL PLANTS... HERETOFORE DEEMED IMPOSSIBLE
1942–1945

> You have built cities where none were known before. You have constructed industrial plants of a magnitude and to a precision heretofore deemed impossible.
> Groves to all members of Manhattan Project, 23 December 1946. Hewlett and Anderson, 1962, 655.

Mobilization and cooperation

In a single day, Pearl Harbor changed the scale and nature of the war. It also changed fundamentally the tempo and outcome of the bomb project. It did so by triggering a massive mobilization of American resources that would have been inconceivable in the absence of a sudden, damaging, unprovoked attack. The nuclear research programme, huge in itself, represented only one small fragment of the mobilization effort. In the bomb project, as in other directions, American effort would overtake and then overshadow the efforts of Britain, Germany, or Japan within a few critical months in 1942.

These developments would have deep consequences for Franz's life and work, and for his family. For more than two years he would lead a split existence of another kind. In Britain, he remained a key figure in the Tube Alloys team and project leader of the gaseous diffusion group. In North America, he would spend about nine months on five separate visits between January 1942 and May 1944, working with the teams of scientists and engineers who were breaking trail towards a new age. In a year-end letter to Lotte on 22 December 1942, he could reflect on this divided life: "Incidentally, I have spent this last year 40% of my time in America, 40% in Oxford and 20% travelling in England!"

The first of these overseas visits poses problems of documentation. When he travelled to North America in January 1942, the family letters stopped. Unlike his earlier American visits, no exact itinerary has been found for this one, whereas after his return to Britain in April he began to keep a different style of diary, a Lefax pocket organizer which he carried about with him and continued to the end of his life. The new diary and calendar list places, travel times, appointments, committee meetings, laboratory and factory visits, and even leisure-time activities, though normally the subject matter of these contacts is not indicated. After Franz's death, Lotte transcribed a substantial portion of the diaries—Franz's original script is microscopic and difficult—beginning from Sunday 19 April 1942. She knew, or had met, a good many of the individuals named there, and later generations owe her an immense debt for making the entries more accessible.

A high priority was the long-delayed family reunion after 18 months of separation. As on later visits, reunion began in New York, where a temporary flat served as a base of operations and as "home" for the time being. Rudolf Peierls recounts that Akers and Simon flew across the Atlantic on this occasion by Clipper via neutral Lisbon, a city "full of spies who watched the transients,"[1] where German and British planes drew up on the airport tarmac side by side. Family legend has it that Akers and Simon spent a few days in transit here, chained securely to locked diplomatic bags, which were unlocked only on arrival in America. Halban, who had heart problems, travelled by sea, while Peierls, whose work in nuclear physics was widely known, was sent by bomber from Prestwick via Newfoundland to Montreal.

"Jointly or separately," Peierls notes, "our delegation visited many of the laboratories where atomic energy studies were going on."[2] These included most obviously Columbia, Chicago, and Berkeley, but research was not yet so concentrated geographically as it would become later. Peierls and Simon went together to the University of Virginia to discuss centrifuges with Jesse Beams. The lack of coordination had advantages. Project scientists were able to move freely, and to exchange information on nuclear developments in both countries that had been restricted from normal publication since 1940.

By 19 April, Franz was back in Oxford, writing to Lotte of his return trip by Ferry Command bomber, 13 hours non–stop from Montreal in thick flying suits at 20,000 feet to "somewhere" in Scotland, using oxygen masks the whole way, and subsisting on frozen sandwiches and chocolate ("But chocolate at -30^0 does not taste for anything!"). Peierls was not yet back, and Akers was delayed again in Lisbon. Despite laboratory visits and extensive exchange of scientific information, the British visitors do not appear to have been told how rapidly American top-level planning was evolving during these months. Only in June, when Akers's deputy Michael Perrin visited North America on other business, did the accelerating American tempo—and its implications for joint development—become clear. In an urgent letter to Akers written just 4 days after arrival, Perrin warned that, except for Simon's work on diffusion and Halban's on the "boiler," the American effort had largely overtaken or surpassed British work: "The time available for making any plans for co-ordinated work between the Americans and ourselves is extremely short and probably less than a month from now."[3]

In Britain, the top decisions lay with Sir John Anderson's Tube Alloys Consultative Council. At its June meeting, the council hesitated to endorse proposals of its own Technical Committee for joint Anglo-American control structures and joint experimental work on U^{235}. At the end of July, after further technical review, the Consultative Council agreed to recommend that a full pilot plant on the British diffusion model be constructed in the United States under joint Anglo-American control. Their reasoning was outlined clearly in Anderson's explanatory minute to the Prime Minister on 30 July 1942:

> It is still considered probable that our method is the best; but it has little chance so long as work on it is handicapped by the limited resources in this country . . . we must now make up our minds that the full scale plant for production according to the British method can only be erected in the United States and that, consequently, the pilot plant also will have to be designed and erected there. The immediate effect of this would be that, whilst certain of the more academic research work would continue to be carried on in this country, we would move our design work and the personnel concerned to the United States. Henceforth, work on the bomb project would be pursued as a combined Anglo-American effort.[4]

After Churchill approved, Anderson wrote to Bush on 5 August, stressing that the pilot and full-scale plants should be built by the same organization and treated in the same way as the other projects for fissile material.

Bush's reply on 1 September was diplomatic but not encouraging. With four different methods already being supported in the United States in the race for fissile material, he was reluctant to add a fifth contender. In a further letter dated 1 October, he outlined the latest, far-reaching American policy decisions and organizational changes. The programme now included top priorities for fundamental experimental work, three pilot plants—diffusion, centrifuge, and graphite pile—a full-scale electromagnetic separation plant to produce 100 grams of U^{235} per day, and a heavy water plant. These plans would be reviewed and upgraded again before year-end as the American programme moved into still higher gear. Instead of a jointly run model, Bush suggested continuation of parallel development with close liaison between them, leaving further integration to be resolved after the results of the diffusion methods were available. He also suggested that Akers come to Washington again to work out details.

Vannevar Bush, a shrewd and capable administrator, lost no opportunity to build, expand, integrate, and streamline the American organizational structure, always with a view to the ultimate objective: delivering a bomb at the earliest possible date. Single-mindedly, he viewed everything, including the scope of Anglo-American collaboration, in strict cost–benefit terms. His progress through 1942 in incorporating new stages of the project and accommodating apparently irreconcilable viewpoints was quite remarkable. In January, Eger Murphree's planning board of engineers had begun to meet, concerned with industrialization of laboratory-scale results and with supplies of needed raw materials. On 9 March, in an optimistic progress report to the President, Bush had raised the issue of involving the Army. The President concurred. On 23 May, a systematic review by the S-1 Committee of the most promising projects for fissile material reported that "there were now at least four on equal footing" and recommended support "at full speed"[5] for all four, on the grounds that a premature choice among them might become a determining factor in the war. On 17 June, Bush reported again to Roosevelt to suggest a division of responsibilities between the office of Scientific Research and Development (OSRD) for the scientific side and the Army for plant sites and construction. On the same day, the Army designated Colonel James Marshall to take over the as yet unnamed project.

The summer months proved disappointing. The new structure had no clear demarcation of responsibilities, and no overall authority to resolve differences. Colonel Marshall was unable to secure the high priority level that the project needed for access to essential materials, and was unsuited by temperament to do constant battle with the military bureaucracy. Though a suitable site for plant construction had been identified in Tennessee as early as April, Marshall held back on acquiring it until the research outcomes could clearly justify the need for a large-scale plant.

A General takes charge

By September, Bush expressed his frustrations at the highest levels. His concerns brought action more rapidly than he expected. On 17 September, another Colonel, Leslie Groves, appeared unannounced in Bush's office and introduced himself as the new Director for the

Army of the entire project. The appointment had been made by the head of Services of Supply, General Somervell, from his own Quartermaster Corps construction team, without consulting Bush. Bush's impressions at this first meeting were not good. He was quite overcome by the brash arrogance of the man, his sharp tongue, contentious spirit, and conspicuous absence of elementary courtesy or tact. Bush protested both the procedure and the specific appointment to Somervell's chief of staff, but without success. He then recorded his dissent in a written memorandum, ending with a final comment: "I fear we are in the soup."[6]

Groves's lack of manners was generally admitted by his superiors, but they also knew him as an officer who could get things done. His current assignment was to oversee construction of the Pentagon, on which construction had begun just 12 months earlier. It would be completed by January 1943. He accepted the new assignment reluctantly. He had wanted an overseas field command, but found some compensation in a promise of promotion to Brigadier General. At his insistence, this promotion had to be implemented before the new assignment was officially announced.

Groves was already familiar with the project, having watched the summer slowdown from a vantage point in the Corps of Engineers in Washington. With new orders signed on 17 September "to take complete charge of the entire [Manhattan] project," he lost no time in getting to work. On 19 September, he signed a directive to acquire the Tennessee site. On the priorities issue, he wrote a letter to himself, cleared it with his Army superior, and took it personally for signature to the civilian chairman of the War Production Board, Donald Nelson, who signed with little resistance. The letter authorized the top–most priority for needed materials, a status that the military bureaucracy had hitherto refused to consider. In two days, Groves solved two problems that had clouded the project throughout the summer. In the same two days, he also moved to buy the 1,250 tons of uranium ore from the Belgian Congo that had been shipped to the United States in 1940 and stored on Staten Island. On 23 September, high-level discussions produced agreement on a new Military Policy Committee to resolve disputes and oversee the project generally. At Groves's insistence, this committee would have only three members, representing the Army, the Navy, and OSRD. The new arrangements were devised for rapid action on very large decisions.

These sweeping organizational changes coincided closely with Franz's arrival for a second visit to North America. He flew from Foynes in Ireland by Clipper, via Botwood, Newfoundland, to reach New York on 14 September, for a stay of some nine weeks. This was primarily a diffusion visit. He was supported by two British specialists in membrane or barrier research, initially by Michael Clapham, whose starting point was from photoengraving processes, and later joined by S.S. Smith, the research manager of ICI Metals Division. In the absence of Wallace Akers, Franz remained in daily touch with the local ICI staff and also wrote extensive reports back to Akers on the rapidly evolving American research scene. He also attended a meeting of the S-1 Executive Committee in Washington on 26 September to report on the British diffusion project. On top of several crowded weeks of meetings, laboratory visits, and individual appointments, there was time for family reunion and for seeing colleagues, friends, and cousins.

In the diffusion project, the most critical problem was to develop an efficient membrane or barrier for separating the lighter U^{235} isotope. The terminology differs here, for British research had begun from metallic sheets with thousands of microscopic holes to filter the gaseous hexafluoride, while American work already included trials of other barrier materials.

By mid-1942 researchers were experimenting with filters formed from powdered nickel. Copper mesh had been rejected as unsuitable. Despite the massive American effort, the British leadership believed that British research was still ahead in some respects, and that full information exchange could only work to mutual advantage. For the diffusion process, the stakes were high. Any small gain in barrier efficiency, in a process involving up to 4,000 stages, held out the promise of saving hundreds of units in the separation plant.

The academic focus of Franz's second visit was Columbia University, where Harold Urey directed the diffusion project for the S-1 Committee, and where much fundamental and applied physical and chemical research was concentrated on the new substances. Franz's diary gives a very exact chronicle of this visit, listing travels, meetings, and even meal times almost hour by hour. During the 40 days he spent in New York, the diary shows 34 morning or afternoon meetings and another 34 individual appointments with colleagues at Columbia, plus another 15 lunch meetings at the Faculty Club, usually with one or more of the various team leaders.

One of the earliest meetings (22 September) brought Franz and Michael Clapham together with Edward Norris and Edward Adler, two researchers who were developing a barrier of metallic nickel. More meetings on barrier problems would follow, with Clapham or Smith or both present when available. By the end of 1942, the type that became known as the "Norris–Adler" barrier would be a much-improved result of developing and testing "literally hundreds of different nickel barriers,"[7] using either metallic or powdered nickel. In September, however, as the official history notes, "Both types of nickel barrier . . . were far behind those employing other metals in terms of process development, and there was doubt that either would be ready before the end of the war."[8] In retrospect, the autumn of 1942 was an ideal time for assessing the experience of both countries with all types of barrier research, since neither country was within sight of an efficient solution.

If barriers were the top problem, the second was developing suitable pumps. Both centrifugal and reciprocating designs were under intensive study, and the problems in either type concerning vacuum seals, lubricants, and corrosion damage from the unknown hexafluoride posed unprecedented challenges to the industrial engineering of the day. By the end of 1942 almost 200 scientists, in five divisions, would be working on the diffusion project centred at Columbia.

The diary, read with care, reveals the range of questions discussed during the days in New York.[9] A few meetings are noted by specific label, such as "Chemical Method Meeting" or "Chem. Group," or "Corrosion Meeting" (twice). A few are labelled simply "General discussion." For most others, and all the one-to-one appointments, much can be gleaned from the Columbia research described in detail in the official history. Discussions with Urey are recorded almost daily, and some other names appear several times. On 20 October, Franz, along with Urey and Ernest Lawrence, had a day-long visit to Princeton, where other important experimental and theoretical work on the diffusion project was under way.

Among the large contractors, the industrial phase of the diffusion project was already at the planning stage. On 25 September, Franz records an afternoon "Meeting with Kellogg and Physicists," the Kellogg Corporation having already been chosen as the main contractor to build both a pilot diffusion plant and the eventual full-scale plant in Tennessee. This meeting, the first of four, was apparently at Columbia; the second, on 7 October, was a visit with Smith to the Kellogg laboratory in Jersey City, where a 10-unit cascade was being built under the direction of Kellogg's chief project engineer, Percival Keith. Two further sessions with Kellogg engineers in the firm's New York office on Broadway followed on 19 and 21 October, and on 3 November the diary shows a talk with Keith himself.

Other industrial contacts during the visit included two visits to the Bell Telephone Laboratories at Murray Hill, New Jersey, where Foster Nix headed a team developing a barrier based on powdered nickel (6 and 31 October). On 8 October Franz travelled overnight to Pittsburgh to visit the Westinghouse Research Laboratory, which had several contracts for the diffusion project, including work on pump designs. A final industrial visit was a two-hour journey to Phillipsburg, New Jersey, on 2 November to Ingersoll-Rand, which had worked on large centrifugal pumps and seals. Early in 1943, however, this firm would withdraw from the pump assignment and be replaced by another contractor, Allis-Chalmers.

In the absence of Akers, the Tube Alloys Chief Executive, Franz accepted certain wider responsibilities. His diary records almost daily liaison by telephone with the British Central Scientific Office (BCSO) in Washington and the ICI office in New York. It shows sizeable periods spent drafting letters to Akers to keep him informed on the fast-moving American scene. One of his earliest acts after arrival in the United States was to head for Washington with Clapham, to visit the BCSO. The next day they had a 45-minute meeting with Bush himself, and then a meeting at the British embassy before their return to New York.

In the following week Franz returned to Washington twice. On 23 September he met his close friend and colleague from the Bureau of Standards, Brickwedde, at a meeting at Columbia, and later had dinner with him "at home." They then returned together to Washington, where Franz stayed with the Brickweddes. In the morning he accompanied Brickwedde to the Bureau, where Franz had an hour with the Director, Lyman Briggs, and a further half-hour with Conant before revisiting BCSO in the afternoon. He returned to New York for his first meeting with the Kellogg group on 25 September, but then caught an early train back to Washington on Saturday 26 September.

This second trip to Washington is documented by a laconic line in the diary: "12.30 Executive Committee, my report. 1.15–2.15 Lunch with Comm[ittee]." Probably this report dealt with British diffusion experience to that date, and Franz's appearance before the S-1 Executive Committee was to support Anderson's proposal to Bush that a jointly sponsored pilot plant on the British model should be built in North America and assessed on an equal basis with the other methods of isotope separation in the race for the bomb. When Bush wrote to Anderson in detail on 1 October, he suggested that consideration of joint efforts be deferred until test results from both countries' pilot units were available. Bush knew that events were now moving at vastly different speeds. The American programme was already backing four different competing entries in the race. He also knew that the newly formed, streamlined, three-man Military Policy Committee at the summit would be impossible to combine with any form of joint control.

As the diary makes clear, Franz attended only his own part of the S-1 meeting. After lunch with the committee, his afternoon was mainly clear. At 6 p.m. he went to the Cosmos Club, where he dined with Conant, Lawrence, and Arthur Compton until 8 p.m. After dinner, there was an extra half-hour to chat with Ernest Lawrence before catching the 9 p.m. train to New York. Had he been invited for the entire S-1 meeting, he would have heard Groves's latest request. The general asked each project leader for "a statement of needs"[10] regarding their eventual full-scale plants—materials, building layouts, site requirements. Nine days after taking charge, he was making plans for major construction on the Tennessee site.

Another part of the Tube Alloys project, the Halban group's heavy water research, was being moved at this point from Cambridge to North America. Recommended by the Maud Committee in 1941, the transfer to Canada was decided on a year later. The essential requirement was proximity to—and collaboration with—American reactor work at Chicago. Anderson proposed the transfer in a letter to Bush on 5 August, 1942, and with Bush's agreement the proposal was taken to the chairman of Canada's National Research Council, C.J. Mackenzie, on 17 August. The Canadian government gave a rapid approval in principle, with details to be worked out in October. On 16 October Franz appeared in Ottawa with Halban and the ICI administrator, Frank Jackson, for a meeting at the British High Commission. Those attending included the High Commissioner, Malcolm MacDonald, his deputy Gordon Munro, and the scientists planning the future Montreal laboratory. Simon's diary lists no purpose for the meeting, though the single word "Telegram" appears mysteriously in both morning and afternoon sessions. He had flown up from New York for a one-day visit to Canada, and probably he signed or witnessed—possibly as Akers's deputy—an agreement or other documents relating to the project. After the session, he flew to Montreal and caught the night train to Boston, for a one-day visit with Harvard colleagues en route back to New York.

Halban, a close friend and fellow member of the Tube Alloys Technical Committee, figures more than a dozen times in the diary during Franz's visit, with meetings from the morning after his arrival (15 September) to the day before his departure (15 November). On 28 September they met at some length in New York, after Halban returned from discussions in Ottawa. They lunched the same day at Columbia with Urey, Dunning, and others. After lunch, they continued discussion with Urey. On 16 October, as already noted, both Halban and Simon were in Ottawa, and after the 27th, when Akers appeared, they met in three-way sessions to brief him. By early November, Halban had moved to Montreal. He would see Simon again when the latter was awaiting his return flight to Britain.

With Akers present and briefed, it was time for Franz to return to Britain. As November arrived, he had a few last meetings and appointments in New York; farewells to colleagues, cousins, and friends; an evening with the family at Rockefeller [Radio City] Music Hall; papers to be sealed at the consulate; a last talk with the Kellogg project's chief engineer, Percival Keith. Then on to Washington for two days with a similar agenda: farewells to friends and colleagues; lengthy, intensive meetings with Akers; visits to the Embassy and BCSO; documents to be sealed; dinner with Dr and Mrs Briggs. On 5 November, he left by overnight train for Buffalo and Toronto. The official mission was complete.

In Toronto, he had 7 days to unwind, enjoy the family, and meet friends. The diary records time to relax, walk with his daughters, go shopping with Lotte, take photographs, catch up on sleep, and eat a few meals "at home" in the Walmer Road flat. They had an evening with the Mendels, and another with the Fischers. Hermann Fischer, an organic chemist at the university, was another exile from Berlin. There was time to see other evacuee children from Oxford; Franz made a point of telephoning their parents after each of his visits to Toronto. He also called on Eli Burton, head of the university's McLennan Laboratory. On 12 November, he left Toronto by the night train to Montreal.

In Montreal, two more days of meetings were planned with Halban and ICI representative Frank Jackson. The group there was beginning to expand. Franz had met Pierre Auger in New York, and now Bertrand Goldschmidt, both originally from Paris but newly recruited in the United States. Both would have senior positions in the Montreal group. On the second day, 14 November, the diary records an afternoon meeting in "the Building," undoubtedly the rented mansion on Simpson Street[11] that served as an interim base while the leaders looked—without success to this point—for suitable accommodation for the new laboratory. On 16 November, after a one-day postponement, Franz's plane left for Gander and Prestwick.

At that point in mid-November, the outlook for the Tube Alloys project looked generally good. Anglo-American information exchange was working smoothly, and to all appearances effectively. Problems were being solved. Occasional negative signs appeared, but they hardly seemed significant. Franz sensed American reluctance to take on an additional pilot plant to test the British diffusion process, and commented on it in his report to London: "I would guess that the real reason is that the Army people want to run the thing 100 per cent American or something similar."[12]

One consequence of Army control and a vastly expanded work force was a much enhanced security system, based on restricting access to information to areas needed for each employee's particular tasks. This policy of compartmentalization, though resented by many scientists and often blamed on the Army, was strongly backed by Bush and Conant, who indeed shaped the detailed rules for its application. Although compartmentalization was barely visible during Franz's visit, the security system set up by General Groves to monitor it was already in place. As Gowing notes:

> Already the United States Army, which was checking calls and visits, was inquiring why Professor Simon had been in touch with Ernest Lawrence at Berkeley since the British were doing no work on electromagnetic separation.[13]

That the two had been colleagues at Berkeley in 1932, and close friends for over ten years, was deemed insufficient grounds for crossing the barrier between sectors. In time, this kind of contact became highly sensitive, requiring advance clearance. It would be left to Akers, however, who stayed on until February 1943, to discover the full implications of compartmentalization for the work and morale of Tube Alloys in Britain.

Around his two extended absences in North America, totalling about five months, Franz fitted the remaining weeks and months of 1942. It was not easy to return to the old patterns. The stress of family separation remained, unrelieved. After the first 1942 visit, he ended a letter of 10 May by admitting the problem:

> My dearest Lo, it is much more difficult to get accustomed to the separation again than I thought, I am restless and longing for you and thinking of you the whole time. But we both know that it is the only reasonable solution and that it would be sheer foolishness to try to change anything – we must take this separation as part of our war effort.

For Lotte, thoughts of return to Britain were seldom far away, and not always amenable to reason.

The main new element in the British diffusion project was the pilot plant envisaged for a Ministry of Supply industrial site in the Rhydwmwyn Valley, near Mold in north Wales. Franz refers to it casually in a long letter of miscellaneous news on 24 May. Nancy Arms, his secretary, who had recently married his former student and research associate Shull Arms, was now pregnant:

She does not yet stop working, but will probably go with part of our group to a place where we are beginning experiments soon. I'll also spend a part of my time there. They are now looking for a house for us – we'll be about 12 people – and I think that from July on I'll be there at least half of my time.

These plans would later be deferred somewhat. Franz's diary and calendar show brief visits to "Valley" and to "the house up there"—as the factory and the residence would be known—in July, August, and September, but longer visits do not appear until December, after his second return from North America.

In other respects, the project continued in Oxford and the scientific staff expanded, spilling over from the Clarendon into the chemistry laboratory of Jesus College. Franz's committee work in London required several days a month. There were the usual meetings with Cherwell, and some extra time in August to prepare a jointly written obituary of Nernst for the Royal Society. Plant visits continued, to Metropolitan Vickers in Manchester, the contractors building the pilot plant, and to firms researching meshes and membranes. The diary mentions a meeting in Birmingham with a Mond Nickel Company representative on 12 May 1942, followed by several more over the next 13 months. The implication seems clear. The search for an efficient barrier is being widened at this point from metal meshes to powdered nickel.

Franz's familial and social life in Oxford appear little changed, with regular dinners in College and occasional Royal Society meetings in London, including "a kind of celebration . . . for the Newton Tercentenary" at Burlington House on 30 November. As 1942 drew to a close there was time to reflect on the year just past. The letters sound a happier note. Thus on 13 December: "I have just been reading our letters of a year ago. Well, 42 has really been a happier year. Or not?" Christmas fell on a Friday, allowing an extra day to look back. In the last letter of the year on 27 December, he writes of a few colleagues who gathered on Christmas night to watch home movies:

> . . . and as I wanted to see our films too, I took them along and we had a long performance, very nice to see the old pictures. To-day I lived in the same world, looking through old photos and making enlargements . . . . To-morrow normal life begins again and day after to-morrow I am going north; this time I'll take my car and leave it there.

The same letter explains the context for his improved outlook:

> Main reason for feeling happier are [sic], of course, the news during these last weeks! On Xmas eve I was playing with the wireless set, which had just been returned from the repairer after a long time, when I heard the oily voice of Goebbels! – I believe I haven't heard him for 5 or 10 years. He always has been talking nonsense or lies, but at least he did it in a way which fascinated his audience. But this time, absolutely empty phrases, repeating himself continually, soaked in sentimentality, appealing in a repulsively larmoyant manner to the mothers! Very encouraging indeed for us! – I continued to play and was soon luckier: I found my beloved "Kleine Nachtmusik" on a Swiss station.

Like other refugee scientists, Franz monitored German scientific periodicals whenever possible for evidence of nuclear activities by his German counterparts. On 10 December, he had written a formal letter to Lord Cherwell—whom he saw on a weekly basis in Oxford—to point out "astonishing" numbers of known and even well-known scientists listed in the *Physikalische Zeitschrift* as killed in action in Russia or in Africa while serving as ordinary soldiers: 16 in 10 months. "I do not know whether one can deduce very much from these facts," he concludes, "but I thought it worthwhile telling you about them."[14]

His intuitions were sound: after long months of Allied disasters, a turning point of the war was at hand. Germany was about to lose 22 divisions as Soviet forces closed in on besieged Stalingrad, and total losses of those killed or captured on the German side in three months of winter fighting in Russia would exceed 500,000 men.

In December and January, and especially over the year end, activity intensified at the northern site. In his first letter of 1943 on January 6, Franz describes to his daughters the scientists' house and its isolation:

> The house where we are living up there is quite an agreeable place in summer, not so much in winter! Definitely too cold for my liking – in addition our present cook is not a genius. The house is quite for itself, the next "inhabited locality", a very small town, is two miles away. One evening we went there to the movie, however. It is quite a funny sensation: one walks for half an hour along slippery country lanes in the complete black-out, hardly notices that one has reached the "town", is led by the experts to a barnlike building, and when we entered, what was on the screen? Corner of Fifth Avenue and 54th Street – it was quite near the place where the dentist had his office! The film quite entertaining, Fred Astaire.

Though nothing appears in these letters about the factory, the diary has clues about ongoing work. On 1 January 1943, it shows a small emotional spark: "First run of 1 stage!" In the exclamation mark, a small cry of triumph. Or, as a possible alternative reading: Finally! At last! For this is the contract placed in May 1941, before the Maud Report, for pilot units expected to be ready by the end of 1941, then for August 1942, and this was only a first, single-stage unit. For 2 January, the diary notes "no run," and for 3 January, "1st run with Fl carbon." In the official history, Gowing notes that using fluorocarbons for lubrication was an idea that came "from the Americans."[15]

The long stalemate

Any brief euphoria after the first Valley runs was about to be abruptly extinguished. The American leaders were about to launch a drastic new security policy, based on detailed rules drafted by Conant and delivered to Akers on 13 January. The "basic principle" was that all information exchange on weaponry and equipment would henceforth be limited to recipients "in a position to take advantage of this information in this war." This meant that interchange could continue on the diffusion process (under the control of General Groves), and also on "use of heavy water in chain reaction" between the Montreal and Chicago groups (but excluding plutonium). For the electromagnetic method, heavy water production, fast neutron reactor, and all "bomb" matters, the rule would be "No further information to be given to the British or Canadians."[16] For other areas, basic scientific information would require direct approval of the chairman of the S-1 Executive Committee (i.e. Conant).

The British leadership reacted to Conant's blunt language with stunned outrage. They felt humiliated and cheated. By any careful reading of the memorandum, the restrictions were shaped to maximize American advantage in basic science areas combined with minimum exchange of information on military or industrial applications. The new rules seemed a direct repudiation of the Churchill–Roosevelt Hyde Park agreement of June 1942, and a sudden reversal of the full sharing of information that had been the rule up to this point.

The American leaders saw no basic unfairness. The central reason for restriction was security, over a vastly increased labour force. Was it unfair to impose compartmentalization on foreign scientists when it now operated stringently at the domestic level? Moreover, the disparity in costs had become huge, roughly half a million pounds annually against half a *billion* dollars. Emphasizing their responsibility to Congress, Bush and Conant were unwilling to concede any possible commercial or industrial benefits to foreign firms for knowledge paid for by the American taxpayer. As a last-minute argument, a new Anglo-Russian agreement for the exchange of scientific information signed in September 1942 was used to win Roosevelt's approval for restriction. The Canadian nuclear historian Wilfrid Eggleston, who describes the Anglo-American rift in relatively balanced terms, sees the treaty issue as "a convenient rationalization for the new restrictive policy."[17] Although Conant and Bush emphasized to British and Canadian colleagues that this policy was based on an order received "from the top," they did not mention that it rested, as the postwar record shows, on carefully reasoned recommendations developed by Conant and submitted "to the top" in December 1942 by Bush himself.[18] In a cold calculation of risks and benefits, the American project leadership contended that changed circumstances justified new limits on interchange regardless of previous commitments.

The Bush–Conant rules produced a long stalemate, from December 1942 to the Quebec Agreement of August 1943, at a critical point in the project and the war. For the British programme, the consequences of the new policy would be devastating. The first British and Canadian reactions were split between angry resistance and temporary acquiescence in hope of future flexibility. Those close to the daily scene, including Akers and the Halban group in Montreal, favoured making the best of a harsh bargain and working for ad hoc improvements. The Montreal group and Mackenzie in Ottawa were only too aware of their dependency on Chicago for both scientific expertise and allocation of raw materials—even for the heavy water expected from British Columbia.

Despite the ban, some informal ties between Montreal scientists and their former colleagues in Chicago continued. On 3 February, Auger and Goldschmidt returned to Chicago and were welcomed by their previous colleagues and shown results from the projects on which they had collaborated. Goldschmidt would long remember the parting advice of Samuel Allison: "I hear that your visit may be the last for a long time. I hope that you will not leave with empty pockets."[19] They returned to Montreal with souvenirs of their earlier work: fission products and four micrograms of plutonium.

At the official level, Anderson and Cherwell leaned towards a harder line. Tempers became frayed. The American leaders continued inviting individuals or small groups to help with specific American problems, but London increasingly discouraged these ad hoc contacts. Thus at a key meeting on 26 January, Groves asked Akers to send him experts for "three or four days" to discuss the design and construction of a diffusion plant, with any further exchange to be dependent on the benefit to the Kellex Company, a new Kellogg subsidiary created to build the gaseous diffusion plant. He also repeated an earlier invitation for Chadwick and Peierls to revisit the theoretical physicists before their isolation under the compartmentalization rules became complete. Akers was cool to both requests.

At the same meeting, Akers noted that, under the old rules of interchange, W.T. Griffiths, a British expert on powdered nickel then in the United States, could have met with his American counterparts to discuss recent British barrier work. Conant replied that Griffiths should discuss only nickel powder, and "not discuss barrier manufacture."[20] In February London refused yet another invitation to Chadwick and Peierls, and in early March an invitation to Halban to meet with Urey and Fermi in New York was vetoed by Anderson after provisional approval by Mackenzie and the British High Commissioner in Ottawa. At issue were discrepancies between Fermi's and Halban's earlier Cambridge research results that were critical for planning the extent of the American heavy water programme. The Canadian leaders were shocked at the refusal, and the Anglo-American rift now became an Anglo-Canadian one as well. On 30 March, in retaliation, the American Military Policy Committee decided not to deliver the first heavy water output from British Columbia to the Montreal laboratory, as had been promised 3 months earlier.[21] The break struck like a thunderbolt. Just 6 months earlier, as Franz's diary shows, Simon and Halban had lunched with Urey, Dunning, and others at Columbia on 28 September, and afterwards continued discussion of matters now forbidden.

In April, with Anglo-American relations in complete disarray, the British leadership reconsidered briefly the option of going it alone. Cherwell suggested construction of a full-scale diffusion plant and a heavy water production plant and reactor in Britain. Churchill asked Anderson to prepare cost estimates in terms of time, money, manpower, and competing priorities for other projects. Such estimates had been made before, and they could easily be updated. But this exercise was futile, for if Metropolitan Vickers was still months behind in its delivery of the first units for the pilot plant, how could anyone seriously contemplate a project hundreds of times larger? Even the most optimistic estimates of the Technical Committee could only project a result for 1946 or 1947, and this based on overriding priorities that Britain could not currently provide. The calculation showed clearly that renewed interchange was the only possible option within the expected duration of the war.

From the beginning of the impasse, the most promising path seemed to be direct negotiation at the very top between the two leaders. Churchill, advised and briefed by Cherwell, accepted the extra burden and pursued the question with a tenacity that justified his bulldog image in the press. At three major conferences—Casablanca in January, a special trip to Washington (the Trident Conference) in May, and finally Quebec in August—Churchill pressed the issue with Roosevelt face to face. In between, he sent letters, telegrams, and emissaries. When these were ignored, he sent others. The full chronology is complex and tedious, but Churchill's determination paid off, for he was prepared to link the agreement on Tube Alloys with much larger issues of high politics, including British agreement on the timetable for a projected landing in France.

On 9 July, Churchill again demanded action from Roosevelt. His adviser Harry Hopkins, when asked for his advice, reminded the President that "you made a firm commitment to Churchill in regard to this when he was here and there is nothing to do but go through with it."[22] On 20 July, Roosevelt directed Bush to "renew, in an inclusive manner, the full exchange of information with the British Government regarding tube alloys."

Much hard negotiation still lay ahead. In mid-July, Bush and Secretary of War Stimson came to London on other business, and while there had frank discussions with Churchill, Anderson, and Cherwell to talk through the issues, clear up misunderstandings, and develop a first outline of a solution around a framework proposed by Churchill. The details were worked out during a visit by Anderson to Washington early in August, and a final text was signed by the two leaders at Quebec on 19 August 1943. It would end the eight-month stalemate and govern international aspects of atomic energy for the duration of the war.

The text signed by Churchill and Roosevelt at Quebec is published in full in the British official history.[23] Its chief institutional change was to create a new six-member Combined Policy Committee of three American representatives, two British, and one Canadian. Its functions were to "keep all sections of the project under constant review," to allocate scarce materials, and to settle questions of interpretation of the agreement (Article 6). The same article called for "complete interchange" of information at the level of the Policy Committee, and "full and effective" interchange among those from each country working in the same sector. In theory, the compartmental principle continued. For the "design, construction and operation of large-scale plants," special ad hoc arrangements, subject to the approval of the Policy Committee, were to be allowable if they "appeared to be necessary or desirable" for the earliest possible completion of the project. This last was a considerable American concession, and it would open the way to effective participation by British experts in several sectors of the project. At this point in 1943, some very basic scientific questions remained unsolved.

The Quebec Agreement came with a heavy price. For months attempts at renewing exchange had foundered on deep-seated American suspicions that British firms (and notably ICI) were overly interested in the postwar commercial benefits of atomic energy. Despite repeated disclaimers, neither side could separate postwar commercial, military, and strategic calculations from the immediate objective of forestalling a German bomb. The eventual breakthrough came in London. When Bush openly expressed the American suspicions in the frank discussions of 22 July, Churchill dramatically renounced any interest in postwar commercial advantage and offered to make a written commitment to this effect. In the end, this concession became Article 4 of the Quebec Agreement, which acknowledges the "heavy burden of production" on the American side, and specifies that "any post-war advantages of an industrial or commercial character shall be dealt with . . . on terms specified by the American President to the Prime Minister of Great Britain." By this clause, the partnership was indeed restored, but also formally recognized as an unequal one.

Throughout the dark period from the Conant memorandum to the Quebec Agreement, those working for Tube Alloys in Britain endured months of painful uncertainty. Franz was close to the leadership but mindful of secrecy. Month after month, his letters to Lotte testify to the pain of hopes repeatedly raised and deferred, pain intensified by a parallel uncertainty concerning family reunion. On 24 January 1943, he writes:

> Saw Alexander twice and am going to see this week one of his colleagues in London, my highest "boss"; rather stormy developments. We expect A[kers] back soon; it is very necessary and we'll have quote an exciting time.

In December, Cherwell had been appointed to the Cabinet as Paymaster General, and Franz henceforth refers to him by his second given name. Akers, still in America, was about to have his "stormy conference"[24] with Groves and Conant on 26 January. In his next letter on 31 January, Franz writes:

> But now the news; Tuesday morning [January 26] to London, the Lord President wanted to see me; had a talk alone with him, nearly an hour.

February brought little change: the same impasse, the same surface optimism. The next 2 months continued the same theme:

> We haven't arrived at any decisions yet; there was a long stalemate, but now things begin to move again. [February 28]
>
> . . .we are still waiting for a decision, which however has to come very soon. [March 21]

> You ask about some advice for your summer planning. Difficult to give from here.... Don't get impatient, Darling, we'll have a decision in the near future without any doubt! [April 23]

By early May, the unending uncertainty was becoming too much for Lotte. As the summer approached, she became depressed, unable to concentrate or plan ahead. "I am rather fed up with everything," she writes on 27 April, "I wish I knew more about your plans. I am always waiting and waiting and waiting . . ." Franz tried to help by pointing out the broader context on 9 May:

> I am so sorry, Darling, to see you in such a state of depression and this depresses me of course too. The price of separation we had to pay was no doubt very high, but very much was at stake too – actually much more than we imagined at the time. Think where we would be now if the Nazis would have been successful! Please bear this in mind. "An insurance premium isn't considered wasted if the house doesn't catch fire". . . . In one way or another we'll know soon.

Three days later, Churchill, Roosevelt, and their respective advisers would meet in Washington to shape "a set of strategic decisions embracing the entire war effort."[25] Among these was the issue of Tube Alloys, discussed on 25 May. When Churchill left Washington the next day for Gibraltar and Algiers, he believed this question settled, and sent Anderson word that "The President agreed that the exchange of information on Tube Alloys should be resumed . . . . Lord Cherwell to be informed."[26] On 29 May Franz writes of a meeting with Cherwell, just back from Washington:

> In the afternoon long discussion with Alexander. Now the decision really can't be very far away . . . . You certainly should do something about your border card soon if you intend to visit Yale, but it also may be of interest to me. I certainly cannot tell you anything definite, but decisions may come very quickly and suddenly.

However, Roosevelt delayed action once again. Two more months of hard negotiations still lay ahead. In mid-July Lotte went to New Haven for three weeks to visit the Cassirers. "I feel very much at home and am very happy to be here," she writes on 16 July.

Through all the emotional highs and lows of the dark period, the work of Tube Alloys continued. Franz attended the usual committee meetings: the Technical Committee, the Chemical Panel, the Diffusion Committee, and other unspecified meetings at Old Queen Street. He continued monthly visits to the factory at Valley, but the scope of work there was limited by manufacturing delays. Following the initial runs of the first unit in January 1943, experiments with the second single-stage unit and the two-stage unit began only in June and August, respectively, while the two 10-stage units, delivered in August and November, became redundant before they could be made operational as a consequence of high policy decisions after the Quebec Conference. In the absence of sufficient manufacturing priorities, the work at Valley effectively ceased to be a wartime project, though its scientific results would be used to good effect later.

More significant was the work being done in Birmingham, at ICI Metals and the Mond Nickel Company, on barrier research. Franz's diary shows nine meetings with Mond Nickel personnel, the first on 12 May, 1942, then on 25 July, and then at approximately monthly intervals from December 1942 to June 1943. These are often combined with visits to ICI Metals at Witton. Seven of these visits are in Birmingham, one in London, and one at the

Clarendon Laboratory. Concerning barrier materials, important basic questions remained unresolved, and the prospect of renewed interchange was well received on both sides. Even Conant, the strongest opponent of wider interchange, conceded privately to Bush that at this point the diffusion programme "needed all the help it could get."[27]

The Manhattan Project in 1943

At the end of 1942, a wave of optimism had pervaded the American programme, fuelled in part by Fermi's success in achieving a chain reaction at the Chicago Stagg Field pile on 2 December. For a time his success seemed to justify the Conant rules of January 1943. The gains in security from restriction appeared to outweigh any losses from diminished flows of information. While the American leadership continued to invite inputs from Britain on specific questions without a guarantee of assistance in return, this was soon stopped by Sir John Anderson as part of his effort to restore reciprocity. As time would reveal, the optimism of December 1942 was misplaced, and the two years from January 1943 to the end of 1944 would amply demonstrate that the secrets of the new world of nuclear energy would be unlocked only slowly, after much frustration, disappointment, and unremitting effort.

By the time that the Quebec Agreement in August 1943 opened the way to renewed interchange, some very large construction steps had been initiated in the American programme, even though some major scientific problems were still unresolved. For the gaseous diffusion project, the Kellex Corporation had signed a first contract in December 1942 to build the giant full-scale diffusion plant in Oak Ridge that would be known as K-25. At Kellex's suggestion, this was followed in January 1943 by a contract with Union Carbide Corporation to operate the plant, so that construction and operation of the plant could be closely integrated. By March, the Kellex group had a basic plot plan for the site but unresolved issues in component design forced delay in plant construction. After much experimentation with pump design, a suitable vacuum-tight prototype was developed and was being tested by the late spring of 1943.

The barrier question was more serious. In January Columbia scientists began building a pilot plant on campus for producing the Norris–Adler type of barrier by a continuous process rather than by hand. By early April the Army contracted with the Houdaille–Hershey Corporation for a plant to produce several million square feet of this barrier by the end of 1943. By summer this company had its own pilot plant in a new building in Decatur, Illinois. By July, the pilot plants had produced enough material for testing, but test results proved disappointing. The materials were too brittle, too prone to corrosion, and too variable in quality even in the same batch. When Percival Keith for Kellex declared on 3 August that no acceptable barrier for the Oak Ridge plant was yet in sight, Urey became deeply pessimistic, reporting back to his Columbia team that the next 4–6 weeks could be critical for the entire gaseous diffusion project.

At this point the electromagnetic method of separation seemed to be pulling ahead of other routes as the most promising path to the bomb. In November 1942 the Military Policy Committee had authorized construction of a full-scale electromagnetic plant at Oak Ridge, to be known as Y-12. In late December and early January, Groves signed separate contracts for plant construction, plant operation, and four more for various types of equipment. The first construction at the site began on 18 February 1943. In aiming to have the first units operating by July, Groves was telescoping additional basic research, engineering design, and plant construction into one continuous process, with the hope of discovering improvements and solving unforeseen problems along the way.

Like the diffusion project, the electromagnetic method was a process of hundreds of stages. The first plant at Y-12, which aimed at 100 grams per day of enriched, fissile uranium 235, was designed as a series of oval "racetracks" for the ion beams, each with spaced gaps and collector tanks around the outer wall to receive slightly enriched product. At a meeting in Berkeley on 13 January, Groves set a first objective of five racetracks, each one with 96 collector tanks for the product. Further laboratory research led to a decision in March to build a second type of magnetic plant adapted to higher stages of enrichment. The Beta units, as they became known, were designed to handle more concentrated product from first-level Alpha tracks more efficiently, with better chemical recovery and less wastage of partially enriched feeder gas.

By May, the experimental units at Berkeley produced samples containing 9 per cent uranium 235. The process was clearly working, and Lawrence, ever confident, began to suggest enlargement of Y-12. In mid-August, in the light of Lawrence's optimism and Urey's pessimism, Groves made a decision to double the size of Y-12 and to reduce the target for separation at K-25 to enrichment up to 50 per cent uranium 235.

If electromagnetic separation was looking better than gaseous diffusion by the summer of 1943, the other two Army-backed paths to the bomb were apparently lagging. The third method for uranium separation—the gas centrifuge—had been formally eliminated by the Military Policy Committee in November 1942. Although Jesse Beams had achieved some separation in his laboratory model at the University of Virginia, efforts by Westinghouse to produce a full-scale, vacuum-tight, industrial machine in the summer and autumn of 1942 had proved frustratingly difficult.

The fourth path, that using plutonium for fission, appeared to have more unknowns and more difficulties to solve: a new chemical extractive process, radiation hazards, buildings requiring operation and maintenance by remote control, and a location further away from settled areas. By January 1943 the eventual site for plutonium production had been chosen on the Columbia River near Hanford, Washington, with the Du Pont Corporation as prime contractor. A more immediate objective, however, was to move a major part of the experimental work at Chicago to a separate part of the Oak Ridge tract. There, on a site designated X-10, Du Pont was to design and build a graphite-moderated experimental pile, a pilot plant for chemical separation of plutonium, and associated laboratory facilities. By the summer of 1943, these research buildings were approaching completion.

During the closing months of 1942, a new idea had emerged and gained strength: the creation of a separate weapons laboratory to centralize and coordinate the problems of military application of the Manhattan Project. The foremost reason was to improve security for the most sensitive parts of the project, but centralization of fast-neutron work from several existing locations also promised greater efficiency. Groves found the idea attractive and commissioned a search for possible sites. The requirements included a setting allowing for maximum security and secrecy, isolation of the scientific personnel, areas suitable for weapon tests, and remoteness from concentrated populations. In November 1942 Groves examined two sites in New Mexico selected by Army engineers, and chose one of them. In later years, the Los Alamos site became the best known part of the entire Manhattan Project, but in its early stages it was known to very few. Its mission could begin only when the arduous struggle to produce fissile material had succeeded.

Initially, Los Alamos had been conceived as a military laboratory, with senior scientists and engineering staff holding temporary commissions in a military hierarchy under direct Army administration. This idea encountered opposition when Groves's choice as director, Robert Oppenheimer, found that some of those he sought for the most senior positions would not serve under these conditions. As a compromise, Groves and Conant drafted a letter, signed on 25 February 1943, undertaking that the new laboratory would be run as a strictly civilian organization, under a War Department contract with the University of California, at least until January 1944. On this basis, recruitment proceeded, construction started, and laboratory equipment was transferred, until by mid-summer 1943 the new site was ready to begin operations. Although the Army had security and engineering responsibilities from the start, the plan to militarize the scientific side was never invoked.

The new laboratory was well equipped. It had, for example, an operative cyclotron. In a secret negotiation between Conant and Groves, Harvard sold its cyclotron to the US Government for $1 and an informal promise to replace it after the war. It was dismantled and shipped to St Louis, to be forwarded to an "unknown destination" for medical treatment of the military. When asked to approve the transfer for the Physics Department, Professor Bridgman's response was classic: "If you want it for what you say you want it for, you can't have it. If you want it for what I think you want it for, of course you can have it."[28]

Restarting cooperation

When the Quebec Agreement was signed by Roosevelt and Churchill on 19 August 1943, a few senior scientists from Tube Alloys were already mobilized to resume the interchange of information. Oliphant, Peierls, and Simon arrived in New York via the Clipper from Foynes, Ireland, on the same day, while Akers, Chadwick, and Halban were already in North America. They would encounter an American leadership unprepared and not yet willing to reopen discussion. Much had changed since 1942. The Quebec text did not change the policy of compartmentalization, but rather left its application to be spelled out by the new joint Combined Policy Committee(CPC) in Washington. Until this committee could meet to examine and delimit specific areas of exchange, the American leaders—Groves, Bush, and Conant—reluctant to consider any exchange wider than absolutely necessary, preferred to wait for authorization from the new body. This authorization would not be available until after the CPC's first meeting on 8 September. With tempers still raw from past frictions, both sides proceeded correctly and cautiously.

Such a timetable left the British team with little to do at official level for the next three weeks. In the first week after arrival, Akers held a four-hour Sunday morning meeting in New York of his full team—Chadwick, Halban, Oliphant, Peierls, Simon—before leaving with Peierls and Simon on a four-day visit to the joint British–Canadian project in Montreal. There, morale was at a low ebb. During the dark period its scientists had been cut off from colleagues doing similar work at Chicago, and deprived of the uranium and heavy water needed for experimental work, though they still had most of the 180 litres rescued from France in 1940. Both refined uranium and heavy water were by now being produced in Canada, but the Canadian government was not disposed to override signed private contracts committing this output to the American Army. Among its other provisions, the Quebec Agreement provided a channel for negotiated allocation of raw materials. Without a fair allocation, the very future of the Montreal project was in jeopardy. After four days of meetings, extended discussions, and a farewell dinner given for C.J. Mackenzie, Franz returned to New York.

The continuing delay provided an unexpected break for Franz, an occasion to enjoy the long-delayed family reunion. A flat in Greenwich Village, rented before the visit to Montreal, became "home" for the next six weeks. For two weekends and the intervening week, he could relax with the family and enjoy the late summer atmosphere of New York. Day by day, the diary portrays the mood: "Off to Staten Island. South Beach. Swimming. Lunch on Beach"; "Walk through Village," "Photo children"; an evening performance of the operetta Rosalinda; "With children stroll town"; "Stroll Central Park, Zoo." He kept in touch with mission matters through daily phone calls. On one afternoon the British colleagues came for tea with the family. There was time to connect with a few refugee friends. What is missing in these entries is any evidence of contact—even a phone call—with the colleagues at Columbia or elsewhere with whom Franz had worked so closely on previous visits. The security issue had become tight and delicate.

The week following was one of high expectations and key decisions. After the Labor Day holiday, the British group met in Washington on Tuesday 7 September, in Akers' office, ready to proceed. Franz's diary briefly records: "Chadwick reports on Council refusal." The next day, Wednesday, the first meeting of the CPC was held at the War Department. Again Chadwick, Oliphant, Peierls, and Simon met for almost 4 hours in Akers's office, but no result came back. From the official history we know that the CPC, unbriefed and with two of its members absent from Washington, decided to set up a tri-country subcommittee, chaired by General Styer and consisting of Richard Tolman, Chadwick, and Mackenzie, to draft specific rules and limits for interchange. Tolman, borrowed from Cal Tech, had been Groves's scientific adviser since January 1943.

When the subcommittee met two days later at the Pentagon, it was presented with an initial plan drafted by Groves and approved by the American Military Policy Committee the previous day. The proposal reflected American priorities: scientific exchange on gaseous diffusion, the centrifuge, and thermal diffusion, but not on electromagnetic separation; on the heavy water pile, but not on the graphite pile. In a long session, Chadwick won a few improvements: for diffusion and the heavy water pile, interchange would extend to production plants. Groves's plan also suggested that some key British scientists—initially Chadwick and Oliphant—might move to the weapons laboratory at Los Alamos. Franz's diary for Friday shows that at 1 o'clock Chadwick and Mackenzie returned from the meeting to Akers's office, and that after a break for lunch the full British team discussed these results for the rest of the afternoon.

Further meetings of the British group continued through the weekend, all day Monday, and Tuesday morning. The possible transfer of senior scientists from Britain raised important new questions. What would be the consequences of their departure for the British projects they left behind? What would be the longer term status of research discoveries made in North America under such conditions? The weekend talks were attended by the department head of DSIR, Sir Edward Appleton, and a DSIR patent expert, Arthur Blok. Chadwick, as a member of the subcommittee, quickly understood the mutual advantages that would result from British participation in what appeared to be the fastest route to the bomb for the Allies.

For the immediate future, two points emerged from these meetings in Washington. In line with Conant's admission that gaseous diffusion could use any help it could get, the way was clear for Simon and Peierls to sit down with their colleagues at Columbia and at Kellex. By contrast, the subcommittee meeting on 10 September had made no decisions for the Montreal laboratory. Franz's diary for Monday afternoon, 13 September, notes "General Discussion (Montreal situation)," an indication that this problem continued.

On Wednesday, 15 September, the diary records that Simon and Peierls joined "Gen. Groves, Keith, Urey, Benedict, etc." for a first meeting on the diffusion question. They had been in the United States for almost four weeks, and now only 17 days remained until their intended departure. The meeting ran from 9 a.m. to 5.15 p.m., including "Lunch together," though Groves was absent in the afternoon. In the days remaining, the diary lists 17 separate meetings, including four full-day sessions at Kellex, fitted into 13 working days. Of the four remaining days, three were spent in Montreal, where high-level meetings were held over a weekend—with Groves again attending—to consider plans for the heavy water laboratory. On the last Sunday before departure, 26 September, there was time for a family excursion upriver by boat to Bear Mountain, and a walk on the slope along the ski trail.

The diary lists interesting details of these meetings, typically listing those attending, topics or problem areas, the time and duration of the meeting, and noting other colleagues seen separately, or over lunch. The names listed are mainly those that Franz knew from previous visits. The topics mentioned give hints of the problem areas: "Visit Pilot Plant, Norris. Disc[ussion] Urey about our place"; "Conv[ersion?] Factor and Memb[rane] Testing ... Pilot Plants"; "Mainly Corrosion"; ". . . leave by car for Kellex NJ . . . Visit Lab"; "Kellex . . . Control Meeting" (an all-day meeting); "Columbia . . . with Boorse's group [pumps] . . . Membr[ane] Meeting in Urey's office"; "Kellex . . . Corrosion membrane Meeting"; "conversion factor" (on the morning of departure). For the membrane meetings at Columbia on 28 September and at Kellex on 30 September, the diary lists an attendance of 10 or more, a small indicator that the barrier question was still unresolved. As the official history shows, the most critical barrier issues lay ahead, just weeks away.

On 1 October, at the moment when Franz was leaving Urey at Columbia to catch his flight from La Guardia to Montreal and on to Britain, the diary has an odd note—"Shake Hands Tolman"—which requires knowing a wider context. Groves, increasingly concerned about delays in developing an effective barrier for K-25, had just asked Tolman to review the state of barrier work in the New York area and then in Britain. Within 2 weeks, Tolman would be in Britain. He would stay until mid-November, visiting British barrier work sites and discussing British participation in other Manhattan sectors. The diary entry suggests that Franz knew, before he left New York, of Tolman's planned mission.

For Groves, the barrier issue reached a time for decision in November. Essentially, the choice lay between the Norris–Adler type, developed and improved in the Columbia laboratories over many months, and a newer, less tested powdered nickel type developed in the Bell Telephone, Bakelite, and Kellogg industrial laboratories. Percival Keith, the head Kellex engineer, thought the second type more promising. At a key meeting on 5 November 1943, Groves heard the arguments for both types, and in the end decided that both sides should continue research with maximum effort until a choice could be made between them. This question loomed large as Tolman toured diffusion sites in Britain between mid-October and mid-November. Facing this momentous decision, Groves was prepared to use every resource available to him, including wider consultation. Even before the 5 November meeting, he had requested London to send a team of British diffusion experts to New York for a full-scale review of the American situation.

Franz, back in Britain on 3 October, faced a formidable round of formal meetings and wider discussions. He had to catch up and also report on results from the American visit. His diary shows an all-day meeting of the Diffusion Committee on 7 October, followed on 13 October by a meeting of the Tube Alloys Technical Committee, which also examined the "proposals" to be presented at a meeting of the Consultative Council the same afternoon. The diary entry also adds: "Cont[inued] Disc[ussion] with Akers, Perrin, Peierls on Am[erican] collaboration." After lunch, discussion continued at the Tube Alloys offices, and at 3.30 p.m. the diary shows a brief meeting there of Dirac, Peierls, and Simon with Niels Bohr, who had been secretly flown to Britain from Stockholm by Mosquito bomber a week before, after escaping from Denmark to Sweden on a fishing boat on 29 September. At 4.30 p.m. Franz had still another appointment, at the London office of Mond Nickel.

By the weekend, Tolman was in Britain to make his tour. Franz expected to introduce him to the various diffusion sites, but the fates intervened in the form of a serious illness. On the eve of Tolman's arrival in Oxford, Franz was already taking aspirin to ward off a cold. As he later explains in a letter to Lotte on 28 October:

> I expected my American visitor the next day and of course didn't want to miss him. We had Dinner in College (quite a nice evening, Lord Samuel with son and grandson, all three Balliol, was there; it turned out that his son who is Chief Censor in Palestine knows my sister). At night I again took aspirin and spent the morning with Tolman and also had lunch with him. In the afternoon I felt rather unwell, went home and when I took my temperature I had over 102. I then cancelled my engagements, went to bed and had 104!

After the fever persisted for several days, Ludwig Frank prescribed sulfa drugs:

> Then the temperature went down, but I had some very disagreeable days – or nights – lying about, sweating, feeling depressed after the M&B [a sulfa drug], not being able to accompany Tolman to see our places as I should have and so on.

Counting several days in bed and then convalescence at home, the illness cost Franz two full weeks of Tolman's month-long visit. He would see Tolman again in London on 2 November, at meetings called "to make the plans for the near future," and again on 15 and 16 November, by which time the plans and decisions on collaboration were largely in place.

In the same time period, and still convalescent, Simon welcomed two other visitors to the Clarendon. As part of their orientation to Britain, Niels Bohr and his son Aage, also a physicist, were also given a tour of sites of the Tube Alloys project. Franz's diary records that, on Monday 8 November, he met Bohr and his son at the Oxford station and took them to the Clarendon. At one o'clock the group went to Ronald Bell's room in Balliol College for a private lunch, followed by discussion there till 4.15 p.m., then a break for tea and a further laboratory visit at the Clarendon. The Balliol location for the lunch was not picked haphazardly. Bell had spent four years at Copenhagen University for postgraduate study in physical chemistry, and he was fluent in Danish. Part of the college was used in wartime by the Royal Institute of International Affairs for a group preparing reports on events in occupied and neutral countries, and Bell himself worked part-time at the Danish desk for this project.

For dinner, Bohr senior was a guest of the President of Magdalen, Sir Henry Tizard, while Aage Bohr dined at Balliol with Simon. The next morning, Franz collected the two guests for a further two-hour visit to the Clarendon before their departure at 12.55 p.m. The visit was strictly a private one. Untypically, Franz does not mention names—only "Monday and Tuesday visitors"—in his letter to Lotte on 14 November. In December, the Bohrs would travel to New York, where they would be welcomed but known in public as Nicholas and James Baker.

The barrier question

By mid-November, specific arrangements for renewed collaboration on Manhattan Projects were largely agreed. They included the diffusion review group requested by Groves. By this time, Franz knew that his next visit to the United States would be different from the others. On 14 November, he writes to Lotte:

> Yesterday and to-day, Sunday, in the Lab, busy with preparations which have to be more elaborate this time as most of my senior collaborators are going to accompany me.

On 23 November a first contingent of scientists—some with their families—converged on Liverpool and embarked on the Andes, a cruise ship adapted as a troop carrier. Rudolf Peierls describes the voyage: "We were a large party, destined for a variety of American laboratories. Some were to stay only for a short period of discussions, the rest were going to join American research groups. The party included Frisch, who was to go to Los Alamos."[29] It also included Chadwick, who became the senior British representative to the Manhattan Project in place of Akers, Peierls, Simon, and several others in the diffusion group. After a stormy, zigzag, 9-day crossing, the *Andes* docked at Newport News in Virginia. Franz's diary records virtually nothing for the days at sea, except "stormy night," "rough weather," and a one-hour talk with Chadwick on the fifth day out. In Peierls's account, his wife Genia was so seasick that she vowed to live out her life in America rather than "to face such a crossing again."

For this mission—Franz's fourth wartime visit to North America—the mandate for the British diffusion team was quite explicit: a full review of the American diffusion project as it existed at that date. It was not to be a reciprocal exchange, and this was accepted by both sides in view of the current disparities. On the diffusion side, the review teams had to put aside memories of the earlier confrontation of Groves and Conant with Akers on 26 January 1943. The context for this December meeting is different: Akers is bringing 15 experts, the barrier question has become more critical, it is 10 months later, and the new Quebec Agreement is in place, though not yet tested.

The passengers from the *Andes* reached Washington on 3 December. After an orientation meeting the next morning, those in the diffusion group continued on to New York, where they would quickly become involved in working meetings almost on a daily basis. Although at first sight these meetings might appear similar to those of Franz's earlier visits, a closer look at the diary reveals noteworthy differences. They begin with a marathon meeting at the US Army Engineers' office on Fifth Avenue on 7 December, a sombre anniversary day for Pearl Harbor. The diary lists "Lunch together . . . meeting continued. 6.00 Dinner together," with no hint of the agenda or those attending. It then notes "Walk with Peierls" from

9 to 11 p.m., a very long, untypical day. Whether Groves was present or not on this occasion, his presence at three later meetings is clearly recorded.

The next several days were spent on more routine activities, including four visits to Columbia, three to Kellex, one to the Bell Telephone Lab, and one "Membrane meeting" at the British group's own office. For Saturday morning, 18 December, the diary lists a meeting at Kellex with General Groves, Colonel Nichols, Urey, Keith, and "our whole group." This meeting, which lasted about two hours at most, seems to have been a preparation for a more important one to follow. Franz attended four more preparatory or planning meetings on 20 and 21 December, at Kellex ("Meeting on Diffusion"), at the British mission office ("plans Wednesday meeting") and two at Columbia ("Separation Situation" and "Conversion Factor"). Then on Wednesday 22 December, he writes "Great meeting at Kellex with General Groves, Conant, Tolman, etc." from 9 a.m. until 1 p.m., followed by lunch with Urey, Taylor, Dunning, Cohen, and others.

The official history provides further detail on this key meeting and its background.[30] It lists by name others attending: "Wallace Akers and fifteen of his experts," four representatives from Kellex, three from Union Carbide, and three from Columbia. Its purpose was a full review of all aspects of the gaseous diffusion project, the most comprehensive review in more than a year. "Armed with these data," the account reports, "the British rolled up their sleeves to study the problems that had haunted Keith, Urey, and Dunning during all those months." The review then adjourned for two weeks over the holidays. At Columbia, the project leadership had just been changed. Harold Urey had disagreed vigorously with Groves's decision in November to continue research on both types of barrier; he feared that this might delay completion of K-25 beyond the end of the war. Groves relieved Urey of responsibility for a project in which he no longer believed by bringing in Lauchlin Currie, an executive borrowed from Union Carbide, as Associate Director. Currie would direct all research except that on the new barrier, which was being directed by Professor Hugh Taylor of Princeton.

Following the meeting on 22 December, the British mission team went to work. Its leaders met at length the same afternoon, followed by a long, formal mission meeting on the 23rd. After Christmas, Franz spent half a day in "formal" and "informal" discussion at Columbia on the 27th. For the rest of that week he notes sessions at the mission office, numerous telephone calls, a talk with Niels Bohr, and some social events. The tone is active but unhurried. After the New Year weekend, the pace quickens. On Monday 3 January, the diary shows a "General Discussion" with Urey, and a lengthy meeting at the mission office with Chadwick and Peierls, followed by another with Akers, who gives "reports on U.C.C. [Union Carbide] and Keith," presumably regarding their positions for the coming barrier meeting. The next day is largely given over to meetings at the British office, a "Mission Meeting (with Chadwick, Oliphant)" in the afternoon, preceded by a morning meeting of the Technical Committee and a second meeting of the same committee after the mission meeting. All of this seems preparatory to the resumption of the diffusion review scheduled for 5 January.

Franz's diary barely mentions this key meeting: "9.00 – 1.00 Meeting at Kellex with General etc." The official history outlines the barrier discussion in detail. The British agreed from the start that the "new"—nickel powder—barrier looked better for the long run, but felt if speed were paramount that the Norris–Adler type had a significant lead. Keith, for Kellex, argued that the new barrier could be developed and produced earlier. Kellex had produced samples in the laboratory by hand processes and if necessary could produce the entire amount needed by using thousands of workers on piece work. While accepting this possibility, the British nevertheless urged continuing work on both types of barrier until a clearer judgement could be made. A measure of consensus then formed around a proposal to continue research on both types but limit the mass production effort to the new barrier alone.

When it then emerged that the Kellex–Carbide scenario would require a radical conversion of the Houdaille–Hershey barrier plant in Decatur from Norris–Adler machinery to new equipment, the British balked, considering such a plan to be audacious and even reckless. They also considered the American time schedule, which called for a plant to produce a still untested barrier in four months, to be all but impossible. The official histories diverge slightly but significantly in summarizing this meeting.[31] The American version calls it "a check on the plan . . . tentatively accepted," but the British account says that British input "undoubtedly helped Groves to make up his mind."

The meeting of 5 January 1944 marked a decisive point in the diffusion programme, but formal meetings and discussions with individuals and small groups continued through the following week. By 13 January Franz was busy writing a report at the mission office, and on Saturday 15 January the diary records "Last Mission Meeting." On the same day, Groves headed west by train for Decatur, to announce personally the stripping and rebuilding of the barrier plant. After this date some members of the British diffusion group returned to Britain; others stayed on for varying periods to help Kellex or other groups on specific problems. Rudolf Peierls stayed on with Kellex for several months, and then went to a senior position at Los Alamos.

Franz continued daily contacts at Columbia and the Bell Telephone laboratories, but on 19 January this work was disrupted by a renewed round of illness. Lotte, who had gone back to Toronto on the 15th, returned to New York on the 21st to look after him. Since his bout of flu and pneumonia in October, this was a third relapse, more serious than the first two and slower to clear. Apart from telephone calls, and a few visitors while he convalesced, he was out of action for 10 days. After his recovery, there was time only for quick visits to Washington, back to New York, then Toronto and Montreal before his scheduled return to Britain.

On the return flight to Britain, he was accompanied by two American colleagues. After the meeting of 5 January, the British mission acted on Urey's release from administration by inviting him to visit Britain. The diary shows that Franz himself went to Columbia on 6 January to see Urey and convey the invitation in person. In the depth of his pessimism the previous November over the barrier issue, Urey had expressed the view that "British design and technology"[32] might offer a better prospect than K-25 for any full-scale diffusion plant. Groves, who had watched British developments and sent Tolman to Britain in October, was open to finding improvements from any source. On 11 January Urey telephoned Franz to say that a visit to England had been approved. He would be accompanied by Eugene Booth, an expert on barrier research who had been among the first to develop multistage test units in the Columbia laboratories. These two joined Franz on 4 February in Montreal, where they had a brief visit to the heavy water laboratory before their flight. After two postponements, their plane left on 7 February, following the usual bomber route from Montreal to Prestwick.

The Urey–Booth visit ran like a small-scale review mission in reverse. It lasted for five weeks, and much of it is documented in Franz's diary. The visitors, usually escorted by Franz himself, were taken to see all the major sites of the diffusion project in Britain. For Franz, who had been away for 11 weeks at this point, this was an opportunity to bring himself up to date on the British project. The travel itinerary included six days in Birmingham in two separate visits; a few days in the north, including visits to "Valley" and "Factory," Winnington, Widnes, Runcorn, and Manchester; a second northern visit to Darlington and Billingham; and three trips to London for meetings at the Tube Alloys office in Old Queen Street. Among the talks and discussions during these five weeks, the diary lists four Saturday sessions with Lord Cherwell, who came to Oxford almost every weekend. After their return to the United States in mid-March, Urey and Booth submitted a report on the visit. The official Manhattan history is dismissive: "Their report on the work there contained nothing to make further interchange seem worthwhile."[33] Whether it was worthwhile from a British perspective seems unexamined.

The lighter, non-secret side of the visit is mentioned in every one of Franz's letters to the family in Toronto from 13 February to 19 March. The sudden transition from New York's comfort to Oxford's ancient stone buildings in February proved a severe thermal challenge:

> ... Harold came up the next day, as he couldn't find accommodation in London. It is very difficult here also; the first night I put him up in Balliol, but he left the next day as he couldn't get accustomed to breaking the ice before washing. (I am wearing two pair of pants for a transition period!). [13 February]
>
> Have quite a busy time with Urey in the Lab and taking him out . . . . Yesterday an interesting afternoon with the President of Magdalen [Tizard]. To-day he is in London. To-morrow we have a meeting here and then we go to Birmingham until Saturday afternoon. Weekend here, then next week on a tour, weekend in London, following week touring again. [15 February]
>
> Often in Balliol, one evening Dinner in Christ Church with Ch[erwell], so far the coldest place encountered – but I slowly get accustomed to the temperatures. [27 February]
>
> On Tuesday I am going away for some days again; by the end of the week we are returning to Oxford and then Harold is going to wait here for his departure. To-night we are having dinner in Balliol. He is staying in Christ Church, Balliol is too cold for him although he got a little accustomed to our temperatures. [5 March]

The challenge of cold weather did not prove insurmountable, however, for Urey would return to postwar Oxford and Balliol as Eastman visiting professor in 1956–7.

The family reunited

Franz's last recorded meeting with Urey before Urey's return to the United States was in London on 14 March. Five days later, a telegram arrived from Lotte: ". . . too busy to write" code words for an imminent sailing, particularly after she had cabled "on short notice" on 8 March. The time for family return was at hand. Passages for Lotte and Kay had been under negotiation for some time, but shipping space was scarce during preparations for landings in France. Dor, just short of 16 years old, was ineligible for return, and had to stay on in Toronto, becoming a boarder at Havergal College. When Lotte's telegram arrived on 19 March, Franz replied immediately with the last of his 277 letters to Canada. His first words reflect the general uncertainty:

> Dearest Dor (and Lo?)
>
> This morning the telegram, it was a great disappointment as I was convinced that Mummy and Kay had already left and I was expecting them here every day. Now the waiting begins again but I do hope that when this letter arrives they have left.

They did not leave Toronto, however, until early April, landing in Liverpool on 12 April. They travelled in a slow convoy from Halifax, on a cargo ship that had a number in place of a name.

Lotte and Kay reached Oxford early the next morning, and over the weekend of 15–16 April Franz helped them to unpack and settle in. They had been absent for almost four years. The next day Franz had a meeting in London, catching the 8.40 a.m. train and reaching the Tube Alloys office in Old Queen Street by 11.00 a.m. The diary then records discussion with the Bohrs, Akers, and Perrin: at 12 noon, "War Cabinet [offices] . . . Cherwell, Appleton, Dale. 12.35–1.00 Meeting Sir J. Anderson, Cherwell, etc." By this date, Anderson was not only in charge of Tube Alloys but also Chancellor of the Exchequer. This non–routine meeting was followed by lunch with the Bohrs and Perrin, then back to Old Queen Street for more briefing before returning to Oxford. Clearly, something unusual was afoot.

Ten days later, and one day after another day-long meeting at Old Queen Street on 26 April, Perrin telephoned Franz with a simple message: "leave for U.S." They left together for Liverpool 5 days later to board the *Mauretania*. In the interval before they left, however, there would be a previously arranged lecture in Oxford by Franz's California colleague, Joel Hildebrand, on 29 April, and a day with the Birmingham group on 1 May. Hildebrand, an early postdoctoral student of Nernst, was in London at this point as American liaison representative for the OSRD committee. The diary for Saturday 29 April shows Hildebrand at Belbroughton Road ("Lunch with us"), his lecture at 2.30 p.m., tea at Wadham College ("Cherwell, Tizard, etc."), a visit to Jesus College (presumably the chemistry laboratory), a dinner at Balliol with several leading scientists, and after the dinner some further discussion with Ronald Bell and others. Another guest at the dinner was Charles McIlwain, the eminent Harvard historian who was the current Eastman Professor.

The passage to New York on the *Mauretania*, from 4 to 11 May, was routine. There were stormy days, fine days, and disembarcation, after a delay and anchorage in fog, on 12 May. The diary lists walks on the deck, talks with Perrin, and an enigmatic "meet Roskill" just after the ship sailed. Three more meetings followed, the last with Perrin in Roskill's cabin. This fellow passenger was Captain Stephen Roskill of the Royal Navy, heading for Washington to join the British Admiralty Delegation. On retirement from active service in 1948, Roskill would go on to write the official naval history of the Second World War and other works on naval history.

Landing in New York early on 12 May, Franz had time only for the briefest of contacts with Urey before catching a 12.30 p.m. train to Washington. On arrival, he proceeded to the office of W.L. Webster, the British co-secretary of the CPC. From there, he joined Cockcroft, Oliphant, and G.I. Taylor for dinner. From Saturday morning until Wednesday evening, five days of intensive meetings followed. The Saturday, Sunday, and Monday sessions were all-day affairs, with breaks for lunch. Most names listed in the diary for these days are familiar: Akers, Perrin, Chadwick, Cockcroft, Peierls, Oliphant, Jackson (the administrative officer), and also Frank Kearton of ICI (who had stayed on with Kellex in New York after the previous mission), and Geoffrey Taylor (a British explosives expert who was consultant at Los Alamos). On the first day, Joel Hildebrand and a colleague joined Chadwick, Cockcroft, Peierls, and Simon for dinner. On the fourth day, Franz notes a 90-minute afternoon appointment with Cherwell's brother, Brigadier Charles Lindemann, at the British Embassy.

On the subject matter of these meetings, the diary is admirably but vexingly discreet. On the fifth day, Wednesday 17 May, the timetable is more diversified and contains two small hints: "short talk Akers, Peierls before Breakfast about new developments... 10.30–11.15 x metal disc[ussion] G.I. Taylor, Chadwick." From 11.20 the regular sessions continued, in two parts, until 5.00 p.m. As this was the last day in Washington, the diary also notes "6-7.35 Sherry party for General, Conant, Tolman & us," followed by more discussion among the "Engl[ish] group" until 8.30 p.m., dinner with Chadwick, Oliphant, and Taylor, and a midnight sleeper to New York.

The diary shows three working days in New York at the British Supply Mission office, where Peierls and Kearton were still assisting Kellex, though Peierls would move to Los Alamos in the summer and Kearton would return to Britain in September. For Franz, there was time to telephone friends, visit his cousin Alfred at the Simmon plant, and spend an evening with the Ureys at Leonia, New Jersey. His schedule in New York was less hurried than it had been in Washington.

On 20 May, he took the overnight train to Montreal, where he rejoined Akers and Perrin. In two days at Montreal, there were brief talks with Halban, Placzek, and Auger, plus a meeting with the laboratory's Technical Committee. By this point, the outlook for the Montreal laboratory was better. The long months of waiting, shortages of needed materials, and deep frustration appeared to have ended, following a decision by the CPC on 13 April 1944 to authorize construction of a heavy water pilot reactor in Canada. The plane left at the time scheduled, and Franz and Michael Perrin were back in London within 24 hours.

This hurried visit to North America gives rise to a question. What would explain this short 12-day visit and a 22-day absence from work and family? Almost haphazardly, a clue in Gowing showed that the answer lay not in the Manhattan Project but in Tube Alloys. Gowing mentions that the Tube Alloys Technical Committee met only three times between January 1944 and August 1945 to produce reports for Anderson's Consultative Council, and two of these meetings were held in North America. One of these, she notes, was "in Washington in May 1944."[34] Cockcroft, the newly appointed Director of the Montreal laboratory in place of Halban, later described his orientation to the new post in a note to Wilfrid Eggleston:

> On 19 May [sic] I took a leisurely train [from Chicago] to Washington to meet Chadwick and Oliphant and also to attend the meeting of the U.K. Technical Committee which Akers had convened in Washington. At this meeting we heard a great deal about the U.S. projects and their relative prospects. There was some pessimism about the diffusion plant project owing to a belief that more effort was required on the electro-magnetic separation project of Ernest Lawrence. We also had a first discussion about future U.K. plans....[35]

Cockcroft, returning to nuclear physics after several years of war work on radar, was understandably impressed by progress reports on the Manhattan Project, but Gowing makes it clear that the prime focus of this meeting was on future British nuclear policy.

Once the purpose of the meeting is understood, the rationale for a Washington venue is clear. Since Chadwick, Peierls, Oliphant, and Cockcroft were already in North America, it was easier for Simon to go there than to bring other committee members to London. Further, because the issue was sensitive for the future of British–American relations, there were advantages in having open discussion, in Washington, with American observers present. The sherry party for the American leaders on 17 May was a social event, but also a diplomatic one.

In addition to the review of American work described by Cockcroft, the Technical Committee was asked to advise on two basic questions regarding future British nuclear policy: what technical direction should it take, and how much should be attempted during wartime? In May 1944 the American project was still backing three routes to fissile material. For Britain, a more modest and selective programme was essential, but none of the American routes was yet demonstrably superior to the others. On the issue of timing, Chadwick insisted repeatedly that nothing should impede the fullest British cooperation and participation in the American project, even if this meant delaying Britain's own plans. His emphasis on this point was key to his exceptionally successful working relationship with General Groves.

Concerning future directions, this meeting recommended against a heavy water plant, hitherto advocated by ICI, in favour of a "compromise solution"[36] proposed by Halban. Though Halban does not appear at the Washington meeting, he had suggested constructing a pile that would use ordinary water plus uranium enriched up to about twice the proportion of U^{235} in natural uranium. This lower cost alternative could then serve as a first stage towards producing either high-separation U^{235} or plutonium, whichever would be needed later on. The committee recommended design and development of such a low-separation plant, but was divided over constructing it during wartime. It was probably this proposal that prompted Franz's pre-breakfast private chat with Akers and Peierls on 17 May about "new developments."

The Washington meeting of May 1944 marks the opening of an extended debate on Britain's nuclear future, one that involved both technical options and basic issues of high policy. It would unfold in the Tube Alloys Technical Committee, in Anderson's Consultative Council, and in two larger ad hoc scientific panels set up in 1945 to plan a new nuclear establishment. The Consultative Council did not endorse the Halban compromise, tilting towards caution in a situation of multiple uncertainties. Since shortages of personnel and materials effectively closed off any large new initiatives, the cost of waiting a little longer was not high. In the end the question of timing—in wartime or afterwards—was resolved by continuing non-decision. Amid the ebb and flow of debate, however, were two areas of wide agreement: that nuclear research should continue as actively as possible with whatever resources remained available in Britain, and that a new and stronger establishment under government control would be needed after the war in place of Tube Alloys.

During this debate over future goals, Franz's work schedule did not change radically. In the Clarendon Laboratory, research, especially on membranes and membrane testing, continued at full strength. The administrative burdens of managing a sizeable work force continued. In each month 5 or 6 days were needed for meetings at the Tube Alloy office in London. Diary entries also show occasional visits to Cambridge and to industrial sites where research was continuing. The absence of several British project leaders in North America led to adjustments in the Clarendon team.[37] Shull Arms moved to Birmingham, where he led a team working on separation of the hexafluoride by thermal diffusion. When he was recruited to the Montreal laboratory in 1945, he was succeeded in Birmingham by G.O. Jones. At Valley, work had been scaled back. Franz visited the "Factory" and Maes Alyn on 25–26 July 1944. The diary for these days lists "Long talk with Manning" on arrival, a meeting with senior staff the next morning at "Factory," lunch in the canteen, a lengthy walk back to Maes Alyn with Heinz London, and dinner there. This would be Franz's last visit during the war. W.R.D. Manning, of ICI Billingham, was resident manager of "Factory."

In one respect, Franz's life had improved beyond measure. When he arrived home from North America on 24 May, his younger daughter Dor was at the station in Oxford to meet him. With other returning evacuees she had travelled from Toronto on a sealed train to New York just as Franz disembarked there on 12 May. Her group would board the *Mauretania* for its return passage to Liverpool. After an absence of almost four years, the family was at last reunited, and home in Oxford.

For the remainder of 1944 the diary has many examples of the pleasures of mundane activities: of short walks, and longer ones—the University Parks, Mesopotamia, Port Meadow, Elsfield, two hours to "Marston, etc."; shopping with Lotte in town; visiting Blackwells, the bookshop; a family visit to a cinema. On 3 July the goods sent from Toronto by freight arrived, and were unpacked. On 11 August, the entry begins: "1st day of 'Holidays at home'." That day was the first of 12 set aside for gardening, reading, resting outdoors "under apple trees," longer walks with Lotte, and only minimal contact with the laboratory and with London. The daily entries suggest a summer idyll. After 4 years, the Simons, normally irrepressible travellers, were discovering felicity in their own garden.

The mood of the summer continued through the autumn to the close of the year. Despite extra meetings in London in September and continuing activity in the diffusion project, the diary shows much after-hours familial, social, academic, and cultural activity. A Balliol concert on 22 October is highlighted in a single word, "Brandenburg." The year-end double festivities for Christmas Eve and Kay's birthday included a performance of "Messiah" on radio on 27 December and— a rare treat—skating on the frozen, flooded Port Meadow on 31 December. Rather oddly, the wider war is barely mentioned during these last seven months, despite the landings in Normandy, the liberation of France, and the unleashing on London of Hitler's vaunted "secret weapons" (flying bombs or V1s in June, rocket bombs or V2s in September). It is as if these pages are an attempt to shut out the war and make up for time lost in four years of separation.

Two exceptions, however, stand out markedly. On 21 July a brief notation reads: "News of revolt in Germany," that is, the foiled assassination attempt, barbarously punished, on Hitler's life. At midnight on 31 December, a second notation reads simply: "Hitler speech." At 12.05 a.m., German time, the Nazi leader emerged from seclusion to address the German people for the first time since 20 July. As a 25-minute New Year's message, it was shorter, more direct, and less flamboyant than its predecessors. He explains the long silence by noting that his work left little time for speeches. Delivered at the height of the German counteroffensive in the Ardennes, he calls for "still greater effort" from all German workers to rebuild bomb-shattered factories by "increased working zeal." There is even a strange recognition of the slave labourers: "[German workers] are more and more being joined – I may say this today – by those thinking peoples of other nationalities who, as workers in Germany, perceive the essence of our social community."[38] He also warns that the war will continue into 1946—unless Germany should win earlier. Franz adds no comment in the diary, but no doubt deduced correctly that occupied Europe's agony was coming to an end.

Thirteen days after Franz's return to England in May 1944, the long awaited Allied landings in France were launched on 6 June. The European war was on the threshold of its last, decisive phase. At this point prospects for any atomic weapon were clearly bound up with the Manhattan efforts in North America, unless, of course—as many still feared—the Germans should get there first. To counter that possibility, General Groves had set up his own intelligence unit, headed by Colonel Boris Pash. This unit accompanied the American army as it advanced through Normandy, and then joined with Free French units to become the first American unit to enter Paris at its liberation on 25 August.[39] That same evening, Pash found Joliot Curie at the Radium Institute. Groves, with his usual foresight, would lose no opportunity to track down German scientists and nuclear raw materials as the Allied forces advanced.

Manhattan: continuing uncertainties

While the Allied armies were advancing through France in June 1944, the situation for each of Manhattan's three remaining paths towards fissile material left much to be desired. The fourth path, separation by centrifuge, had been abandoned in November 1942. For the electromagnetic route, four Alpha tracks were in operation at Y-12 in Oak Ridge but were producing only a fraction of the material expected from them. The early history of these units, in operation since December 1943, had seen several months of breakdowns, repairs, and hopes for improved results as more tracks came on stream. The gaseous diffusion plant, K-25, was not yet operative, though its first units were installed and ready for testing. For plutonium, the giant production complex at Hanford in Washington had six plants under construction, with one pile and one separation plant near completion, another similar pair half complete, and a third pair barely begun. The weapons laboratory at Los Alamos, operative since mid-1943, had yet to receive any significant supply of plutonium or U^{235}.

If the situation looked difficult and uncertain in June 1944, worse days lay ahead in the summer and early autumn. In electromagnetic work, the first two higher enrichment Beta tracks were operative by the end of June, and so were the first Alpha II units, but their performance was one of continuing breakdowns, insulator failures, assembly and operator errors, and negligible output of U^{235} product. In the initial operative stages,[40] no more than four per cent of the potential U^{235} was reaching the Alpha receiver boxes and only about five per cent of this first amount reached the receiver boxes in the Beta units.

In gaseous diffusion, the 1944 summer outlook was worse. By August, Houdaille–Hershey had shipped a few thousand barrier tubes to the Chrysler plant in Detroit for assembly into converter units, but these tubes failed to meet minimum standards for separation. Even at this date the barrier question remained in doubt.

In the plutonium project, a serious new problem surfaced in July. Although early research at Los Alamos had investigated both gun and implosion techniques for triggering a bomb, most of the research had then focused on gun devices as a simpler and better known technology for ordnance work. However, closer laboratory work with plutonium samples revealed the presence of a new radioactive isotope, Pu^{240}, with strong tendencies to spontaneous fission. When this was confirmed experimentally, the gun method had to be ruled out as unworkable in mid-July. Only a successful implosion technique could save the plutonium project and the whole vast Hanford complex. With all uranium enrichment still in trouble and the plutonium route still dependent on developing a procedure to avoid premature detonation, an earlier timetable for having a plutonium bomb ready by late 1944 was unsustainable.

In the midst of these uncertainties, Groves reached out for help from an unexpected and almost forgotten source. On 28 April 1944, he had received a letter from Oppenheimer that reminded him of the work of Philip Abelson for the Navy on liquid thermal diffusion.

Groves himself had seen Abelson's work on a visit to the Naval Research Laboratory at Anacostia on 21 September 1942, just 4 days after taking "complete charge" of the Manhattan Project. At the time, he had discounted this in his own mind: "My first impression . . . was not favorable, primarily because of the evident lack of urgency with which it was being prosecuted."[41] The naval research was focused primarily on low-level uranium enrichment for possible development of nuclear-powered submarines. A further complication was that President Roosevelt, temporarily at odds with the Navy, had explicitly directed Bush in March 1941 not to involve the Navy in S-1 affairs.

Abelson had studied uranium isotope separation since 1940, experimenting with thermal diffusion of the liquid hexafluoride under pressure in long vertical pipes. The principle was simple. If a steam-heated hot pipe was placed inside a larger cooled pipe, some separation could be achieved by simple convection, with improved yields from longer pipes and from greater temperature differences. Abelson's work, backed financially by the Navy, was only in sporadic contact with the S-1 Committee of OSRD. When Groves visited the Anacostia laboratory in September 1942, it had only minor results to show. By December, however, the Lewis Review Committee noted that Abelson had achieved "small but significant" separation of uranium isotopes. This led to a visit by Bush on 14 January 1943. Bush was impressed enough to recommend continued funding by the Navy, but in the light of the presidential directive he refused a request from the Navy laboratory to provide data on nuclear constants needed by Abelson and rumoured to be obtainable from the Chicago group.

In the late summer of 1943, while the gaseous diffusion teams agonized over the unsolved barrier impasse, Conant requested another review of Abelson's work for the S-1 Executive Committee. The Review Committee found little change from the situation in February, and its findings were reported to the Navy on 15 September.

Conant sent the bad news to Admiral Purnell in a firm but polite letter. He would consider it most unfortunate if the Navy drew away from the Manhattan project any of the scientists now engaged in S-1 work. No additional supplies of uranium hexafluoride would be given to the Navy for its experiments, but the Navy was requested to exchange any material enriched at Anacostia for the hexafluoride of normal concentration.

But Abelson would need more of the hexafluoride. His patient experimentation was about to pay off significantly. When the Navy requested more of the hexafluoride in October, Groves at first refused. The Navy then reminded Groves that it was Abelson who had developed the original hexafluoride process, and that the Navy had given it to S-1. Reluctantly, the Army filled the order. "On this sour note, all exchange of information between the two projects ended."[42]

Even in the grey summary of an official history, the tone of Conant's Navy letter seems more "firm" than polite. Its effect was remarkably similar to that of his peremptory memorandum of January 1943, which precipitated the eight-month rupture between the American and British projects. Now, barely a month after the earlier rupture was repaired by the Quebec Agreement, another Conant letter precipitated a second hiatus of roughly equal duration, this time between the American Army and Navy projects. When Groves received Oppenheimer's reminder of Abelson's work in April 1944, the general was quick to sense—and grasp—a possible opportunity. If all three of his contenders for fissile material were still struggling uncertainly, some help might be available by entering another horse in the unfinished race.

During the seven months of this second dark period, Abelson obtained Navy approval in November to build a 300-column liquid diffusion plant at the Philadelphia Navy Yard, which had ample supplies of steam. Construction of the first 100 48-foot columns was begun in January 1944, and these were the units inspected by the Manhattan Project Review Committee in June as soon as Army-Navy contact was restored. Groves was impressed by the committee's favourable report and moved swiftly to action. On 27 June he signed a contract to construct a large-scale liquid diffusion plant at the Tennessee site adjacent to the K-25 powerhouse, which would have ample excess steam as long as K-25 remained unfinished. The plan was simple: to make 21 exact copies of Abelson's 100-column plant in Philadelphia. The contract specified completion within 90 days, or sooner if possible. Although this timetable might seem hopelessly unrealistic, the first rack of pipes was completed and ready to operate just 69 days after construction began. It would not be completed, however, until the end of the year.

The troubles of the summer of 1944 continued into the autumn. The plutonium predetonation crisis of July forced a revision of the schedule for bomb production, which since 1942 had aimed at a first bomb by the end of 1944. On 7 August, Groves informed General Marshall that several implosion plutonium bombs "would be available between March and the end of June, 1945."[43] The first uranium bomb would be ready by 1 August, with one or two more by the end of 1945. This meant that, if the Allied advances in Europe continued at the same rate, both types might be too late for the European war.

At the uranium separation plants, patient efforts continued, month after month, with the engineers and technicians seeking unnumbered incremental improvements wherever they could. In September and October, the electromagnetic Alpha tracks at Y-12 were still producing only a thin stream of enriched material, and there were still no satisfactory barrier tubes for the gas diffusion units at K-25. By December, after several weeks of improvements at Y-12, nine Alpha and three Beta tracks were in production with improving outputs, and the S-50 liquid thermal plant was near completion though still mediocre in performance. More significantly, Chrysler in Detroit was receiving larger numbers of barrier tubes that met specifications, and the assembled compressors were installed and tested in K-25 as soon as they were ready. On 20 January 1945, the first uranium hexafluoride was introduced into the initial operative units of K-25, and by late February, after similar trials and improvements in S-50 and Y-12, all three plants were ready for effective production.

At this point a strange transformation occurred. As early as September 1944, when outputs were still doubtful or minimal, a production control committee had been set up to plan the most efficient use of the three plants. By March the result was a system of lowest level enrichment in S-50, followed by further treatment in the first operating units of K-25 or in the Alpha units of Y-12, and completed by high-level enrichment in the Beta tracks of Y-12. The optimum proportions at each level, calculated mathematically, were modified as new units of each plant came on stream. Thus the three contenders for the fastest route to the uranium bomb, including the last-minute "dark horse" entry copied from Abelson, ended up no longer competitors but harnessed together as a team, with each plant contributing in the range where it produced to best advantage. On this basis, the output for March 1945 was more than double the combined total for all previous months. At regular intervals, relays of special couriers escorted containers of refined product by car to Knoxville, train to Chicago, another train to Lamy, New Mexico, and by car again to Los Alamos. Similarly, from

February onwards the first plutonium shipments were driven in a small truck from Hanford to Los Alamos.

By March 1945, after more than three years of prodigious effort, the quest for fissile bomb material was demonstrably successful. As a military man, General Groves could look ahead. His strategy was to phase out the less efficient processes—liquid thermal diffusion and Alpha electromagnetic units—and to multiply the more efficient ones—gaseous diffusion and Beta electromagnetic units—as construction continued. On 31 March, with his superiors' consent, he authorized a second gaseous diffusion plant (K-27), and two days later a fourth Beta plant. The giant K-25 plant, which had begun first operations by enriching material to 1.1 per cent, had added more units until, by 10 June, it could supply material enriched to seven per cent directly to the Beta tracks at Y-12. The two new plants were scheduled for completion by February 1946, for a war expected to continue until July of that year.

With fissile material ensured, the fate of the Manhattan project would depend on the weapons work at Los Alamos: chemical work on plutonium, purification techniques, bomb design, bomb construction, gun design, and an implosion mechanism for a plutonium bomb. There were theoretical and practical problems to be solved in many directions. As the official history notes with understatement, "nothing went very smoothly."[44] The most persistent difficulties involved the special qualities of plutonium and its requirement for detonation by implosion. In August 1944 this problem forced a postponement of the expected date for a bomb from the end of 1944 to the second quarter of 1945, and late in 1944 a second postponement to "the second half of 1945."[45]

One result of these delays was that, in May 1945, when American military planners turned their full attention to the Japanese war after the capitulation of Germany, many questions concerning nuclear weapons still remained unanswered. When would either bomb be available? Would either work? Should they be used, and, if so, how? Even the first test in New Mexico was scheduled only for July. With so many unknowns, the Joint Chiefs of Staff reviewed conventional alternatives: continued air raids and naval blockade; a front with the Chinese on the Asian mainland; or landings on the main Japanese islands. Timing was important, and complex interests of four major powers—the USSR, China, Britain, and the United States—were at stake.

On 25 May, a directive from the American Joint Chiefs set 1 November as a target date for a landing on Kyushu, the southernmost of the main Japanese islands. The plan and date were endorsed by President Truman at a briefing on 18 June and reaffirmed on 24 July by the British and American Combined Chiefs of Staff at a meeting between Truman and Churchill. By this date, however, the plutonium bomb had been tested, successfully, on 16 July, on a site south of Los Alamos in central New Mexico.

Germany: the Army's withdrawal and its consequences

Two days before the Japanese attack on Pearl Harbor in December 1941, the German Army Ordnance Department announced a general review of the German nuclear project, to be completed by the end of February 1942. Under tightening constraints from a deteriorating war economy, the department could no longer support projects unlikely to find military application in wartime. In nuclear fission research, continuous Army support since September 1939 had produced no successful separation of U^{235}, no chain reaction, and no production of element 94. This review resulted in a long and thorough report, written by Army scientists, which assessed the importance of the project and the results to that point. The report emphasized the importance of nuclear energy for Germany's future, both for weaponry and for industrial power, but it also recognized that major research gaps and key material shortages—heavy water, enriched uranium, element 94—stood in the way. These obstacles would have to be overcome before industrial development could proceed. The report's conclusion was that nuclear energy was "certainly possible, but not close at hand."[46]

At Army Ordnance, General Schumann and his superiors were less sanguine than the authors of the report. Early in February, Schumann met with Albert Vögler, a steel industry magnate and President of the Kaiser Wilhelm Society, to consider a possible transfer of the nuclear project from the Army to a civilian body. Both parties were agreeable. On 27 February 1942, Vögler, for the society, formally accepted responsibility for a project that the Army considered beyond its expected time horizon for winning the war, estimated at one or two more years.

There is a curious contrast here between the German and the American programmes. Ten days after the transfer from the German Army, Vannevar Bush asked permission from Roosevelt on 9 March to *involve* the American Army in the work of the S-1 Committee. After three months of intensive planning and research following Pearl Harbor, Bush felt that the moment for large-scale work and Army involvement was at hand. On 11 March, Roosevelt gave his approval. By June, the American Army was firmly committed. The German Army would never reconsider the 1942 decision to withdraw. In weaponry, it had other priorities to pursue.

After the Army gave up control, the tracking of German nuclear development becomes more complicated. For Germany there is no comprehensive, officially sponsored history comparable to those for Britain and the United States. On the research side, the activities of the several scientific teams and laboratories continued as before, with similar funding, and this aspect is well covered in Mark Walker's (1989) careful study. What seems to be less complete is a similarly detailed study of the administrative side, an account that would trace the tangled lines of formal authority and informal interventions by other ministers or their officials. Paradoxically, the final years of Nazi rule appear not so much an iron dictatorship as a collection of fiefdoms and satrapies, based on raw power, in which scientists gained protection and access to resources through the favour of individual ministers and officials.

The state of nuclear research at the time of the transfer can be assessed from two conferences held in February 1942. General Schumann had scheduled a research conference for scientists before the transfer decision was taken. This became a three-day meeting, with 25 scientific papers presented, held from 26 to 28 February at the Kaiser Wilhelm Institute. Parallel with this meeting, the Army and the Reich Research Council jointly sponsored a one-day conference of popular lectures on nuclear energy for an invited wider audience. The Ministry of Education and the Reich Research Council had been forbidden to continue sponsoring nuclear research in October 1939, but their leaders saw an opportunity at this point to re-enter the field. The three principal figures of the ministry were the same as in 1939: the physicist Abraham Esau at the Council, the top-ranking official in the Education Ministry, Rudolf Mentzel, and Minister Bernhard Rust. This was the same Minister Rust who had boasted after 1933 of having removed more than 1,000 teaching academics from German universities. Among those ousted were an estimated 106 physicists and 86 chemists,[47] of whom seven were Nobel Prize winners and eight more were future Nobel laureates.

The popular lectures on 26 February were delivered to invited representatives of the Nazi Party, the German state, and industry. The invitation underlined the importance of nuclear power for the military and economic future of Germany. Those attending included the Minister himself, Vögler for the Kaiser Wilhelm Gesellschaft, and members of the Reich Research Council. As speakers, the programme listed General Schumann (wearing his academic hat as Professor Doctor Schumann), Esau (for the Reich Research Council), five leading nuclear scientists—Hahn, Heisenberg, Bothe, Clusius, and Harteck—and one other senior experimentalist, Hans Geiger. In his address that day, Heisenberg spoke of basic neutron theory, of using uranium machines for power, of their potential for producing a "new substance, element 94," of easier access to an "unimaginably powerful explosive" from this element rather than from the less accessible uranium 235. He also stressed the work that lay ahead, and the financial and institutional support needed, to realize these goals.[48]

Rust had high hopes that these lectures would increase awareness of nuclear power in official circles and build the support needed for its success. In this he was successful, though eventually beyond his expectations. Within a few days, he moved to appropriate the nuclear project for the Reich Ministry of Education. If Army Ordnance could appropriate the Kaiser Wilhelm Institute for Physics in 1939, his own department could move in again in similar fashion when the Army withdrew. In this way control of the nuclear project was transferred from President Vögler of the Kaiser Wilhelm Gesellschaft to Abraham Esau of the Reich Research Council, who worked under the direction of Rudolf Mentzel in the Education Ministry.

After the Army withdrew, there remained the question of a director for the Kaiser Wilhelm Institute of Physics, which had been directed by Army physicist Kurt Diebner. The prewar director, Peter Debye, still formally on leave, was teaching at Cornell University. Just before Christmas in 1941, the Army generals had proposed Walther Bothe as director. In January 1942 a lobby of senior scientists spoke out in favour of Heisenberg for the post, on the grounds of Heisenberg's greater stature as a theoretical physicist, and Bothe's difficult personality. Heisenberg was also well established in the institute as a leading researcher. Vögler, for the Gesellschaft, nevertheless backed the Army's choice of Bothe. In the first week of March, however, word came that Bothe's appointment had been blocked by the Reich Research Council, and when Vögler queried this, Mentzel replied that the council now had complete control.

Two months after Bothe's appointment was blocked, Heisenberg was appointed director instead. The convoluted background to this appointment begins in 1937, and it offers a revealing glimpse of Nazi practice regarding academic appointments. On 15 July 1937, as the Deutsche Physik group grew bolder in its attacks on "Jewish" theoretical physics, Heisenberg had been subjected to a vicious personal attack by Stark in the SS organ *Schwarze Korps*, labelled as a politically unreliable "White Jew" and defender of relativity theory. He realized at once that if the charges stood his career would be in jeopardy.

To fight back, he called on a longstanding connection between the Heisenberg and Himmler families in Munich. He asked his mother to visit Himmler's mother to plead his case. In a postwar interview, Heisenberg recalled his mother's account of this visit:

She said that the elderly Mrs. Himmler said immediately "My heavens, if my Heinrich only knew of this, then he would immediately do something about it . . . . I will tell my Heinrich about it. He is such a nice boy . . . . So if I say just a single word to him, he will set the matter back in order."[49]

The elderly widow was as good as her word. Following her advice, Heisenberg wrote to the SS chief directly on 21 July, mentioning the maternal encounter and stating his case for protection against further attacks.

From this point the file moved slowly. Himmler replied in November, with a brief formal note requesting Heisenberg to respond to the specific charges in Stark's article. Heisenberg complied immediately in a letter of 7 November. Friends and colleagues also intervened on his behalf. The SS and its security arm, the SD, investigated the case at length. He had supporters within the SS itself, but the matter remained on hold for several months, awaiting Himmler's personal decision. Finally, on the anniversary of Heisenberg's first letter, Himmler wrote him a letter of exoneration dated 21 July 1938:

> Because you were recommended by my family I have had your case investigated with special care and precision. I am glad that I can now inform you that I do not approve of the attack in 'Das Schwarze Korps' and that I have taken measures against any further attack against you.

After this year-long inquiry, Heisenberg could count on protection and support from one of the highest Nazi ministers. Yet Himmler's letter ends with a postscript of warning:

> P.S. I consider it, however, best if in the future you make a distinction for your audience between the results of scientific research and the personal and political attitude of the scientists involved.[50]

Stark's attack on Heisenberg and Heisenberg's fight for exoneration were just one engagement in a larger battle: a protracted fight to succeed Heisenberg's renowned teacher, Arnold Sommerfeld, in the Chair in Theoretical Physics in Munich. Sommerfeld reached retirement age in 1935, and Heisenberg had been widely regarded as his natural successor. But the faculty's first nomination of him in July 1935 was blocked by the Education Ministry, apparently in response to local party influences and local leaders of the National Socialist University Teachers League, or *Reichsdozentenbund*. This resistance continued and intensified after the 1937 Stark attack, until Deutsche Physik supporters could put forward candidates of their own. By the time that Heisenberg was cleared by Himmler in July 1938, a new agreement between the Education Ministry and party headquarters in Munich left final authority for assessing the political reliability of candidates for chairs to Deputy Führer and Party chief Rudolf Hess. On this basis, the Munich chair was filled from 1 December 1939 by one Wilhelm Müller, who was not a theoretician, nor even a physicist, but an aeronautical engineer and published critic of relativity theory as a "specifically and typically Jewish affair."[51] It was a stunning victory for Deutsche Physik.

Although Heisenberg's candidacy for the Munich chair had support from both Rust and Himmler by mid-1939, neither was prepared to challenge Hess and Party headquarters on the issue. After discussions with Heisenberg himself, his candidacy was withdrawn in exchange for a promise from Himmler to find Heisenberg an equivalent post in the future. On 7 June 1939, Himmler wrote to Heisenberg to mention an upcoming chair in Vienna, but this plan failed when the universities were closed temporarily at the outbreak of war. The vacancy at the Kaiser Wilhelm Institute for Physics, carrying with it a professorship at Berlin

University, thus represented the next available opportunity. With Himmler's continued backing, the appointment was passed through Mentzel and the Reich Research Council without difficulty in April 1942. It was recognized by both the ministry and Heisenberg himself as fulfilment of the earlier promise. As Walker (1989) observes, "Heisenberg certainly deserved the two Berlin appointments on scientific grounds, but his professional competence was not why he received them."[52] The appointments also had no specific linkage to the nuclear projects, which at that point were headed by Abraham Esau under the Education Ministry following the takeover by that ministry.

For Heisenberg, the double appointment was gratifying, but insufficient. On 4 February 1943, he addressed an extraordinary two-page letter to Himmler[53] to remind the SS chief that the plan developed earlier for the official restoration of Heisenberg's honour was twofold: first, a call to a suitable Chair in Physics, and then publication of an article by Heisenberg reflecting his position in the *Zeitschrift für die gesamte Naturwissenschaft,* a journal "which has made many attacks on theoretical physics." Regarding the first stage, he reminds Himmler of his accession to the chair at the Kaiser Wilhelm Institute in May 1942, "and I thank you for the reestablishment of my honour bound up with this."

The second stage, however, had run into difficulties. The intended article had been completed and sent to the SS Security Office on 26 June 1942, for forwarding to the journal. On 17 November, Heisenberg had received a message from the journal's editor—Bruno Thüring, who was not an SS member—that "such an article would not be accepted in the journal." Heisenberg had written to the SS Security Office again on 12 December to ask where the article stood, but had received no reply. "Since it is impossible for me to obtain an explanation of this matter," he writes to Himmler, "I am afraid I must ask you to order the explanation yourself. I am extremely sorry that I must burden you at this most difficult time with this unpleasant matter. Heil Hitler!"

Heisenberg's letter was answered on 15 February by a member of Himmler's staff, who undertook in Himmler's absence to look after the matter on his next visit to the Security Office. "I assume that your request will find a solution without a need for the Reichsführer-SS to be involved again himself, since the preliminary steps for publication of this article have been taken already. Heil Hitler!"[54] With SS and SD backing, the article appeared in the October issue of the journal. It was cleverly written, and mindful of Himmler's original warning to Heisenberg to separate the science from the scientists. By comparing physics to geography, Heisenberg reviews modern theoretical physics including relativity and quantum mechanics while carefully avoiding any credit to Einstein: "America would have been discovered if Columbus had never lived . . . . So too undoubtedly relativity theory would have emerged without Einstein."[55] By means of this article, widely read, Heisenberg completed, to his own satisfaction, the recovery of his honour. He was clearly untroubled by the means needed to achieve this end. In wider context, the article marked a clear setback for proponents of Deutsche Physik.

The higher profile of nuclear research after the February 1942 conference led to further consequences. By stages, the work came to the notice of other, more powerful ministers. One of these was the ambitious young Albert Speer, Minister of Armaments and Munitions, newly appointed in February 1942. After first hearing of the project in April, he arranged a major briefing meeting for his military and civilian staff for 4 June 1942.[56] It was attended by about 50 persons: Speer and his staff, Vögler, and 16 prominent nuclear scientists. Speer was deeply impressed by the future potential for nuclear energy, but the meeting left him with mixed impressions. To improve the situation, he obtained a decree from Hitler, dated 9 June, that transferred the Reich Research Council, together with Mentzel, Esau, and the nuclear teams, from Rust's Ministry of Education to the ministries headed by Hermann Göring. Rust had proved a weak minister; Göring was one of the strongest. For the nuclear scientists, backing from powerful ministers such as Göring and Speer could enhance access to money, scarce materials, military exemptions, and more, but it would do little, given the Nazi power structure, to shape a more integrated nuclear project.

Though no minutes were kept of Speer's briefing meeting on 4 June, the discussion that day has been carefully pieced together by Thomas Powers (1993) from the memories of those who attended. Heisenberg, speaking for the scientists, talked of shortages of materials and money, of young scientists drafted by the military, and of the postwar promise of nuclear reactors, but he avoided direct discussion of element 94. Speer and the generals had other concerns. They raised direct questions about bombs. Heisenberg's reply, as Speer recalled it, was that a bomb was theoretically possible, but "the technical prerequisites for production would take years to develop, two years at the earliest, even . . . given maximum support." As Heisenberg remembered it, he had said "yes, in principle we can make atomic bombs and can produce this explosive stuff, but . . . it would take perhaps many years and . . . expenses of billions if we wanted to do it." Given Germany's successes in the war to this point, even a two-year wait seemed over-long. When a general asked how big a bomb would be needed to "reduce London to ruins," Heisenberg cupped his hands in midair: "About as big as a pineapple."

Under further questioning about bombs, Heisenberg became evasive. When Speer asked how his ministry could help, Heisenberg and Weizsäcker named budgetary figures so low that Speer found them ridiculous. Although he would continue to back the project with increased funding and access to scarce materials—including an expensive underground laboratory for the Institute of Physics in Berlin and approval to build Germany's first cyclotron—he left the meeting of 4 June with no expectation of any nuclear application to weaponry during the war.

This was not the first occasion for Heisenberg to evade an outside offer of help. Just three weeks earlier, he had received a letter dated 12 May from a research physicist at the giant Krupp firm. Having heard of the fission project through Vögler, the writer asked if the firm could be of assistance in developing nuclear energy at an industrial level. On 15 May Heisenberg replied negatively. He acknowledged that his group was doing research on nuclear fission, but, as this research was secret, he could not discuss it. He also expressed doubt that Krupp could help in this way in wartime, though after the war the situation might be different. Krupp did not pursue the question further.[57]

This double refusal of outside aid—whether from the public or the private sector—is consistent with Heisenberg's declarations and behaviour on other occasions. The contrast here between Britain and Germany is striking. While the Maud scientists were talking meaningfully with ICI executives as early as January 1941, Heisenberg and his team were still keeping their work strictly to themselves after almost three years of research.

Despite the decision of Army Ordnance to withdraw, research continued through 1942 and into 1943. Heisenberg had been working with his colleague Robert Döpel in his Leipzig laboratory on experiments to increase neutron multiplication, using layers of powdered uranium metal with a moderator of heavy water. On 23 June 1942, something went awry. The machine overheated, exploded, and burned for two days, wrecking the Leipzig laboratory. Heisenberg and Döpel narrowly escaped injury, but the experiments to that point appeared to show the possibility of a chain reaction if the dimensions were enlarged. After moving to the Institute in Berlin-Dahlem in July, Heisenberg began to plan for larger scale trials, but this time using uranium in cast metal plates rather than in powder form. His specifications for machine cutting of these plates were demanding. An official of the Auer company informed him that, barring a higher priority, plate production would have to be delayed until March 1943. In the event, successive delays would stretch this delay to January 1944.

Heisenberg's move to the directorship in Berlin led to further complications. Kurt Diebner, the former institute director from the Army, found himself isolated and even ostracized by Heisenberg's team, which included Carl Friedrich von Weizsäcker and Karl Wirtz. Abraham Esau therefore arranged for Diebner and other Army physicists to transfer their work to the laboratory of the Reich Physical-Technical Institute—of which Esau was president—in Berlin-Gottow. There Diebner, who was also a capable nuclear physicist, directed a younger team in a series of experiments on neutron multiplication, at first using uranium oxide with paraffin as moderator, then later with uranium metal cubes suspended in a lattice design in heavy water. These experiments, continuing over the next year and a half, gradually worked out a clear hierarchy for successful reactor design: a lattice of cubes was superior to rods, which in turn were superior to plates, though rods were easier to control. The Gottow experiments were producing the highest neutron multiplication reached so far in Germany. By November 1943, Heisenberg's associate Karl Wirtz wrote to warn him that the still-delayed plate experiment "might turn out poorly."[58] By this time Heisenberg was almost alone in defending his plate design, but in the long interval of waiting for the plates he had turned to other, more theoretical work. He also undertook several government-sponsored visits as a cultural ambassador to countries under German occupation: the Netherlands (October 1943), Poland (December 1943), and Denmark (February and April 1944). In December 1944 he would lecture in neutral Switzerland on a similar basis.

The recurrent friction between the Berlin-Dahlem and Berlin-Gottow groups was not over trivialities. Since raw materials and manpower were limited, rational allocation of resources was important. In March 1943 Esau, as Göring's plenipotentiary for nuclear physics, ruled that 600 litres of heavy water, which had been stored in the basement of the Berlin institute pending completion of the underground laboratory, were to be delivered to the Gottow group for experiments there. The transfer prompted Heisenberg to call for a meeting to discuss priorities in backing experiments and allocating scarce materials. One was called for 7 May, with Esau presiding but Vögler absent. The question of plates and cubes was vigorously debated, but in the end Esau accepted a compromise proposed by Heisenberg: to perform the plates experiment first, and later cut the same metal into cubes for use in a lattice pattern. It did not help in these questions that the two principals were working under the patronage of different ministers, Heisenberg through Vögler to Speer, Diebner through Esau to Mentzel to Göring. By December, Speer would persuade Göring to replace Esau with another plenipotentiary more acceptable to the Heisenberg group.

Notwithstanding frictions and shortages, the outlook for nuclear research in Germany remained optimistic during the first half of 1943. In March Bothe received the magnet needed to complete his cyclotron. On 6 May, a second series of popular lectures on nuclear physics was delivered at Göring's Air Force Academy, with talks by Esau, Hahn, Clusius, Bothe, and Heisenberg. As before, the purpose was to emphasize the utility of physics and the need for its support by government. Heisenberg's talk, unlike his 1942 lecture, avoided amy mention of "bombs" or even "explosives," stressing instead the utility of "uranium burners" for ship propulsion. In the interval since the 1942 conference, the regime had been promising "wonder weapons" or "vengeance weapons" as retaliation for increasingly successful Allied air attacks on German cities. By 1943, German nuclear scientists and their managers were careful not to feed Hitler's obsessions on this point. To mention any new weapon that could not be delivered in the timeframe available was to invite extreme personal danger.

Half a year after Fermi's achievement of a self-sustaining chain reaction in Chicago, Mentzel could forward Esau's progress report to Göring's personal assistant, Fritz Görnnert, with a comfortable assurance of continuing German nuclear superiority:

SECRET July 8, 1943

Dear Party Brother Görnnert:

Enclosed I send you a report by State Counsel Prof. Dr. Esau, Plenipotentiary for Nuclear Physics, on the present state of the work with the request to inform the Reichsmarschall about this. As you can see from the report the work has progressed rather considerably in a few months. Though the work will not lead in a short time towards the production of practically useful engines or explosives, it gives on the other hand the certainty that in this field the enemy powers cannot have any surprise in store for us.[59]

On this last point Mentzel is committing a cardinal error, a surprising one in the light of his long association with Education Minister Rust. He is forgetting that for over 10 years Germany has forcibly exiled scores of its most eminent physicists and chemists, and uncounted numbers of promising science students. He does not know that many of these exiles are working, with many others and more resources, on the same problems as their colleagues who stayed in Germany.

Germany: the war closes in

The situation deteriorated sharply with intensified Allied bombing of German cities in the second half of 1943. In the summer, plans were developed to move the Kaiser Wilhelm institutes in Berlin to small towns in south Germany, in Baden-Württemberg south of Stuttgart. By autumn, with Hamburg bombed heavily, the Harteck group moved south to Freiburg, then to nearby Kandern, and in 1944 northward again to Celle, near Hanover. The Diebner group at Berlin-Gottow and the Clusius group in Munich were each relocated just outside those cities. Part of Heisenberg's Institute of Physics moved south in 1943, but the new underground bunker laboratory, with reinforced concrete walls two metres thick, was safe from air raids. Work would continue in this laboratory until January 1945, when the remaining scientists of the Heisenberg and Diebner groups were ordered south to keep clear of approaching Russian forces. On the night of 15 February 1944, Otto Hahn's Kaiser Wilhelm Institute of Chemistry was destroyed in one of the raids, after which he too relocated southward to Tailfingen, close to the physics group at Hechingen. In Heidelberg and Göttingen, university towns, the research teams continued without disturbance.

The end of 1943 brought another administrative switch. Minister Speer, increasingly at odds with Plenipotentiary Esau, persuaded Göring—through Mentzel—to replace him. Esau was removed, given another post as compensation, and a successor was appointed effective January 1944. The new nuclear plenipotentiary was Walther Gerlach, a respected experimental physicist at Munich University. From the start, Gerlach treated his appointment as a double mission. He would steer the nuclear research programme to achieve the best results possible given diminishing resources and deteriorating conditions of work. Of equal importance was his professional loyalty to science, and a determination to preserve modern physics and a younger generation of physicists for a postwar reconstruction in which both would be needed.

The question of military exemption for scientists became acute at the end of 1941, when the Army was reviewing support for the nuclear programme amid mounting losses on the Eastern Front. Simon, watching Germany from Oxford, noticed increased casualties among recognized scientists in 1942 and pointed this out in his letter to Cherwell. Under urging from leading scientists, Göring was moved to action in July 1943. He established a planning board for research priorities and scientific manpower in the Reich Research Council, headed by an SS scientist, Werner Osenberg. By December, military authorities had agreed to withdraw some 5,000 scientists from combat service, but the project moved slowly, "strongly opposed"[60] by Speer but approved by Göring, Bormann, and Himmler. A postwar British inquiry into this policy found that, of 6,000 scientists selected for release, some 2,000 were already dead or untraceable.[61]

On 14 June 1944, as conditions worsened, Hitler issued a decree for total mobilization of resources. On 12 July Speer issued regulations for its execution that left his favoured nuclear projects basically intact. In the same month, Himmler, as head of the SS, blocked a plan to induct some 14,600 hitherto exempt scientists and engineers by issuing a peremptory order to his immediate deputy, SS General Hans Jüttner: "I direct you to cease this inductment operation in the military research sector forthwith, as I consider it madness to dismantle our scientific research."[62] On 3 September, Martin Bormann, who also had close access to Hitler, secured a further decree providing that scientists could not be forced under the total mobilization decree into any special service outside their own immediate neighbourhood. Gerlach himself, in becoming nuclear plenipotentiary, had demanded and received the right to cancel military call-up notices. On 16 December, he wrote to Bormann to complain that Heisenberg, von Laue, and others in Hechingen had been drafted into units of the Volkssturm, the last-stand "people's militia," and hence liable to service outside their own area, in contravention of the September decree.[63] Apparently this complaint was successful, and research continued.

The other side of military exemption was that nuclear work towards a self-sustaining reactor must continue at maximum possible speed. By 1944, the centrepiece of this effort was Heisenberg's underground laboratory at the Institute of Physics in Berlin. Following a series of experiments up to the end of January 1945, it was ordered to be dismantled and removed by lorries to Haigerloch, just west of Hechingen, on Gerlach's orders, when the Russian advance moved close to Berlin. By late February it was rebuilt and ready for the next experiment in the Berlin series, B-8, this time using a lattice of uranium cubes instead of Heisenberg's stubbornly defended plates. Neutron multiplication this time was the highest yet achieved, but not yet critical. Heisenberg estimated that about 50 per cent more uranium and heavy water could make the pile self-sustaining. On 1 March, Heisenberg notified Berlin of this by telegram, but to bring the extra material to Haigerloch was no longer possible.

Beyond the push for a reactor, other experiments were pursued with similar zeal. Although the Army—and Heisenberg too—had earlier written off uranium isotope separation as too difficult, others had not given up attempting it. The two methods most favoured were Erich Bagge's isotope sluice and Paul Harteck's work with centrifuges. Bagge, in a series of attempts, first in Berlin, then after relocating to Butzbach north of Frankfurt, and to Hechingen in August 1944, developed a model that successfully produced a few grams of enriched uranium hexafluoride. His first two attempts in Berlin were destroyed in air raids, but his apparatus was reconstructed in Butzbach and then rebuilt on a larger scale in Hechingen. Harteck's team had worked on a centrifuge design and then on an ingenious double centrifuge, which they developed during successive relocations to Freiburg, Kandern, and in September 1944 north again to Celle. Despite the loss of his early models in a bombing raid in July 1944 on the factory producing them in Kiel, Harteck by early 1945 had a bank of double centrifuges operating in Celle, attaining an enrichment of 15 per cent more U^{235} than in normal uranium.[64] If heavy water were unobtainable, sufficiently enriched uranium might not need it. Still another separation method pursued to the very end in Hechingen was work on liquid thermal diffusion. Horst Korsching, a young colleague of Karl Wirtz at the Berlin institute, had been working on this process since 1939.

As nuclear plenipotentiary, Gerlach marshalled his scientific forces like an army general even as enemy troops overran his country. He personally believed, and assured his superiors in all sincerity, that Germany still had the lead in nuclear research. More naively, he saw this presumed technological lead as a potential bargaining chip. If Hitler could be overthrown, he confided to Paul Rosbaud on 24 March 1945, "a wise government"[65] might negotiate a more satisfactory peace treaty on the strength of it. The scenario that Gerlach dreamt of was roughly parallel to the second part of Cherwell's dreaded nightmare: ". . . or reverse the verdict after they had been defeated" (see Chapter 5).

As late as 19 December 1944, Minister Speer had written to Gerlach to stress the "exceptional significance" of nuclear research, and the "great expectations"[66] the Minister held for it. Like the captain of a doomed ship, Gerlach stayed with his command. In January he organized the dismantling and dispatching southward of the Berlin underground laboratory and reactor, travelling personally with Wirtz and Diebner in a convoy of motor vehicles. When he heard the results of the B-8 experiment in Haigerloch from Heisenberg's telegram of 1 March, Gerlach believed—mistakenly, for his grasp of nuclear physics was tenuous—that the problem of a chain reaction had been solved.

In mid-March he returned to Berlin and met Martin Bormann in an underground bunker, announcing "We've got the chain reaction going. Our progress in Hechingen means we'll win the peace." He spoke with the same excitement to his friend Paul Rosbaud on 24 March: ". . . we know something of extreme importance which others don't"; with a post-Nazi government this "could perhaps get better conditions"[67] for peace. Rosbaud, less naive scientifically and politically, tried in vain to persuade him that such a tactic would not work, and could prove personally dangerous. Gerlach remained in Berlin until 28 March, and then headed south for last meetings in Hechingen with Heisenberg, Hahn, and von Laue. He then returned to Munich to await the end of the war.

The round-up of nuclear scientists

As General Groves's Alsos staff prepared to enter Germany with the first Allied troops, they still lacked certain specifics. Where were the teams and institutes that had left the main cities to avoid the bombing? Interviews with scientists in Rome after its capture in June 1944 had yielded very little hard information, and in liberated Paris the result was similar. The first major breakthrough came when Allied forces entered the University of Strasbourg, where Weizsäcker had held the Chair in Theoretical Physics since 1942. Among his hastily abandoned institute files, Goudsmit uncovered in early December a treasure trove of documents and correspondence relating to the entire German research programme, complete with current addresses and even telephone numbers.

On the Western Front, Allied forces crossed the lower Rhine and entered northern and central Germany before they could advance into southern Germany. By the end of March, Alsos team members were in Heidelberg, where they found Walther Bothe, his cyclotron, and his colleague Wolfgang Gentner, who had worked with Joliot's cyclotron in Paris during the occupation. On 8 April, Stadtilm in Thuringia was captured, and by 12 April the team found more documents there, the remnants of Diebner's pile in a schoolhouse basement, but not Diebner. On 17 April, they flew to Göttingen, where they discovered the Reich Research Council's files on scientific manpower compiled for Werner Osenberg. On the same day, other team members found the Harteck centrifuges in Celle, but not Harteck.

By 20 April, with Russian troops storming Berlin, resistance was fading and the time was at hand to reach for the top prize, the laboratories in the Hechingen area. On 21 April, Colonel Pash led a special unit of American Engineer Corps troops from the nearby town of Horb into Haigerloch, where two technicians opened the small laboratory building and a cave behind it containing Heisenberg's reactor pit. That evening a mixed assemblage from Alsos and other intelligence groups watched as Michael Perrin, the only person present who had previously seen a pile—Fermi's in Chicago—supervised the lifting of the lid of the improvised apparatus.[68] Perrin travelled from London, on short notice, for the occasion.

The next day Pash's troops moved on to Hechingen and seized several buildings housing the rest of Heisenberg's institute. The round-up of scientists ran smoothly. Some 25 scientists and technicians were segregated and questioned. Among these, von Weizsäcker, Wirtz, Bagge, and Korsching were detained. They reported that Heisenberg had left on 20 April to join his family in Urfeld in Bavaria. Also detained at Hechingen was Max von Laue, as was Otto Hahn the next day in Tailfingen. The trail next pointed to Bavaria, where Gerlach was found in his laboratory in Munich on 1 May, Diebner outside Munich on 2 May, and Heisenberg in Urfeld on 3 May. With nine scientists in custody, the only figure still sought by Groves's team was Paul Harteck, and he was soon found in Hamburg.

The round-up of scientists, important as it was, was only one part of the Alsos mission. Its objectives also included the seizure of German nuclear materials and documentation. The original motive for Alsos had been a widespread fear of a German project more advanced than Manhattan. Once this fear had been overcome, the objective shifted to preserving Manhattan's security and secrecy for a few more months while the war continued. To this end, the German detainees were held incommunicado, and documents, apparatus, and raw materials were either carried away or destroyed. Just before the Hechingen operation, a combined intelligence operation carried out a daring raid that recovered and removed from Germany about 1100 tons of oxide—the bulk of the uranium still unaccounted for—from Stassfurt, a town south of Magdeburg in the future Soviet zone. A month earlier, Groves himself had secured a massive 600-plane raid against the Auer plant in Oranienburg, just north of Berlin, before it could fall into Russian hands. With an equal distrust of the French—and Joliot's Communist connections—he ordered similar scorched-earth tactics for Hechingen and Haigerloch. After all apparatus was dismantled and removed, the cave and reactor pit were blown up.

The six scientists detained in the Hechingen area were driven to Heidelberg on 27 April. The next day, joint Allied custodial arrangements for them were worked out at Eisenhower's headquarters in Reims. On 2 May, in the custody of Major T.H. Rittner of British intelligence, they began a difficult peregrination that included five days in Reims, and four days at a regular camp for German prisoners at Versailles—where segregation was not possible and anonymity was in serious jeopardy. This was followed by 24 days at a large house at Le Vesinet, near St Germain, and 29 more at a chateau at Huy in Belgium. In all of these places a longer stay was unavailable and conditions for anonymity were less than ideal. The original six were joined by Heisenberg and Diebner after the first week, Harteck on 11 May, and Gerlach on 14 June.

The legal status of the 10 detainees posed a problem. Though detained on General Groves's orders, they were all civilians, accused of no crime, offered no explanation for detention, and could not be held in the United States under American or international law. In the face of this dilemma, Lieutenant-Commander Welsh, supported by his superior R.V. Jones in British scientific intelligence, urged that they be relocated to Britain. There, under wartime regulations, they could be detained "at his Majesty's pleasure" for up to six months—long enough for purposes of secrecy—without a hearing and without any need for explanation. On 20 May Major Rittner was informed that Washington was being requested to agree to this proposal, and on 14 June Welsh telephoned him to report that American permission had been given.

Rittner then went to England to inspect the new location, a country house near Cambridge named Farm Hall. It was owned by British intelligence but not needed after European hostilities ended. The displacements for the detainees ended with a flight from Liège airport to England on 3 July. In his preliminary report to Perrin and Welsh dated 14 July, Rittner declared the operation "successful to date." In 10 weeks, there had been no security leak, and "The professors . . . have, with considerable difficulty, been kept in a good frame of mind."[69] The Farm Hall arrangement solved an awkward legal problem for the Americans. The Alsos mission had recovered practically all the German raw materials and German documents for direct shipment to the United States, but the "guests" detained at Farm Hall represented a considerable resource for British scientific intelligence.

In the new location, the scientists were closely monitored and recorded by hidden microphones for information bearing on their wartime nuclear work. From the recordings, edited transcripts were sent simultaneously to the Tube Alloys office and to Groves's representative in London. The importance of the information at issue overcame any hesitation about procedure. The scientists, occasionally suspicious, usually ignored possible listening devices and talked freely. On 6 July, Diebner asked: "I wonder whether there are microphones installed here?" Heisenberg laughed. "Microphones installed? Oh no, they're not as cute as all that. I don't think they know the real Gestapo methods; they're a bit old fashioned in that respect."[70]

The results would justify the monitoring effort. On 6 August, when the scientists first heard—with total incredulity—the BBC news reports of an atomic bomb dropped in Japan, their comments would capture a cardinal moment in twentieth-century history.

The monitoring technology used at Farm Hall was primitive. The recordings were made on shellac-coated metal discs, which were recoated and reusable. The parts deemed relevant were transcribed in German and translated. The translated and edited material was then incorporated in regular reports to Welsh and Perrin which ran from 3 July to 31 December 1945. Despite pressure from historians, the transcripts were not declassified until February 1992. Before that point, the official position had been that the Farm Hall transcripts did not exist, a position harder to maintain after Groves himself quoted extensively from them in his 1962 memoir *Now It Can be Told*. The appearance of a British edition in 1993 and an American one in 1996 revealed raw materials for dramas rivalling any play of Shakespeare. For present purposes, these editions are an essential source for understanding the complex, uncoordinated evolution of Germany's nuclear effort.

From April 1945 onward, echos of the round-up of German scientists appear in Franz's diary. The key figure is Michael Perrin, who by now is prominent in nuclear intelligence by virtue of his position as Deputy Director of Tube Alloys. On 12 April, writing from newly captured Stadtilm, where Diebner's group from Berlin-Gottow had been working, Fred Wardenburg, an Alsos scientist from the Paris office, sent an urgent note by courier to Goudsmit, which read in part:

Sam,
... After three hours here, it is obvious we have a gold mine ... [he lists events, files, apparatus]
I think you should get here post haste. Mike Perrin should also be here. We will certainly learn the broad outlines of the whole project here and in the south fill in the technical details.
See you soon,
Fred[71]

Franz's diary for this day, however, shows Perrin in London at a two-day meeting of the Tube Alloys Technical Committee, in the midst of which the diary lists a half-hour discussion with Perrin alone on "Ch.'s letter," which is not further explained.

The reference to "Ch.'s letter" may refer to Churchill's letter to Roosevelt, dated 1 April, which questioned Eisenhower's apparent southerly shift of plan and re-emphasized the strategic, symbolic, and political importance of getting to Berlin as an Allied objective: "from a political standpoint . . . should Berlin be in our grasp we should certainly take it."[72] Roosevelt, however, died suddenly in Georgia, only a few hours after this discussion between Simon and Perrin. When the Alsos action shifted towards the Hechingen area, however, Perrin was ready to be there on short notice.

On 1 May, Franz went to London for meetings. At midday, the diary records: "1.10–2.30 Lunch with Perrin at R.S. [Royal Society] Club. P. tells his exp[eriences] in Germany." Perrin's week away from London—21–28 April—had given him quite a lot to tell about: of leading the group of intelligence experts into the Haigerloch cave and opening the lid of Heisenberg's pile; of lengthy questioning of Weizsäcker and Wirtz in Hechingen; of discovering containers of heavy water hidden in petrol cans in a Haigerloch cellar; of wielding a shovel himself as they dug up buried uranium cubes from a field outside the village; of documents in a sealed canister hidden in a cesspit; of a report drafted in Hechingen by Perrin and Welsh, coded, transmitted to London by radio, and forwarded through Akers to Sir John Anderson; and finally of the road journey with the detainees back to Heidelberg.[73] After lunch, Perrin and Simon attended the afternoon meeting with Akers and others in the Tube Alloys offices, after which Franz returned to Oxford. At bedtime that evening, he notes in the diary: "News of Hitler's death. Listen."

For the war years, Simon and Perrin had been in contact by telephone or in meetings almost daily, but these private lunches at roughly two-week intervals from April to early July, along with occasional sessions in the Tube Alloys offices noted as "Perrin disc[ussion]" or "Perrin alone," indicate that more confidential matters are involved. Franz also records contact in this period with Perrin's assistant, David Gattiker, who had been active in April in the round-up of missing uranium ore in Belgium and in planning the raid on Stassfurt. Gattiker's superior in tracking raw materials, Sir Charles Hambro, who accompanied Perrin when the Haigerloch reactor was opened, is also mentioned in the diary entry for 1 May. The Tube Alloys offices became an assembly point for German nuclear information. On 17 May, in London for meetings, Franz records intervals both morning and afternoon to "read German reports."

The "German scientists" question also involved sporadic contacts with others. On 14 April, Franz received a visit to the Clarendon by Major E.W.B. Gill, an older Oxford don and Fellow of Merton College, who was working on an intelligence report on German scientists during the war. They would talk again on 3 August, when this report was apparently near completion, followed the next day by more discussion on "Germany, later Physics" at Gill's home in Boar's Hill, with Gilbert Murray also present.

A more enigmatic diary entry is a brief note: "Look up W. Groth" on Sunday evening, 22 April. Wilhelm Groth, a member of Harteck's team, had been found and questioned, but not detained, three or four days earlier in Celle, where the centrifuge separation apparatus was again operating after multiple migrations. Whether by accident or design, the diary entry is sandwiched between two telephone calls from Cockcroft, the director of the Montreal laboratory, who was visiting Oxford at this point. Cockcroft had spent the previous evening at Belbroughton Road with the Simons and Shull Arms. They talked "about Montreal," presumably about the heavy water laboratory. Though Harteck was one of Germany's leading experts on heavy water, Groth, his colleague, had been working mainly on their centrifuge project since the autumn of 1941. Though not detained by the Alsos team, Groth was picked up soon after by T-Force, a rival British intelligence team rounding up a wider range of German scientists, and flown directly to London for interrogation. After 1945 he returned to Hamburg University, and in 1950 to a Chair in Physical Chemistry at Bonn. During the war he wrote half a dozen reports on uranium isotope separation, mainly by ultracentrifuge. By 1953 he was prominent in a project to export centrifuges for uranium enrichment and supporting equipment from Germany for Brazil's nuclear programme. Blocked initially by the Allies, the shipment was sent in 1956.

Awaiting the end

After the excitement of the spring round-up, July was quieter. The detainees, after 13 weeks of improvised lodging, settled comfortably into Farm Hall. From their monitored conversations pertinent information was extracted and translated. From 14 to 28 July Franz was away on holiday with Lotte and Kay, a rain-plagued fortnight on the north Devon coast. While they were away, the long-awaited test of Manhattan's plutonium design—Trinity—took place on Monday 16 July, the Potsdam Conference on the future of Germany began on 17 July, and Churchill resigned as prime minister on 26 July, as soon as results from the general election of 5 July were released. At Potsdam, the eastern boundary of Germany, still unresolved, would be in the hands of Stalin and two new Western leaders, Truman and Attlee.

While Franz was on holiday, Michael Perrin sent him a brief, undated note on a slip from a small writing tablet:

> You will be glad to know that everything went off according to expectations last Monday.
> This is very particularly much for your own personal information at the moment.
> Hope you are having a good holiday.
> Yours ever
> M.W.P.[74]

Just 35 words, no salutation, but enough to be clear beyond doubt to Franz.

The morning after Franz's return to Oxford on 28 July, Michael Perrin telephoned him about a visit to Oxford that same afternoon of two American officers. Franz met them in Christ Church after lunch, and later took them to meet Ronald Bell. The only one identified in the diary, Major John Vance, a chemist in the Corps of Engineers, was an expert on tracing and rounding up nuclear raw materials.

The rest of this first week back from holiday seemed typical, apart from extra post to be attended to. The further contacts with Major Gill on 3 and 4 August have been noted above. One small difference in the diary is more frequent mention of radio listening from 1 August onwards. The news from Potsdam or Downing Street was interesting enough, but senior scientists in Tube Alloys may have known of Groves's revised time target of a bomb by August.

The first Monday in August was a public holiday, marking bank holiday weekend. In 1945, the holiday fell on Monday 6 August. For the Simons, that weekend marked the twelfth anniversary of the family's arrival in England in 1933. The diary records a leisurely day, with two shortened periods at the laboratory for "tidying," a holiday dinner at midday, an early return home, gardening, and "After supper short walk with Lo." The diary then continues: "8.45 home. Daily Mail rings. 9.00 News. Truman, Churchill's statement. Press Association rings. 10 listen. 11 listen. Kurti rings. 12 to bed . . ."

A few miles away, at Farm Hall, the German scientists were listening to the same BBC news and official statements. Forewarned by Major Rittner just before dinner that evening, their initial reaction to the 6 p.m. news had been total disbelief. As they listened intently to the longer Truman and Churchill statements at 9 p.m., their disbelief slowly ebbed away.

The family documents found in the house include official statements from Tuesday's press. The *Daily Telegraph* for 7 August devotes most of its front page plus more inside to the two official statements, some background on development of the bomb, and related stories previously held back by censorship. The Truman text gives figures on manpower and expenditures, mentions Oak Ridge in Tennessee, Richland in Washington, and "an installation near Santa Fé." It also promises a statement from Secretary Stimson giving further details, and repeats the Potsdam ultimatum to Japan of 26 July. Churchill's statement reviews the history of the project, describes American–British cooperation, and pays tribute to Roosevelt and the American achievement. It also names the leading scientists and institutions in Britain that were active in Tube Alloys. For the detainees at Farm Hall, the magnitude and scale of Manhattan is astonishing. The manpower figure, Harteck remarks, "is a hundred times more than we had."[75]

For Franz, the week that had begun with a leisurely Monday holiday quickly became very crowded. Both the British and the American statements promised more information to follow, and both countries agreed to full disclosure of the underlying science, while withholding details on technological processes and military applications. A book-length American report, known as the Smyth Report, was ready by July and released on 12 August, but the British equivalent material was incomplete at the time of the official statements. Franz, in London at Old Queen Street on 7 August, read both reports "cursorily," but then became involved in writing, revising, and editing material for the unfinished parts. Working under pressure, a Tube Alloys team had a 33-page British statement ready for release by the same date as the American one, 12 August. A shorter Canadian statement was released a day later. These three official statements are printed together, with a combined index, in the fifth and later printings of the Smyth Report, *Atomic Energy for Military Purposes*, a volume that documents the contribution of scientists and other experts from many countries to the Allied nuclear projects.

Apart from this extra editorial work, Franz faced a crowded schedule of other matters after the Monday holiday: on Tuesday, two hours to "talk to all our employees" about the project itself and the announcements, followed by the afternoon in London at Tube Alloys; on Wednesday, a doctoral examination with an external examiner from Cambridge; on Thursday, another visitor to the laboratory. The diary notes "Farkas. Heavy Water, Dead Sea. Cherwell+Farkas. Cherwell alone." Franz took Farkas home to lunch, and later—back at the laboratory—joined Farkas and Kurti for tea. Ladislaus Farkas, a Hungarian physical chemist from Berlin, had moved in 1935 to the Hebrew University in Jerusalem. He and his younger brother, both physical chemists in Germany, and both refugees in 1933, had each worked on heavy hydrogen. On Friday 10 August, Franz had a full, busy day in London. When time allowed, his writing, editing, and checking for the British report continued, and this carried over into Saturday and Sunday. Overall, the week of 6 August had left him few moments for reflection.

It was his sister Ebeth who had the better view. Looking on from a distance, and seriously ill herself, she was better placed to see their lives, their fates, and their achievements in wider perspective. On 10 August, she wrote Franz a letter of congratulation, mixed with jubilation that 12 long years of evil were at an end:

Jerusalem, August 10, 1945

Dear Franciscus,

This is to say our congratulations and admiration – you really have not spent your time on trifles! I suppose people all over the world were breathless – I cannot remember anything having made such an impression in all my life. VE-day was nothing compared with it; strangely that was rather a flop here, owing to this piecemeal release and indiscretion. And people here being not different from people all over the world, were of course particularly proud that some of their brand were among the chief brains. Well, we do sincerely hope that we hear a few details (not scientific, as we wouldn't understand them probably) from mouth to ear . . . .

The rumours have just started, that Japan has surrendered, but it does not yet seem to be the real thing, but we believe that it will be a matter of days. But to imagine that you have played a part in such an achievement is really breathtaking. Much as I am glad that Papa has not lived to see this world and the happenings of the last 12 years, it is a pity that he hasn't seen you doing <u>this</u>.

It is a long time since we sat in the sand-pit talking about the future, and though I remember some of my plans, I cannot remember in the least yours, but I suppose they didn't reach the reality, one which is really given only to a very few people. Still, it would be nice to know how you thought about things then. You were quite a bit older, so I suppose they were not very similar to mine.

Give our love to Lotte and the children.

Yours Ebeth

The study of physics, which Ernst Simon had deemed an unprofitable profession (*brotlose Kunst*), had been worth something after all.

7

WHY MANHATTAN?

This six-year chronicle of the struggle for nuclear energy poses one central perplexing question: Why? In the larger mirror of twentieth-century history, why did Germany, a country pre-eminent by 1900 in scientific discovery—and especially in quantum physics—fall so far short of Manhattan in the nuclear race? The question is important, and the answers may have lessons for science projects at other times and places.

This chapter is an essay in comparative analysis of the five nuclear programmes that were active during the war years, with a view to understanding the factors that were conducive to success or failure. For convenience of analysis, it will establish a few simple categories and consider each in turn: (1) scientific personnel; (2) administrative organization and management; (3) military and industrial priorities; (4) raw materials; and (5) effects of enemy action. In wartime, this last factor can be serious, even decisive. By its nature, this chapter will be less concerned with individual careers than with the working environments in which they evolved.

Prewar developments

A good place to begin is with the scientists themselves: their numbers and quality; their availability; their motivation. For Germany, the initial phase is one of serious brain drain, beginning in 1933. The first and largest loss, Minister Rust's dismissals from the universities in 1933, have been traced earlier. Anti-Jewish measures became more systematic with the Nuremberg laws of 1935, which closed off the Hindenburg exemption for front-line war veterans, and targeted families with one Jewish spouse or those of partly Jewish background. Some of the latter were allowed to remain in Germany but forbidden to teach, examine students, or work in the public sector. As German power spread over Europe, the outflow of scientists expanded, with more refugee scientists from Austria, Czechoslovakia, France, and elsewhere. With some exceptions, replacements for the ousted scientists were less able, or less experienced. In time, the Nazi Party would control all appointments to university chairs, and political orthodoxy would replace academic competence as the dominant criterion for appointments.

When Jewish scientists had been effectively banished from the universities, the attack continued against what traditionalists labelled "Jewish" ideas, particularly in physics. The exponents of Deutsche Physik fought for a science of demonstrable public utility, an applied science dedicated to the well-being of the German Volk. In their view terms such as Jewish, cosmopolitan, theoretical, liberal, and speculative became almost synonymous, and all pejorative. What they did not stop to consider was that this speculative, theoretical world might bring tangible benefits—in war or peace—for societies that could understand it.

Under this concerted attack, theoretical physics in the universities was seriously damaged between 1933 and 1938. In the summer of 1938, an anonymous report on the situation noted that 11 of the 35 theoretical chairs existing in 1933 were "either unfilled, inappropriately filled, or eliminated by 1938."[1] The attack on theoretical physics had major consequences for student enrolments and for the future of the subject in German universities. When staff vacancies arose, courses were not offered. When offered, fewer students were attracted to a field so obviously out of favour with the regime. Those who did enrol found their discipline "harassed . . . continually"[2] by the official organ of the Reich Student Organization. These attacks seriously undermined academic physics in numbers and in quality.

If we seek defining events of this attack on the new science, five events stand out. The first would be the Einstein affair in March 1933. Einstein, the most widely known scientist in the world, refused to return to Germany in 1933 after the Nazis seized power, and promptly resigned from the Prussian Academy of Sciences. The Nazi state reacted by confiscating Einstein's German property, and pressuring the Academy to condemn Einstein publicly. With a statement on 1 April, the Academy complied. A second marker, a few weeks later, would be the bonfires of books—including university textbooks—by Jewish authors or co-authors, proclaiming with pride the barbarism of the new regime. A third would be the humiliation and resignation in 1935 of Gustav Hertz, who left his university chair to work—still in Berlin—for the Siemens company. The fourth was Stark's attack on Heisenberg in July 1937 as a "white Jew," as described earlier.

Fifth, and finally, there is the celebrated experiment of the chemists Hahn and Strassmann on uranium in December 1938. Although this work took place in Hahn's chemistry institute in Berlin, it was Hahn's long-time associate, Lise Meitner, and her nephew, Otto Frisch, who first grasped the significance of Hahn's strange result and named it "nuclear fission."[3] Both were now living in exile, and their history-making meeting occurred not in Berlin, but in the snowy forests of Sweden near Gothenburg, during the Christmas break.

The name of Gustav Hertz appears several times in the Farm Hall transcripts. On 6 August, it will be recalled, the detained scientists moved from initial incredulity regarding the Hiroshima bomb to acceptance after they heard details from the 9 p.m. BBC news and Churchill's official statement. Along with initial surprise and shock, the statement touched off a lively debate on how the bomb had been achieved, continuing in smaller groups until 1.30 a.m. At one point Hahn and Harteck went to comfort Gerlach, who had been emotionally shattered by the news and had withdrawn alone to his room. There, after a discussion of various problems in the German programme, Harteck turned to a more general point:

> Of course, we didn't really do it properly. Theory was considered the most important thing and experiments were secondary, and then almost unintelligible formulae were written down. We did not carry out experiments with sufficient vigour. Suppose a man like Hertz had made the experiments, he would have done it quite differently.[4]

In the presence of Gerlach and Hahn, there was no need to elaborate. Harteck then lists specific experiments, along lines pioneered by Hertz and Otto Stern, that could have been done.

> But such experiments were not made, or rather they wanted to persuade you against it.
> Hahn: Hertz did that.
> Gerlach: Yes. He had all the material he could find.

Hahn: When was that – in 1944?
Gerlach: Yes, the end of 1944. But he had measured the emission of heat already two years before. I just went to Hertz and said: "Look here, Hertz, let's discuss the uranium business." He said: "I know nothing about it," so I told him all about it. Then he told me that Schütze had made such heat experiments and then we discussed it and decided that that really was the best thing.[5]

Here and elsewhere, amid Gerlach's imprecisions and possible loose translation, one can regret the absence of the original German transcript, but these lines seem enough to indicate (1) that Hertz was working on uranium during the war, as early as 1942, (2) that Gerlach and Hahn, and probably Harteck too, knew of this work, and (3) that Hertz was unwilling to talk about it but willing to mention experiments by his co-worker Werner Schütze, a doctoral student.

Hertz's name recurs on 18 August, the day that Sir Charles Darwin visited Farm Hall. That evening, after the detainees had been shown the British White Paper on the bomb, their conversation returned to isotope separation. In the transcript, Weizsäcker remarks: "They have decided to do it by gaseous diffusion." Harteck responds: "Hertz tried that and even he didn't succeed."[6] Apparently Hertz was either not yet successful or unwilling to admit to success with diffusion. He would be more successful a few years later, but in the Soviet Union, and Schütze would join him there.

For the scientists forced into exile, Germany's losses became corresponding gains for the receiving countries. Unlike many refugees who had few resources after moving, senior scientists brought readily marketable skills. Physicists, chemists, and mathematicians were a part of the much larger, aptly named "Hitler's gift"[7] that enriched the scientific and cultural life of Britain, the United States, and several other countries. In the sciences, these newcomers brought priceless assets. The more eminent among them represented the pinnacle of research achievement in a country that still ranked among the top scientific countries of the world. Familiar with the German mentality and social conditions, they monitored developments in their former homeland throughout the war years. With direct experience of the Nazi regime, they recognized early the inevitability of war and sounded the alarm as best they could. Finally, unlike their former colleagues in Germany, most refugee scientists felt no moral ambiguity concerning Hitler. After their expulsion they owed no loyalty to a regime that stripped them of their property and citizenship. Earlier than most, they knew that a German victory would spell annihilation for Europe's Jews and another Dark Age for the world.

One could make a plausible case that the race for nuclear supremacy was lost irretrievably by Germany even before 1939 as a result of forced Jewish emigration. The losses to German science from 1933 onwards are clear, and refugee scientists who moved to Britain or America widened the gap on the positive side as well. On the other hand, Germany entered the war with some notable advantages. After several years of clandestine rearmament, these included a powerful war machine which moved quickly to place nuclear research under Army Ordnance control in October 1939. Comparatively, Germany made a very fast start on nuclear weaponry.

The development of nuclear research from 1939 to 1941 under Army Ordnance control has been described already in Chapter 5. As in Britain and the United States, most research in these years was carried out in university institutes and laboratories. Following a string of German military successes, the war seemed likely to be of limited duration; "neither Army Ordnance officials, nor the scientists, were under great pressure."[8]

In one key respect, however, Germany differed from Britain and the United States. It had a well-developed weapons project, older, larger, and with more influential backing: the self-propelled missile. Before the war ended, the Air Force's "cherry stone" flying bombs and the Army's A-4 ballistic missiles—better known as V1s and V2s—would be allocated resources roughly comparable to the Manhattan Project in proportion to the German war economy. By a combination of human decisions and circumstances, German nuclear research would remain a junior partner to this missiles programme throughout the war.

The missile alternative

The Army Ordnance interest in rockets had its origins before the Nazi seizure of power. By 1936 a test site at Peenemünde on the Baltic coast was acquired and shared between the Army and the Air Force. In January 1939, before war was declared, a plant was planned for mass production of missiles, to be constructed over the next four years. In September, with war declared, the target date for missile production to begin was advanced from 1943 to September 1941, as a project "particularly urgent for national defense."[9] This schedule proved totally unrealistic, collapsing in a sea of changing and ineffective priorities for steel and other materials. As long as blitzkrieg victories continued, Hitler remained cautious on rockets, "lukewarm at best."[10] Even a face-to-face meeting of the Führer with the project leaders in August 1941 produced no authorization to proceed to mass production.

The turning point came, gradually, in 1942. After unsuccessful launches of A-4 test models in June and August, a third attempt on 3 October led to a clear trajectory of 190 kilometres in distance and about 80 kilometres in height, breaking all previous records. The successful test launch confirmed the earlier optimism of Armaments Minister Speer, but several more months of development on launching and guidance systems were needed before production in quantity could begin.

During the winter of 1942–3, while the new plant was taking shape under Speer's guidance, Hitler moved from hesitation about rockets to dangerous infatuation. On 11 June 1943, after another demonstration of both the V1 and the V2, Hitler ordered the highest priority in armaments production for A-4 missiles. On 7 July, at a personal audience with Peenemünde project leaders in his East Prussian headquarters, he demanded the impossible: a 10-ton warhead and a production rate of 2,000 missiles per month. When General Dornberger tried to explain what was possible, Hitler reacted angrily: "But what I want is annihilation – annihilating effect (*vernichtende Wirkung*)!"[11]

Behind the shift to large-scale missile production lay two main reasons. In V1s and V2s Hitler could visualize the "wonder weapons" or "vengeance weapons" so often promised in official propaganda to shore up civilian morale. In the face of mounting Allied air raids on industrial cities, such a boost was needed. A second, longer range factor was the regularly repeated warning from the scientists of Germany's need to keep ahead of foreign competition, especially of American rocket research suspected by German intelligence sources. Michael Neufeld notes the paradox of reciprocal misinformation concerning the American and German projects:

> The Army rocket program had become an ironic mirror image of the Manhattan Project. While the Germans were racing a virtually nonexistent American missile program, the Americans (with British and Canadian help) were racing a virtually nonexistent German atomic bomb effort.[12]

Ten weeks after Hitler's order for maximum production, but before the new plant could begin operations, Peenemünde was hit by a 600-plane RAF raid. The raid of 18 August 1943 was only moderately destructive. The production plant itself escaped major damage, but the warning was clear. The Allies knew of the site and the project, and more raids were predictable. At Himmler's suggestion, Hitler ordered A-4 production moved underground, and by 26 August an underground location was chosen in the Harz Mountains near Nordhausen in Thuringia. Between September and December, by a massive effort, and using thousands of slave labourers in barbaric conditions, the manufacturing plant was installed inside a series of underground tunnels. The Peenemünde site remained in use for research and development.

The new production plant, designated simply Mittelwerk, shipped its first missiles for testing in January 1944, but poor quality workmanship and unresolved technical problems delayed their first use as weapons for several months until 7 September. The V1s, also delayed, were first launched against London on 13 June. From these dates to the end of March 1945, about 3,200 V2s were launched against British or continental targets, killing an estimated 5,000 civilians.[13] A more exact figure, issued by Colonel John Elting of the United States Military Academy, lists 6,184 civilians killed in Britain by V1s and 2,754 by V2s.[14] The same source lists deaths from aircraft bombing at 51,509, almost six times as great. A fatal weakness of both V1s and V2s was the primitive quality of their guidance systems. The average error of the V2s in flight was "at least twenty times worse than Dornberger's unrealistic 1936 goal of less than a kilometer."[15] Beginning only while Allied armies were closing on Paris, they did not and could not sow the terror in Britain that Hitler had demanded of them. The cheaper, noisier, and far more numerous V1s caused more disquiet in Britain in June and July of 1944 until effective defences could be worked out to intercept them.

Like the nuclear scientists, those working on missiles worked feverishly, under deteriorating conditions, to the bitter end. At Peenemünde, development teams worked on desperation projects: a glider missile modified from the A-4 design when the target distance became too great for a normal V2 trajectory; a launch platform towed by submarine, to allow missiles to target New York; projects to develop two different anti-aircraft missiles. For all these last-minute efforts, time ran out. On 31 January 1945, under the sound of approaching Russian artillery, an order came to transfer development personnel, and whatever equipment could be removed, from Peenemünde to the Mittelwerk area in Thuringia. The move was completed in early March.

At the Mittelwerk plant, missiles continued to be produced and shipped, under deteriorating conditions of starvation and mass executions of slave labourers, until 18 March. They were launched against London until 27 March. On 1 April, as American troops neared Thuringia, General Dornberger and Wernher von Braun, with about 500 key Peenemünde staff, were ordered to move to the Bavarian Alps. In Bavaria, the top leaders gathered in a mountain hotel at the Austrian frontier and remained there for almost a month before surrendering to the Americans on 2 May, one day before Heisenberg was found and detained at Urfeld.

The rocket scientists soon discovered that they held a bargaining chip far more valuable than that held by Gerlach and his nuclear scientists.[16] Von Braun was allowed to select an entire team of specialists to take to the United States and continue rocket research, initially near El Paso in Texas, but after 1950 at Huntsville, Alabama. Other experts from Peenemünde were recruited by the Soviet Union, Britain, and France. By these migrations,

Peenemünde's A-4 missile became the progenitor of the postwar Redstone, Jupiter, and Saturn rockets in the United States and the Semyorka series of the Soviet Union.

The postwar history of missiles, however, is not the issue here. The two relevant questions are whether the A-4 missile programme was the right decision for Germany in 1942, and whether the decision to produce missiles in quantity had negative consequences for Germany's nuclear programme. On the first question, the answer is relatively simple. The V2 proved to be a weapon of limited destructive power. Its weakness was a primitive, inaccurate guidance system. The money spent on A-4s could have been used more effectively elsewhere. When Speer later blames Hitler's intoxication with "wonder weapons" for the faulty decision on missiles, he fails to acknowledge his own part in converting the Führer from cautious scepticism in 1941 to impossible fantasies about "vengeance weapons" by 1943.

The second question needs a more careful answer, because it has been obscured by later revisionism or faulty memory. In June 1942 the recently appointed Armaments Minister faced two crucial meetings. On the evening of 4 June, Speer and his assembled staffs met senior nuclear scientists at Harnack House to be briefed on nuclear research (see Chapter 6). After the lecture, under questioning about prospects for a bomb, Heisenberg became negative and evasive, stressing two or more years of development, massive costs, Germany's lack of cyclotrons, and refusing even to assure the audience that a chain reaction would not become uncontrollable. When invited to budget for more funding, the scientists asked only for insignificant increments. As Speer later remembered the encounter, "I had been given the impression that the atom bomb could no longer have any bearing on the course of the war."[17]

Nine days later, Speer attended another gathering of Nazi leaders in Peenemünde to watch the first attempt to launch an A-4 missile. That launch, only partially successful, did nothing to diminish the rocket scientists' conviction that their project could be decisive for the war. On the basis of the information at hand, Speer then made a decision to support and expand the project of the optimists while also offering the pessimistic nuclear group all of their modest budgetary requests and even additional money, which they declined. He also arranged the transfer of nuclear research to a more powerful minister, Göring.

During a long imprisonment for war crimes from 1946 to 1966, Speer had ample time to reflect on these decisions. In time, he came to build alternative scenarios around them that were plainly at variance with earlier realities. By 1953, in his Spandau diaries, he could represent rockets and nuclear weapons as basic alternatives, both available from early in the war:

> Thanks to the insane hate of the leadership we allowed ourselves to lose a weapon of decisive importance. If, instead of backing the – in the final analysis – ineffective rockets with hundreds of millions, we had devoted them to supporting atom research from the start, it would have been more useful for the war.[18]

In mentioning support "from the start," Speer appears to forget that it was Army Ordnance, not the Ministry of Education, that ran nuclear research directly from 1939 to 1941 and was responsible for initial modest levels of support.

After his release from Spandau in 1966, with more documentation at hand, he returned to the question more fully in his memoirs. In five revealing pages,[19] Speer touches on many points: his first contacts with the nuclear project; his recollections of the Harnack House meeting; the scientists' "modest requests" and refusal of more ample funding; his brief, guarded report on this meeting to Hitler; Hitler's lack of response to an idea that "quite obviously strained his intellectual capacity"; and Hitler's occasional disdainful references, inspired by Philipp Lenard, to nuclear physics as "Jewish physics."

Speer also returns in these pages to the question of rockets versus atom bombs as alternative policy choices:

> Perhaps it would have proved possible to have the atom bomb ready for employment in 1945. But it would have meant mobilizing all our technical and financial resources to that end, as well as our scientific talent. It would have meant giving up all other projects, such as the development of the rocket weapons. From this point of view, too, Peenemünde was not only our biggest but our most misguided project.

At the end, he concludes that even this would have proved inadequate because of the "superior productive capacity" of the United States and "increasing air raids" in Germany:

> At best, with extreme concentration of all our resources, we could have had a German atom bomb by 1947, but certainly we could not beat the Americans . . .

Speer's own references to cyclotrons, chain reactions, and an "energy-producing uranium motor" in these pages and elsewhere raise doubts about his own understanding of the underlying science.

There is a further revealing moment in Speer's published diaries for 2 December 1962, when he writes of discussing occasional "innocuous" ministerial acts unauthorized by Hitler with his fellow prisoner Rudolf Hess. Among examples, he cites "the switch (*Umschaltung*) of our atomic research to a uranium-powered motor because Heisenberg could not promise to complete the bomb in less than three to five years."[20] Speer apparently did not know, though the scientists did, that a uranium "motor" could be a route *to* a bomb, indeed the *only* route open if uranium isotope separation were ruled out as too difficult.

Inside Farm Hall: a version for the world

Speer's three volumes of memoirs leave an unforgettable record of life at the higher levels of Nazi Germany, but they fail to answer the critical questions on nuclear research. The Farm Hall transcripts provide a verbatim close-up of the nuclear story, which, despite occasional flaws and gaps, remains untainted by the passage of time or selective memory of the participants. This makes the transcripts a useful corrective to the historical revisionism that grew up around the German project, and even documents the first stages of that revisionism. Among the high points in the transcripts, both as drama and as evidence, is Report no. 4, which records in some 35 pages the reactions of the detainees on 6 and 7 August, as they first reject outright, and then slowly accept, the broadcast reports by the BBC on the Hiroshima bomb.

The detainees, left on their own to talk freely after hearing the official statements, continue their earlier discussion on how the bomb may have worked, but as they absorb the dimensions of the Allied project they begin to compare it with their efforts in Germany. Their first thoughts are both dramatic and revealing:[21]

> KORSCHING: That shows at any rate that the Americans are capable of real cooperation on a tremendous scale. That would have been impossible in Germany. Each one said that the other was unimportant.
> GERLACH: You really can't say that as far as the uranium group is concerned. You can't imagine any greater cooperation and trust than there was in that group. You can't say that any one of them said that the other was unimportant.

KORSCHING: Not officially of course.

GERLACH: (Shouting). Not unofficially either. Don't contradict me. There are far too many other people here who know.

HAHN: Of course we were unable to work on that scale.

HEISENBERG: One can say that the first time large funds were made available in Germany was in the spring of 1942 after that meeting with RUST when we convinced him that we had absolutely definite proof that it could be done.

BAGGE: It wasn't much earlier here [in Britain] either.

HARTECK: We really knew earlier that it could be done if we could get enough material. Take the heavy water. There were three methods . . . they kept arguing . . . because no one was prepared to spend 10 millions if it could be done for three millions.

HEISENBERG: On the other hand, the whole heavy water business which I did everything I could to further cannot produce an explosive.

HARTECK: Not until the engine is running.

HAHN: They seem to have made an explosive before making the engine and now they say: "in future we will build engines".

HARTECK: If . . . an explosive can be produced either by means of the mass spectrograph – we would never have done it as we could never have employed 56,000 workmen . . . . If we wanted to make ten tons [of heavy water] we would have had to employ 250 men. We couldn't do that.

WEIZSÄCKER: How many people were working on V1 and V2?

DIEBNER: Thousands worked on that.

HEISENBERG: We wouldn't have had the moral courage to recommend to the Government in the spring of 1942 that they should employ 120,000 men just for building the thing up.

WEIZSÄCKER: I believe the reason we didn't do it was because all the physicists didn't want to do it, on principle. If we had all wanted Germany to win the war we would have succeeded.

HAHN: I don't believe that but I am thankful we didn't succeed.

In just over one printed page—with a small deletion of technical material—the detainees touch on eight different problems in their work: (1) insufficient cooperation (Korsching—Gerlach); (2) parsimonious funding under Army control (Heisenberg); (3) indecision over shortages of key materials (Harteck); (4) inability to imagine an explosion (Heisenberg, corrected by Harteck); (5) labour shortages (Harteck); (6) comparison with missiles (Weizsäcker–Diebner); (7) absence of moral courage (Heisenberg); and (8) a first airing of the "We-did-not-want-to-succeed" theory (Weizsäcker, with immediate dissent by Hahn). In the published transcripts, 14 further pages of equally rich material trace the rest of the evening of 6 August, until the last "guests" went to bed around 1.30 a.m., where most spent "a somewhat disturbed night."[22] Over the following several days, the arguments continued, changing, adding new points, taking new forms, until in time the comparison of German and Allied research would be overshadowed by more immediate and more pressing issues of future employment and the future of physics in Germany.

Under Army control from 1939 to 1941, scientists proceeded routinely with their research in university institutes and other laboratories with few problems beyond wartime delays in obtaining manufactured equipment. The situation changed when German military advances on the Eastern Front came to a halt. The Army's decision to abandon nuclear research and its enhanced need for manpower spelled the end for automatic military exemptions for nuclear scientists. As noted earlier (see Chapter 5), even Harteck and von Weizsäcker received draft notices in January 1942. After much difficulty, these call-ups were reversed, but conscription of younger scientists became a continuing problem for the rest of the war. As signalled in Franz's letter to Cherwell of 10 December 1942 (see Chapter 6), reports of deaths of physicists on the battlefield became noticeable in late 1942.

The notion that German scientists failed to develop a bomb because they did not *want* to succeed was apparently first floated by Weizsäcker in the passage cited just above. By the next morning, 7 August, his rationalization of their conduct was polished and ready for future use:

> History will record that the Americans and the English made a bomb, and that at the same time the Germans, under the HITLER regime, produced a workable engine. In other words, the peaceful development of the uranium engine was made in GERMANY under the HITLER regime, whereas the Americans and the English developed this ghastly weapon of war.[23]

In the hands of some postwar writers, the "spin" from Farm Hall took on extra colouring. The popular account by the journalist Robert Jungk, published in English in 1958 as *Brighter than a Thousand Suns*, and written when documented studies were still scarce, relies heavily on postwar interviews with participants—including Weizsäcker, Heisenberg, Gerlach, Hahn, and Korsching. Jungk takes up the Weizsäcker thesis and pushes it much further:

> It seems paradoxical that the German nuclear physicists, living under a sabre-rattling dictatorship, obeyed the voice of conscience and attempted to prevent the construction of atom bombs, while their professional colleagues in the democracies, who had no coercion to fear, with very few exceptions concentrated their whole energies on production of the new weapon.[24]

On the other side of the argument, Weizsäcker's proposal did not go unchallenged at Farm Hall. Hahn, as shown above, voiced his disagreement the moment Weizsäcker first proposed it. Later in the evening of 6 August, Bagge would be more emphatic: "I think it is absurd for Weizsäcker to say he did not want the thing to succeed. That may be so in his case, but not for all of us."[25] The Farm Hall transcripts, incomplete as they are, offer key evidence on this central, long-debated question.

At another point in the same evening, Bagge reminds Harteck, Wirtz, and Weizsäcker of the fateful discussion at the Army-sponsored meeting on uranium in September 1939, when the question of a bomb had arisen:

> Someone said "Of course it is an open question whether one ought to do a thing like that." Thereupon BOTHE got up and said "Gentlemen, it *must* be done." Then GEIGER got up and said "If there is the slightest chance that it is possible – it must be done."[26]

On that occasion, Jungk's hypothetical voice of conscience had evidently counted for little.

The scientific evidence for the "could-not-succeed" theory is compelling: in Germany the Army-supported nuclear teams achieved neither significant separation of U^{235} nor any self-sustaining chain reaction during more than five years of effort. On isotope separation, Heisenberg apparently decided relatively early that it could not be done with any existing technique. Some younger colleagues, notably Korsching and Bagge, continued to work on separation to the very end, despite a lack of encouragement from the top. "When I think of my own apparatus," Bagge says at one point, "it was done against Heisenberg's wishes."[27]

On nuclear "engines," despite all efforts of the Heisenberg and Diebner teams working independently, a sustainable reaction remained out of reach, though hopes of a breakthrough remained high down to the last desperate experiments in improvised laboratories in Hechingen and Stadtilm in early 1945.

A closer study of Heisenberg's interventions in the discussion that followed the first BBC announcement on 6 August shows how far he was at that point from thinking about either a nuclear explosion or a reactor. In the first conversation over dinner, Heisenberg's first scattered comments emphasize total disbelief:

> Did they use the word uranium in connection with this atomic bomb? . . . Then it's got nothing to do with atoms . . . All I can suggest is that some dilettante in America who knows very little about it has bluffed them . . . I don't believe a word of the whole thing . . . I am willing to believe that it is a high pressure bomb and . . . a chemical thing where they have enormously increased the speed of the reaction and enormously increased the whole explosion.[28]

Gradually, the theme shifts. If there really was a nuclear bomb, how could it have been done? Heisenberg now passes from scepticism into confusion:

> I still don't believe a word about the bomb but I may be wrong. I consider it perfectly possible that they have about ten tons of enriched uranium, but not that they can have ten tons of pure U^{235}.

When Hahn suggests "I thought one needed only very little 235," Heisenberg, replying, thinks and speaks of *slightly* enriched uranium in an engine. Hahn persists, taking an example of "let us say, 30 kilogrammes of pure 235, couldn't they make a bomb with it?" Heisenberg replies: "But it still wouldn't go off." Under Hahn's further questioning on the amount needed, he retreats: "I wouldn't like to commit myself for the moment . . . ."[29] Here, evidently, he has wandered into unknown territory. He is thinking, perhaps for the first time, of *pure* 235, and he lacks essential data for it. In his initial shock and confusion, it is easier to think of an exploding reactor than to imagine the impossible—a bomb of pure U^{235}.

When the others began to discuss possible methods to obtain U^{235}, Heisenberg joins in again, recovering some of his normal sense of authority:

> There are so many possibilities, but there are none that we know, that's certain . . . If it has been done with uranium 235 then we should be able to work it out properly. It just depends upon whether it is done with 50, 500 or 5,000 kilogrammes and we don't know the order of magnitude. We can assume that they have some method of separating isotopes of which we have no idea.

Wirtz then remarks: "I would bet that it is a separation by diffusion with recycling." Heisenberg replies: "Yes, but it is certain that no apparatus of that sort has ever separated isotopes before."[30] He then shifts to a more philosophical plane:

> There is a great difference between discoveries and inventions. With discoveries one can always be sceptical and many surprises can take place. In the case of inventions, surprises can really only occur for people who have not had anything to do with it. It's a bit odd after we have been working on it for five years.

Weizsäcker, however, is not to be sidetracked. Citing the example of Clusius, he steers the issue back to isotope separation:

> Many people have worked on the separation of isotopes and one fine day CLUSIUS found out how to do it. It was just the question of the separation of isotopes which we neglected completely partly knowingly and partly unknowingly, apart from the centrifuges.

Heisenberg then reiterates his position one more time:

> Yes, but only because there was no sensible method. The problem of separating "234" from "238" or "235" from "238" is such an extremely difficult business.[31]

Harteck then closes this exchange by lamenting the vast research terrain left unexplored and their lack of personnel to explore it: "That would have meant employing a hundred people and that was impossible." The transcripts then cease for an unspecified period between the dinner hour and the detainees' return at 9 p.m. to hear the official statements that would dispel their remaining doubts.

These responses of Heisenberg, shown here extracted from the full dialogue, are worth close attention because they are relevant to the much-debated motivation issue. These are not the words of someone trying to hide what he knows, but of an expert vexed at not understanding something he feels he should have known. For Heisenberg, the official statements had not resolved the problem of how it had been done. While others have by now accepted uranium isotope separation as officially announced, Heisenberg cannot.

Three days later, on 9 August, the transcripts show Harteck and Heisenberg in a serious technical discussion. Heisenberg asks: "Well, how have they actually done it? I find it a disgrace if we, the Professors who have worked on it, cannot at least work out how they did it." Harteck suggests that it would be "technically possible" to produce bombs with protoactinium, and for three more pages of transcript they explore this alternative in depth. Towards the end, Heisenberg remarks: "Perhaps the others [in Manhattan] have used protoactinium: this is almost easier to imagine than all other methods."[32] Protoactinium, element 91, was identified by Hahn and Meitner in 1918 and renamed protactinium in 1949.

Later that day, when these two join the others, Heisenberg is ready to make a general pronouncement in his usual style:

> It is still not clear which method has been adopted. Either they have taken pure proto-actinium or they have really separated the isotopes, or they have made element 94 with machines.[33]

The word "really" speaks volumes. Heisenberg clings to the inseparability of uranium isotopes as tenaciously as he had earlier insisted on uranium plates over cubes in designing a reactor. Both positions were costly mistakes. In the postwar years, it would be easier—and less damaging professionally—to go along with the Weizsäcker thesis: that the scientists had not really *wanted* to build a bomb.

On 7 August, the transcripts show, the "guests" spent most of the morning on the newspapers which they read "with great avidity."[34] They were irked and worried by press references to German work on the atomic bomb, "and said they hoped it would be possible to prevent the newspapers from continuing to make such statements."[35] When they were told that such controls were not possible in Britain, they accepted a suggestion from Major Rittner that they should prepare and sign a collective memorandum of their own on the work that had been done in Germany. The finished memorandum, dated 8 August, was drafted by

Heisenberg and Gerlach, and signed by all 10 detainees. The younger ones, initially reluctant to sign, were eventually persuaded to do so by Heisenberg.

The transcripts contain the finished memorandum (in German and English) and a summary of the "considerable discussion on the wording"[36] that went into its preparation. Diebner reported that he had "destroyed all his papers," but his superior, General Schumann, "had made notes on everything." Gerlach wondered if Vögler, as head of the Kaiser Wilhelm Society, "had also made notes." It was recalled that the Post Office "had also worked on uranium, and had built a cyclotron at Miersdorf." Gerlach said that "the Schwab Group also had some uranium." Gerlach and Harteck could each recall approaches from the SS, seeking heavy water. Finally Wirtz reminded the others that a patent for the production of an atomic bomb at the Kaiser Wilhelm Institute for Physics had been taken out in 1941.[37]

The safe navigation of these shoals was clearly a major challenge. Beyond this, the "discussion" has an interest of its own. It contains points not even hinted at in the smoother final text. Samuel Goudsmit identifies Schwab as an SS general who directed the SS's own technical laboratories. In one of these they had their own expert, assigned to work on heavy water. The 1941 bomb patent, presumed to be a secret military application, and dated 28 August 1941, has been tracked by Paul Rose, but he found no surviving text.

The "cyclotron at Miersdorf" was not discussed or explained further, but the Post Office had a large research and development laboratory at Miersdorf, and the postal minister had funds for grants in related areas. One recipient was Manfred von Ardenne, a brilliant inventor who had his own well-equipped private laboratory in Berlin-Lichterfelde. In late 1940 he was awarded a grant from Post Office funds to start work on a cyclotron. After 1940 this project dropped out of sight, and the vagueness of the discussion at Farm Hall suggests that none of the internees had seen the laboratory or the project. In 1945 when Berlin fell, von Ardenne joined with Gustav Hertz and two others in a pact to surrender to Soviet forces. On 10 May a small Soviet group that included Georgy Flerov visited Ardenne at his laboratory, liked what they saw, and soon arranged for all the equipment—including the "60 ton cyclotron"—to be loaded in "about 750 boxes" and transported to the USSR. In the USSR von Ardenne became the head of an institute and leader of a group focused on electromagnetic separation of uranium isotopes.[38]

The finished text of the memorandum contains five short numbered paragraphs, each backed up by corresponding paragraphs after the main text that add scientific and technical detail.[39] These paragraphs deal in turn with: the discovery of fission in 1938; the wartime project in outline; the Norwegian heavy water plant and its destruction; Harteck's experiments to "obviate the use of heavy water" by using slightly enriched uranium with ordinary water; and experiments using "existing supplies of heavy water" to develop "a power producing apparatus." The memorandum contains nothing directly contradictory to established events, but the text is remarkable for a tendency to play down the wartime effort and to highlight its shortfalls and failures. More centrally, it is remarkable for what it leaves out.

Its handling of the war effort in the second paragraph begins blandly: "At the beginning of the war a group of research workers was formed with instructions to investigate the practical applications of these energies." Only in the last of the technical paragraphs is mention made of "extremely small" funding from the "Ordnance Department and later the Reichs Research Board." There is no hint of why these amounts were small, of the scientists' refusal of larger amounts, or of larger support offers from Armaments Minister Speer from 1942 to 1945. The new standard for small or large funding grants is now "compared to those employed by the Allies." Similarly, paragraph 2 hides the central failure to achieve isotope separation in broader language: "it did not appear fassible [feasible] at the time [1941] to produce a bomb with the technical possibilities available in Germany. Therefore the subsequent work was concentrated on the problem of the engine . . ."

The annex to paragraph 2, after briefly listing the chemical research and the experiments to produce a chain reaction, concludes with a firm declaration: "With regard to the atomic bomb the undersigned did not know of any other serious research work on uranium being carried out in Germany." Here, obviously, the operative word is "serious." The technical annex to paragraph 4, on efforts to isolate U^{235}, admits to "no certain positive result" except for slight enrichment in the ultracentrifuge, but it also declares: "No separation of isotopes on a large scale was attempted." Here, too, "large scale" seems to mean Manhattan-sized.

The Heisenberg–Gerlach memorandum is notable for weakness in tracing the role of human agency. Things happen. Experiments take place. Funding appears, though minimally. We learn little or nothing of the human element, the controls by a barbaric regime, threats, fears, Gestapo investigations, strivings, scientific rivalries, ministerial jealousies, the Byzantine Nazi power structure, the Uranium Club, the takeover by Army Ordnance of the Kaiser Wilhelm Institute for Physics, and much more. This manoeuvring and manipulation will be examined in the next section.

The memorandum is also notable for its neutrality, its absence of praise or blame. This was doubtless a necessary price for consensus during the "considerable discussion on the wording" that led to a text acceptable to all 10 detainees. In the confined setting of Farm Hall the detainees developed a cohesion—founded on common interest, professionalism, and concern for future employment—that had been far from evident during the war. Further, the memorandum provided them with a sanitized, jointly agreed document on their wartime work.

Organizational factors

If much depended on the quality of the scientists themselves, much else depended on how that work was organized, facilitated, managed, and coordinated. The organizational side has two possible vantage points. The first is the view from the top down, from the higher Nazi regime down to the teams of working scientists. The second concerns working relations among the scientists themselves. Obviously, too, this review must keep in view the general context of a totalitarian society founded on rigorous ideology, surveillance, and fear.

German nuclear research operated under a top Nazi leadership marked by continuous struggles for power, in this context measured by ease of access to Hitler. The principal power bases included the top leadership of the Nazi Party, the state bureaucracy, the military, and the ever-present SS and Gestapo—Himmler's growing state within a state. This quicksand power structure was far from haphazard. It reflected Hitler's preferred modus operandi. Speer, newly appointed as Minister of Armaments, soon learned of Hitler's distaste for clear lines of jurisdiction. "Sometimes he deliberately assigned bureaus or individuals the same or similar tasks," he writes, on the principle that "That way, the stronger one does the job."[40] Such dispersion also served to undercut potential challengers to the Führer's own authority.

Chapter 6 and this chapter have described the evolution of nuclear research under this system of deliberate fragmentation of power at the top. Ministerial control shifted from Army Ordnance to Education to the Göring group. The supervisory official shifted from General Schumann, the Professor of Military Physics, to Rudolf Mentzel, an SS General, old-line Nazi, and physical chemist. Other changes included shifting the nuclear "plenipotentiaries" from the ardent Nazi, Abraham Esau, to the respected non-Nazi physicist, Walther Gerlach; the replacement of Diebner by Heisenberg at the Kaiser Wilhelm Institute for Physics in 1942; and irregular interventions from Himmler and Speer that initiated these changes. There are parallels here: the changes in control over nuclear work reflect shifting power positions at the top. "The failure of German nuclear physics," Goudsmit wrote, "can in large measure be attributed to the totalitarian climate in which it lived."[41] The most obvious point is that the German programme never had an equivalent to General Groves, and possibly could not have produced one in such an environment.

Apart from the administrative structure, there was ever-present police surveillance. In every laboratory and on every public occasion, one or more spies would be present to observe and prepare reports on the work and loyalty of every prominent scientist. By 1945 all of the senior detainees at Farm Hall had had one or more encounters with the Gestapo or other suspected agents before their arrival in England.[42] Gerlach had been denounced to Himmler by Esau over his allocations of uranium and heavy water. Hahn had had problems with an informant in his institute when "the non-Aryan Fraülein Meitner was still there." Laue was "severely reprimanded" by Mentzel for having mentioned Einstein in a lecture given in Stockholm. Heisenberg, although protected earlier by Himmler personally, faced another SS summons for alleged defeatist remarks made during his 1944 visit to Zurich. Harteck had been denounced by a Nazi colleague in 1943, but forewarned by one of his mechanics who was also a paid police informer.

Nuclear scientists were not being singled out for harassment. Surveillance applied everywhere, and at every level. In the rocket programme, Wernher von Braun, already an SS officer himself, was arrested and jailed on trumped-up charges in March 1944 after he declined to join in Himmler's unsuccessful attempt to take over control of the rockets programme from the Army.[43] After two weeks, he was conditionally released on Hitler's order. Erich Schumann, the Army General and "mediocre physicist"[44] who first directed nuclear research, was also a subject of Gestapo reports that questioned his fitness for the high post that he held. Even Minister Speer became a target for Himmler's intrigues, but he had strong survival skills and lived long enough to expose Himmler's intrigues in full measure in his *Infiltration* (original title *Der Sklavenstaat*). Under Nazi rule, the German state reverted to a Hobbesian state of nature.

The climate of fear in Germany was hardly conducive to innovative thinking or action. Late in the evening of 6 August, the last few Farm Hall detainees about to retire were reflecting on Manhattan's success. Bagge comments:

> We must take off our hats to these people for having the courage to risk so many millions . . . If the Germans had spent 10 milliard marks on it [about $2 billion, the cost of Manhattan] and it had not succeeded, all physicists would have had their heads cut off.[45]

In Britain and the United States, a different fear factor was operative, but this one prompted bolder action and risk-taking. In these countries the great, fearful unknown was whether Manhattan's bomb might come too late.

If direction from the top was shifting and uncoordinated, relations among the scientists themselves were hardly better. In previous chapters we have seen the limited administrative role assigned to Diebner by Army Ordnance at the Kaiser Wilhelm Institute, Diebner's subsequent ostracization after Heisenberg became director in 1942, the removal of Diebner's group from Berlin-Dahlem to the Army laboratory at Berlin-Gottow, and the later evacuation of both Berlin groups to separate sites at Stadtilm and Hechingen. The two group leaders had seldom met, except as rivals for needed materials in short supply. After Diebner's capture in 1945, Goudsmit reports, he was kept briefly in Heidelberg, "living in the same house with members of the Heisenberg clique. Their conversations with him were limited to monosyllables."[46]

There is irony here, for internment would produce closer relationships among the detainees than had existed at any point during the war. As early as 15 June 1945, a fortnight before the move to England, Major Rittner reports on Heisenberg's eagerness to cooperate with Allied scientists:

> He also said that they suspected that their potential value was being judged by the documents found at their institutes. He said that these did not give a true picture of the extent of their experiments which had advanced much further than would appear from these documents and maintained that they had advanced still further as a result of pooling of information *since their detention*.[47]

Another split divided the junior scientists from the seniors—the professors—at Farm Hall. Korsching's row with Gerlach on hearing of the Hiroshima bomb on 6 August earned him the displeasure of the senior detainees and also of his British custodians, who labelled him a "complete enigma."[48] Nevertheless his sentiments were endorsed in private conversations later that evening by Wirtz and by Bagge, juniors on opposite sides of the Dahlem–Gottow divide.[49] The flare-up attracted attention because it was unusual, a sudden challenge to a system that was normally hierarchical. Goudsmit recorded the gulf he found at the time of the round–up in Hechingen:

> ... we also picked up two of the younger men, because they had done some novel research work on isotope separation. This puzzled von Weizsäcker very much; he evidently thought that the young fellows were not important enough to be interned. "What kind of selection is this!" he complained.[50]

Bagge, Korsching, and Professor von Weizsäcker, though different in rank, were all born in the same year, 1912.

Late in the evening of 6 August, Diebner and Bagge, alone together, talked of the new situation, seeking answers, looking for those responsible for German failure. Bagge says:

> Gerlach is not responsible, he took the thing over too late. On the other hand it is quite obvious that Heisenberg was not the right man for it. The tragedy is that Korsching is right in the remarks he made to Gerlach. I think it is absurd for Weizsäcker to say he did not want the thing to succeed. That may be so in his case, but not for all of us. Weizsäcker was not the right man to have done it . . . . (Pause) You can't blame Speer as none of the scientists here forced the thing through. It was impossible as we had no one in Germany who had actually separated uranium. There were no mass-spectrographs in Germany.[51]

Diebner agreed: "They all failed."

Though both were Nazi Party members, neither one on this occasion thought back to the unbending policies of their Führer after his accession to power. On 16 May 1933, Max

Planck, as president of the Kaiser Wilhelm Society, had been granted an audience with the new Chancellor to protest the losses for Germany from the wholesale dismissals of scientists in the universities. It did not go well. Rather than listening to Planck's arguments, Hitler flew into an incoherent rage that literally forced Planck to leave the room, a rage of the sort that became familiar during the war. His reputed response to Planck on that occasion had included an ugly threat:

> Our national policies will not be revoked or modified, even for scientists. If the dismissal of Jewish scientists means the annihilation of contemporary German science, then we shall do without science for a few years![52]

The exact words are debated, but the priorities are unmistakable.

Raw materials and priorities

The organizational weaknesses noted in the previous section reappear prominently on questions concerning raw materials, their processing, and their allocation. Uranium, and especially uranium ore in the limited amounts requested, was not a major problem for most of the war. After initial shortages in 1939, the German attack on Belgium in May 1940 secured ample quantities for shipment to Germany as needed, the same ores that had been deemed unimportant by the British Admiralty (see Chapter 5). When Brussels was liberated in September 1944, the Alsos team could track the shipments from the Union Minière to Germany, to the Auer Chemical Company in Oranienburg near Berlin for preparing powdered oxide, and to a subsidiary in Frankfurt, Degussa, for uranium metal.

In the wartime economy, special factory orders could incur unpredictable delays. Thus Heisenberg's machine-cut uranium plates, ordered in the summer of 1942, would not become available until late in 1943. The "outmoded large-scale uranium plate"[53] experiments on them in the underground bunker (B-vi, B-vii) were not completed until late in 1944. Once again the explanation lies in deference shown to Heisenberg, who persisted in the plates design long after the superiority of cube design was evident to his colleagues. By the time that the plates experiments were completed, further experiments were possible only under primitive conditions in the cave at Haigerloch.

For heavy water, the situation was very different. Scarce, expensive, and vitally needed, it became the focal point for some of the most dramatic episodes of the war. These included: the French deal with Norsk Hydro for its total available stock in early 1940; Halban's removal of this stock from Paris to Bordeaux to London when France fell; the second transfer of this stock with Halban from Cambridge to Montreal early in 1943; the German seizure of Norwegian plants in 1940 and their measures to increase heavy water output; the disastrous British commando attack on Rjukan in November 1942; a substantial sabotage there three months later by Norwegian volunteers; a bombing of the repaired Rjukan plant by 140 Flying Fortresses in November 1943; and the sinking of the Norwegian ferry *Hydro* with its cargo of partially concentrated heavy water en route to Germany on 20 February 1944. The bombing in 1943, and the plant's vulnerability to further raids, led to a German decision to transfer all production of heavy water to Germany. On 11 August 1944, a party of German mechanics arrived without notice at the Norwegian plants. In a single day they packed up the 27 high-concentration cells for heavy water and sent them off to Germany in military trucks.

There is no need to revisit these high points; it is more important to trace developments on the sidelines. As early as 15 January 1940, Harteck had written to Heisenberg, stressing the importance of ample supplies of heavy water and asking "who – if anybody – is working on the production in Germany of heavy water?"[54] Harteck had worked on heavy water before the war, and one continuing project of his institute was to find more efficient ways to produce it. On this occasion, he was rebuked by Army Ordnance for writing to Heisenberg directly.

While still working under Army Ordnance rule, Harteck ran into another, unexpected, obstacle, which he recalls with bitterness when the issue of heavy water came up at Farm Hall:

> When I first talked with Herr Basche I wanted to bring Dramm into it and it appeared that Dramm could not take part because he was a quarter Jewish. He left Ruhrchemie because he had a row – the whole senior Staff was changed – and he was one who had followed the whole construction through. Just think if we had had him! He would have built the thing for us, with him we could have built it.[55]

Dr H. Basche, a civil servant in the Ordnance Department, was General Schumann's deputy and Diebner's immediate superior. That one solitary, faraway Jewish grandparent could block a project of this significance—an echo of the Gustav Hertz case—is all too typical of Nazi bureaucratic behaviour.

The issue of heavy water production in Germany eased a little after the seizure of the Norwegian plants in 1940, but it resurfaced forcefully after the bombing of Vemork and subsequent decision to move all production to Germany.[56] The year 1944 was marked by difficult on-and-off negotiations with I.G. Farben, which turned into angry conflicts over patent rights, costs, plant location, and time schedules. In response, Harteck and Gerlach explored alternative production methods with other firms, and ways to use partially concentrated water rescued from Norway. By early 1945, however, the outlook was bleak, and no adequate replacement for lost Norwegian production could be expected before the summer of 1946.

Per Dahl, who has traced the role of heavy water in the seven countries active in nuclear research during the war years, provides a quantitative comparison.[57] Production of German-controlled heavy water in Norway up to the bombing raid of November 1943 totalled 2.5 tons. A further estimated 0.5 tons was lost in the sinking of the ferry *Hydro* in February 1944, for a total of about 3 tons. Had production continued at Vemork at the same rate without an air raid, Germany would have had about 5 tons by June 1944, the projected amount for a successful heavy water reactor. By failing to replace the lost Norwegian material with equivalent production in Germany, the German project was unable to achieve a chain reaction by the heavy water route. The Manhattan output, by contrast, reached "55 tons by late 1945," which included 32 tons from Trail in British Columbia—where production began in mid-1943—and 23 tons from three new plants in the United States constructed in that year.

The gap between the German programme and Manhattan was widened by differences over graphite as an alternative moderator.[58] Walther Bothe in Heidelberg had investigated graphite in 1940 and reported disappointing results in January 1941. The problem was later traced to impurities in the graphite samples he had used. As Irving notes, "Bothe's regrettable error was not challenged by his peers," which meant that in Germany heavy water became the only moderator with any prospect of success. In the Manhattan Project, the Chicago scientists, including the Fermi group relocated from New York in 1942, chose to work with graphite as a moderator, because heavy water in sufficient quantity was not yet available. By positioning 6 tons of uranium oxide in 250 tons of graphite, Fermi's team achieved the first sustainable chain reaction on 2 December.

Harteck's nuclear research had a second main direction: uranium isotope separation by centrifuge. Since his files survived the war largely intact, both of his projects reveal the role of plunder and confiscation in the German wartime economy. For the high-concentration heavy water apparatus built at the Berlin institute in mid-1944, the generator "was shipped in from Russia, and parts of the Norwegian Hydro high-concentration installation found their way to Berlin-Dahlem as well."[59] In his travels to Norway early in 1945, Harteck discovered that this concentration apparatus had been removed "without any 'private agreement' having been signed" (i.e. confiscated). On a previous visit in February 1944, he had found that Norsk Hydro had received no payment for heavy water shipped to Germany up to that date, and that Germany also had refused to pay anything towards air raid damage to the plant. On each visit, he pointed out to his superiors the negative impact of such behaviour for German–Norwegian relations, but the result, if any, of his intervention is unknown.[60]

On the centrifuge project, the scientists working in relocation at Freiburg and Kandern in 1944 faced critical shortages of tool-making machinery and materials. The military authorities in Freiburg referred them to a "booty camp" (*Beutelager*)[61] in nearby Strasbourg, one of several officially designated depots where machinery and materials looted from occupied Europe were distributed to those needing them. In Strasbourg they located what they needed in March, but were informed that tool-making machines had to be requested through the military bureaucracy. In October, an official of the War Production ministry informed Harteck—by then removed from Kandern northward again to Celle—that these machines, which had been set aside for him in Strasbourg, "had to go through regular channels before he could receive them. Apparently he never did."[62]

Nevertheless, the "booty camp" was not a Nazi invention. In Franz's photograph album there is a picture from 1915 of an improvised *Beutelager* at Radymno in Galicia, with a pathetic assemblage of peasant belongings. Harteck's experience in 1944, however, offers telling evidence of the state of the German war economy at that point, of the low esteem of the military for nuclear research, and of bureaucratic paralysis of a regime in collapse.

Effects of enemy action

A final difference between Manhattan and the German nuclear programme concerns the effects of enemy action, and specifically the effects of Allied bombing raids on Germany's projects. The subject has become controversial, and for this reason a fair assessment requires an appreciation of its complexity. One should weigh the evolution of bomb and missile effectiveness during the war years; the impacts of specific raids on Vemork and other targets; the indirect effects, such as relocation of research institutes to safer locations; and the state of military intelligence on either side.

Bombing capacity, and the accuracy of bomb delivery, increased rapidly as the war continued. The quantity of bombs dropped by the RAF over Germany rose from about 16,000 tons in 1940 to a threefold increase by 1941. After 1942 combined British and American totals rose rapidly to over 40 times the 1940 total by 1944. Bombing tactics also evolved from British night raids alone in 1940 to RAF night raids combined with daylight operations by American Flying Fortresses in formations after August 1942. Up to June 1943, the top Allied priority was defensive, mainly attacks against submarine construction yards and bases, because U-boat attacks on Allied shipping posed a critical threat to Britain's supply lines. After this point, emphasis shifted to the objectives set for the British and American bomber commands at the Casablanca conference in January 1943: "Your primary objective will be the progressive destruction and dislocation of the German military, industrial, and economic system."[63] From December 1943, daylight bomber formations were supported by long-range fighters. These combinations engaged the Luftwaffe directly, and acquired mastery of German air space by the spring of 1944. In turn, air superiority, combined with improved bomb delivery techniques and tactics, would bring the German economy close to paralysis in the closing months of the war.

In earlier night raids against strong defences, accurate bombing was not possible. Targets were typically large urban or industrialized areas. With daylight raids, increasing tonnages, and improving techniques, the targets could be specific industries or individual plants. During 1943 and 1944, systematic attacks targeted plants producing ball-bearings, aircraft frames, steel, tanks, armoured vehicles, and trucks. In the early stages, industrial raids had a limited effect. Plants could be rebuilt, and production machinery often suffered less damage than buildings, as in the raid on the unfinished rocket production plant at Peenemünde in August 1943.

Once the Luftwaffe was defeated and the Western front was assured after the Normandy landings, the primary target became a massive assault on the synthetic oil industry. Its 13 production plants were the major source of Germany's fuel for aviation and tanks. The largest of these plants, at Leuna in Saxony, although strongly defended, was attacked—and quickly repaired—22 times between May and December of 1944. The net result was a plant output reduced to about 9 per cent of capacity during that period. By May 1944 overall oil consumption exceeded production levels, and by early 1945 tank units on the Eastern and Western Fronts were immobilized when fuel supplies ran out on the battlefield.

Even more effective was systematic bombing of the railway network. In 1944, Allied strategists realized that Germany, as an economy heavily dependent on coal, would be vulnerable to attacks on its efficient railroad and canal system. The canals carried about one-quarter of total freight movements, and once damaged they were difficult to repair. In September and October of 1944 a vastly increased attack was launched on railway marshalling yards, bridges, and moving trains, coupled with attacks on key water routes linking the Ruhr with northern, eastern, and southern Germany. This brought a precipitous decline in rail shipments of coal and a complete stoppage in water shipments from the Ruhr.[64] Between July and December of 1944 the country's monthly coal shipments by water fell by 86 per cent; by rail they were down in the same period by 35 per cent, but in February 1945 they were down by 73 per cent from the previous July. The repercussions of these declines spread through the entire German economy.

The nuclear projects, dispersed among several institutes, were seldom a direct target for bombing. A few institutes were hit during larger raids on cities. On 3 December 1943, Heisenberg's institute in Leipzig and his home there were destroyed in a night raid, though he and his family were far away in Berlin and Bavaria respectively. On 15 February 1944, Otto Hahn's institute for chemistry in Berlin-Dahlem was destroyed in a similar raid, but Heisenberg's institute for physics nearby escaped with minor damage. Strangely, General Groves made an odd claim in 1963 that this focus on the Dahlem district had been undertaken "at my request to drive German scientists out of their comfortable quarters."[65] If this is true, its timing was too late. By this date Hahn was already away in Tailfingen arranging the transfer of his institute, and part of the institute of physics had been operating in Hechingen for several months. Further, the physics institute's underground bunker laboratory continued to operate, in spite of air raids, until early 1945. As a general rule, however, institutes doing nuclear research were not targeted as such. Those in Heidelberg or Göttingen, university towns like Oxford or Cambridge, remained undisturbed.

The more visible targets for bombing were plants involved in industrial-scale manufacturing. The most obvious target for bombing or sabotage was the plant at Vemork in Norway, with its crucial output of heavy water. As we have seen, however, the fatal heavy water problem arose from a combination of enemy action and managerial deficiency. As the scientists at Farm Hall realized, this failure was traceable not to Gerlach in 1944 but to the earlier failure of Army Ordnance to identify the problem and devise a sufficient alternative source.

Beyond the loss at Vemork, several projects had apparatus destroyed or delayed at plants contracted to build it. Harteck's centrifuge project was seriously delayed by a raid on the Anschütz Company in Kiel in July 1944, which destroyed the section of the plant that was producing centrifuge equipment. Bagge's isotope sluice was twice destroyed by bombing in Berlin during its construction, twice rebuilt, evacuated to Butzbach, and later moved again to Hechingen, where Goudsmit found it of sufficient interest to remove it and detain its inventor. A massive air raid on the Merseburg-Leuna area on 28 July 1944 destroyed the I.G. Farben hydrogenation works and also its pilot plant for heavy water, leaving the plant manager with a distinct distaste for the heavy water project. At an earlier stage, Heisenberg's cherished experiments with layered uranium plates ordered from the Auer firm in Oranienburg were delayed so long in production that even before delivery they were known to be less effective than lattices of small cubes.

A larger, though indirect, effect of enemy action lay in the evacuation of nuclear research teams from larger cities. The raids on Hamburg in August 1943 led to the move of Harteck's team south to Freiburg and Kandern before the end of the year. Heisenberg had begun thinking of a move to south Germany by June 1943, and over the summer Hechingen was chosen as his institute's new site. The Diebner team at Berlin-Gottow and the Clusius group at Munich each moved outside their respective cities. Diebner would move twice more when the Russian armies moved westward. These moves were made with explicit support from Speer, Vögler, and Gerlach, with the deliberate aim of protecting the future of German science. Once decided on, however, these moves encountered serious shortages of construction labour and materials, more primitive research conditions, and insufficient experimental materials—most notably heavy water. Though we cannot measure these difficulties in any exact way, they clearly meant losses in efficiency from the moment of first displacement, and in the final weeks of the war they became insurmountable. They also point to an obvious question of the "what if?" variety. Since the Groth centrifuges were complete and ready for export to Brazil by 1953, how much earlier might they have been ready without the July 1944 raid on the Anschütz plant in Kiel and the three forced moves of the research site from Hamburg to Freiburg to Kandern to Celle?

One side of enemy action that remained largely hidden from the German nuclear scientists was close Allied surveillance of any external manifestations of their work. The purpose and methods of Groves's surveillance are best seen in his own postwar account. "Early in 1943," he explains, he set up a contract with Union Carbide to report on "the various deposits of uranium and thorium throughout the world."[66] By December of that year he had a small intelligence mission in London working alongside British colleagues. Their watch list soon covered people, raw materials, and enterprises: about 50 known German nuclear scientists; the shafts and tailings piles of the Joachimsthal mines in Czechoslovakia (sole source of new uranium under German control); a long list of laboratories, industrial facilities, metal refineries, metal dealers, and manufacturers of possible components, all of which were suspect until they could be checked and cleared.

The watch list turned up two alarming discoveries. After Paris was liberated, it was found that the entire French stock of thorium had been seized in 1940 for transfer to Auer Chemical, and in November 1944 air reconnaissance showed a burst of slave labour activity around Hechingen and three similar industrial plants under rapid construction. When checked out more fully, the panic receded. The new factories were for producing fuel from oil-bearing shale, and Auer's thorium stock had been earmarked for the postwar commercial market—for a sparkling, radioactive toothpaste. The main point, however, is that Germany's academic research in laboratories posed no military threat, but any evidence of industrial-scale activity would have called for an immediate Allied response.

Among the last acts of Allied nuclear intelligence in Germany were efforts to obliterate, as far as possible, all evidence of Germany's nuclear programme. With the first test of the Allied bomb still weeks and months ahead, Groves made every effort to protect the secrecy of his project. These efforts were particularly sweeping in the future Soviet and French occupation zones. Manhattan had not yet been officially mentioned to Stalin, and Groves suspected that Joliot would try to use the leverage of the Soviet connection to gain for France a larger share in the American–British partnership. The result was a policy to dismantle and carry off whatever could be removed—raw materials, Bagge's and Korsching's apparatus, stocks of uranium metal and ore—and to blow up or destroy what could not be carried away—the cave at Haigerloch, and the Auer uranium processing plant in Oranienburg by a massive air raid on 15 March 1945.[67]

The Soviet troops, on entering Berlin, were not slow to strip the Kaiser Wilhelm Institute of Physics and its underground laboratory of all usable fittings and wiring, but they did not fully understand what they found there. When Goudsmit visited Berlin in late July, he found the building empty save for one equally uninformed American officer, who told him of a basement "swimming pool" and "some junk which we dumped in the backyard." Allowed to look around and inspect the place "thoroughly," Goudsmit describes his tour with some emotion:

> The backyard "junk" contained various pieces of equipment for nuclear physics as well as blocks of pressed uranium oxide. There were also some notebooks indicating the type of research that had been going on. The sub-basement was the bomb proof "bunker" laboratory of which the Germans were so proud. It looked as if it had once been excellently equipped. The "swimming pool" was the pit in which the pile had been constructed. Metal containers and frames for the arrangement of the uranium cubes were still standing near by.
>
> I remembered the primitive setup with which Enrico Fermi had started in a basement room at Columbia University. By contrast, this Berlin laboratory, even empty, gave an impression of high-grade achievement.[68]

Goudsmit stayed a little longer, alone, in the stripped, unlit remains of the bunker laboratory. For him the laboratory was a "physicist's symbol of the defeat of Nazism,"[69] and his visit a culmination, a defining moment in the Alsos mission.

Nuclear projects in the Soviet Union and Japan

In 1942, the war on the Eastern Front did not go well for the Soviet Union. Although Soviet forces regained some territory during the winter months, the Germans opened a sweeping offensive in July, eastwards and south towards Voronezh, Stalingrad, and the oil fields of the Caucasus. The "dormant period" for nuclear work carried over from 1941. Georgii Flerov, who had failed to convince his colleagues in December 1941 (see Chapter 5), continued his campaign to reinstate nuclear research.[70] He wrote a 13-page letter to Kurchatov explaining how a nuclear explosion or bomb might work. Stationed near Voronezh, he visited the library of the evacuated university before the town's capture by the Germans in July. He looked up "the American physics journals," for any comment on his discovery of spontaneous fission reported in July 1940. In the journals, he found no response to his own note, nothing at all on nuclear fission, and nothing by leading nuclear physicists on other subjects. The German situation was similar, and alarming. He then sent a letter to warn the Defence Committee's plenipotentiary for science, Sergei Kaftanov. When this drew no response, he addressed in April a letter directly to Stalin, in a last attempt to penetrate "the wall of silence."

By this date, scientists with access to foreign journals were well aware that something special was afoot in Germany, Britain, and the United States. The Soviet government had an additional information source: incoming intelligence reports from its embassies abroad. By March 1942, sufficient information about these foreign projects was at hand for Beria to recommend an evaluation by senior nuclear scientists. His memorandum led to several months of consultations with scientists, a proposal to Stalin from Kaftanov and his deputy for a new centralized organization for nuclear research, and Stalin's assent to this proposal. The new body, known simply as Laboratory 2 to mask its purpose, was sanctioned by the State Defence Committee in February 1943. The post of Scientific Director fell to Igor Kurchatov, 40 years old, not yet an Academician, who had previously headed the nuclear department of Joffe's institute in Leningrad. Molotov made the final decision, passing over two senior scientists—Kapitsa and Joffe—whose attitudes he found less positive: "I was left with the youngest, Kurchatov, who was not known to anyone . . . . I summoned him, we had a talk, he made a good impression on me."[71] The appointment was confirmed on 10 March 1943.

From this date, under Kurchatov's direction, and to the extent that conditions allowed, work began in several directions. At Laboratory 2 in Moscow, a cyclotron was begun, using parts retrieved from the unfinished machine in Leningrad. Plans were started for a uranium–graphite pile, and for an eventual heavy water pile. Another group was set up to begin work on bomb design. By April 1944, with a staff of 74, including 25 scientists, Laboratory 2 moved into larger premises northwest of Moscow. A continuing problem, however, was acute shortage of necessary raw materials for experimental work. In 1943 Soviet authorities requested shipments of uranium metal and uranium compounds—a few hundred kilograms in all—from the Lend Lease Administration in Washington, and these were mostly approved by General Groves to avoid suspicion about Manhattan. The systematic search for uranium ores in the USSR, repeatedly urged by Vernadskii as early as 1940, was approved in principle but accorded low priority even after the renewal of research in 1943.

The acute shortage of experimental materials eased somewhat after the collapse of Germany in May 1945. The surrender unleashed a race for that country's stocks of uranium. A Soviet counterpart to the Alsos mission rounded up about 100 metric tons or more of uranium oxide, but it missed the 1,200 tons of uranium ore that Groves's group removed from the Stassfurt area before the Soviets occupied it. The Soviet government also concluded a secret agreement in March 1945 with the Czechoslovak government in exile for the total stock and future output from the Joachimsthal mines, the main European source up to this point. The mission in Germany also recruited several capable physicists to work in the USSR, including Gustav Hertz and Manfred von Ardenne, but their numbers were modest compared with Soviet scientists already at work in the project. Germany also yielded machinery. Among many sites, the Auer metals plant in Oranienburg, which General Groves had tried to destroy by a bombing raid in March 1945, was dismantled and shipped to Russia, accompanied by the company's research director, Nikolaus Riehl, who built a uranium metal refinery near Moscow that began production by the end of 1945.

Over the two years from early 1943 to the German surrender, the Soviet project remained at the level of theoretical research and laboratory experiment. Kurchatov, increasingly aware of the widening gap between the Soviet Union and the West, did not hesitate to complain to the leadership about the gap. In September 1944 he addressed a strong letter to Beria on the "completely unsatisfactory" situation of the projects, particularly in relation to raw materials and "questions of separation." In May 1945, with Germany collapsing, Kurchatov and his party superior, Mikhail Pervukhin, addressed a joint letter to Stalin, urging again that the project be speeded up. Neither letter produced any known response from Stalin or anyone else.[72]

When Stalin, at the Potsdam conference on 24 July, first heard President Truman's casual mention of "a new weapon of unusual destructive force," his response was remarkably muted. According to Truman's memoirs, Stalin expressed his pleasure and "hoped we would make 'good use of it against the Japanese'."[73] Other eyewitness accounts report an even briefer response. Stalin's reaction at this point suggests two things: that he already knew something of the bomb project, but that he did not yet recognize its global significance. If Truman's version is correct, Stalin at this point saw it only in military terms, as another weapon.

If the two explosions over Japan were felt around the world, they also focused Stalin's full attention on the Soviet nuclear programme. For the first time, he understood the link between the bomb and the world balance of power. That balance had been disturbed, and had to be restored. To do so, the Soviet Union must have its own nuclear weapons at the earliest possible date.

About mid-August, Stalin had discussions with Kurchatov and the munitions commissar. On the 20th, a decree of the State Defence Committee set up a new committee on uranium energy, still under Beria, with Kurchatov as Scientific Director, and a new directorate to manage the atomic industries. Within the next month, the quest for foreign nuclear intelligence was stepped up, the search for uranium ore in central Asia was intensified, German scientists—most of them unassigned since their arrival in May—were set up in two institutes on the Black Sea to work on isotope separation. Stalin told Kurchatov: "Ask for whatever you like. You won't be refused."[74] These post-Hiroshima developments, however, will be treated in chapter 9.

As early as 1940, as noted in Chapter 5, the Japanese military recognized the weapons potential of nuclear fission. Several months before Pearl Harbor, it authorized and financed Professor Nishina's uranium project at the Riken research institute in Tokyo. After a prompt start in the summer of 1941, the project moved slowly, with little money and a mere handful of trained researchers, first exploring basic nuclear properties. As in Germany, Japanese scientists faced no great pressure until the tide of war turned negative, as it did against Japan in the autumn of 1942. In August, the Riken project was taken over by the Army Air Force, made officially secret, and given the priorities of an army munitions laboratory.

Nishina received a bigger budget and a dozen more staff with military deferments. He assigned one assistant to obtaining "uranium gas" (hexafluoride) and another to studying separation methods for U^{235}. Even with larger numbers, these younger scientists faced formidable obstacles, some of them self-imposed by the secrecy of their project. Interviewed after the war by Robert Wilcox, Kunihiko Kigoshi described having to learn how to get fluorine from potassium fluoride, and metallic uranium from uranium nitrate, before he could combine them to make uranium hexafluoride. In the end he succeeded, but this self-taught exercise had taken him almost a year.[75]

The quest for a separation method for U^{235} was similarly slow and difficult. By the spring of 1943 Nishina's team had studied the separation literature and chosen gaseous thermal diffusion, on the principle of the Clusius–Dickel tube. In April an internal report prepared for General Yasuda maintained that an atomic weapon was feasible, that one kilogram of U^{235} could produce an explosion, and that separation was feasible by thermal diffusion. At this point the Japanese War Minister spoke of ongoing "atomic bomb projects of the U.S. and Germany,"[76] and called for Japan not to fall behind. Nishina received a larger budget, and the project was henceforth to be controlled directly from Army air force headquarters rather than through Army laboratories.

Developing an adequate uranium supply, however, was more difficult than had been expected. In the summer of 1943 Japan made two urgent appeals to Germany for supplies of Czechoslovakian "pitchblende for research purposes." The German Economics Ministry was at first uncooperative. The Japanese Embassy, pressed for a reason for the request, eventually obtained a reason by wire from Tokyo: "Uranium oxide is used as a catalyst in the manufacture of butanol." After accepting this convenient fiction, Germany promised two tons of ore, to be shipped by submarine. Neither side was willing to make further disclosures.[77]

Building the separator, a five-metre vertical column comprising a heated tube inside a cooled one, posed numerous problems, but after a year of effort a pilot separator was complete by March 1944. Further adjustments and an accident delayed effective operations for several months, and operational samples for testing drawn from the lighter gas at the top were not available until February 1945. In the meantime, Lieutenant Colonel Suzuki, the Army physicist who had prepared the first feasibility report for General Yasuda in 1940, had been reassigned to the Nishina project in May 1944 after a stint on another war project. His mandate was to accelerate work on the bomb. With Nishina overseeing projects in several areas, Suzuki assumed a larger role in daily operations. Despite increased funding and repeated official urgings, morale was sinking owing to continuing delays and disappointments.

When the first four test samples were measured in February 1945, the result was either no enrichment at all, or at best "under one percent."[78] Nishina was acutely disappointed, but his team members were willing to try various improvements and continue the work.

Suzuki had a wider view. He had concerns about reliance on a single separator, in a vulnerable location, at a time of increasing enemy air raids. He also had ideas for improving the design and materials of this first trial model. With Army approval, he designed a set of five large new separators, 20 metres high instead of five, with steel inner linings for strength, and arranged for their installation in the well-equipped Physics Department of Osaka University. Given the highest construction priority, they were built in just a few months by the Sumitomo Company at a plant just outside Osaka. Still at the factory site, they were ready for installation early in 1945.[79]

From this point onwards, enemy action played a large role in Nishina's project. On 12 March, an air raid on Osaka cut off the electricity and water systems of the university area allocated for the new separators. After consultation with Sumitomo, the firm agreed to have initial tests and experiments carried out at the plant where they had been built. On 13 April, a massive raid on Tokyo spread fire and destruction through the Riken district. Nishina's laboratory survived unhit, but the next morning it blazed up and was totally destroyed while the last flames in adjacent buildings were being extinguished. The fire consumed Nishina's separator and the entire stock of the hexafluoride gas. Though the team producing the hexafluoride had earlier been evacuated to a small town north of Tokyo to continue their work, supplies of uranium ore and output of the hexafluoride had dwindled. The new separators, still at the Osaka factory, would have no gas to process.

Though his situation after the fire might have appeared utterly hopeless, Nishina ordered his scientists to relocate and continue working. In January 1945, the high command had decreed a state of siege: "Industries that cannot be moved will go underground; those that can be moved will be moved to Korea."[80] In preparationn for a final battle to defend the home islands, the Korean peninsula was being turned into a vast industrial complex, "an enormous supply base for the Japanese Army."[81]

In spite of Nishina's orders to continue working, the best hopes for nuclear success now lay with the Navy, which had decided to investigate nuclear energy in the spring of 1942. Although some sources refer to two separate Navy projects, the first was not a research programme as such but an appointed committee of experts, similar to the Maud Committee or the American S-1 Committee, formed to study the resources available to Japan and the possibilities of successful engines or bombs within the expected timeline of the war. The committee met with naval technical officers at monthly intervals between July 1942 and May 1943. At the start, it was aware that "Research in this field is continuing on a broad scale in the United States, which has recently obtained the services of a number of Jewish scientists, and considerable progress has been made."[82] It was also noted that Germany's efforts might fall short "because of the expulsion of all its 'Jewish scientists'."[83] Seemingly, Japanese intelligence was remarkably informed on these questions.

As the discussions developed at monthly meetings, the scientists explained how difficult the project would be for any country in the time available, while patriotic naval officers urged that anything was possible given sufficient effort. After about a dozen meetings the Navy convened a meeting for 6 March 1943, asking the committee for a "definitive conclusion" on the bomb question. The committee agreed that a bomb would undoubtedly be built at some point, but members doubted whether the United States or Germany or any belligerent could have one ready during the war. Further, they underlined Japan's dependence on imported uranium ores. The committee was then dissolved, but a senior officer attending for the Navy backed up the committee's finding: "The best minds of Japan... came to a conclusion that can only be regarded as correct."[84]

Even as the committee of experts was being phased out, the ordnance branch of the Navy was taking an interest in another nuclear physicist who was unconnected with the study committee or with the Army's project. Like Nishina, Bunsaku Arakatsu had gone to Europe for studies in the 1920s, first with Einstein in Berlin, then at Zurich and Cambridge. After his return he became one of Japan's foremost nuclear physicists, building a strong department at Kyoto University. Arakatsu, when initially approached in 1942, had been doubtful; he "had replied that producing the bomb would be impossible in time for it to be used during the war. The navy countered blandly that in that case it could be used in the next war."[85] His early funding was modest, but in May 1943, under deteriorating war conditions, the grant was increased 100-fold, the project was given an official Navy designation, and he was told to move quickly from theory to bomb development.

Arakatsu, who had visited Nishina's thermal separator, instead chose to try separation by ultracentrifuge. He worked on the design, called in two Tokyo industrial firms to develop and improve this design, and selected the Sumitomo firm for its construction. By the autumn of 1944, with the war situation now desperate, the Army was prepared to cooperate with the Navy project, more money was available, and priorities for the project were again upgraded. The Army team shared information on making the hexafluoride, but finding sufficient uranium ore for an industrial process remained a difficult challenge. The long-held assumption that adequate amounts of minerals could be developed in Korea proved to be more myth than reality.

For Robert Wilcox, "the most important year in the Japanese atomic bomb efforts was 1945."[86] As primary evidence, he describes reports and documents found in his own research of almost a dozen attempts to find and extract uranium in this last phase of the war. The sites include the Ishikawa prefecture (the sole source on the main islands), several places in Korea (but mainly in the north) and other locations in Manchuria, Inner Mongolia, northern China, and Malaya. Many of these were ores or sands of low grade, but the operations ranged in scale up to 1000 workers and "300,000 tons of the uranium ore" mined in the Anshan area south of Mukden. The Navy prepared its own inventory of ores and extractive materials needed, dated 1 April 1945.[87] Disappointingly, records for the experimental situation are far less clear. After the end of 1944, the planned ultracentrifuge disappears in a cloud of conflicting rumours—never built, bombed at the Hokushine plant, destroyed en route to Kyoto, or built but recorded as something else in the postwar scientific inventory.[88]

A similar gap appears in tracking the Army project. Although the destruction of the separator and the hexafluoride supply in the Riken fire seems documented clearly enough, the five large separator columns ordered as back-up by Colonel Suzuki, each 20 metres tall and difficult to hide, were in working condition but still awaiting installation early in 1945. They are not mentioned again.[89] Since they were never traced or found by postwar investigators, Wilcox speculates that these columns may have been removed, along with the factory where they were built and other plant machinery, from the Osaka area to Korea.

As in Germany, the final months of the war were marked by a frenzied effort to mobilize the entire Japanese population against the expected landings on the home islands. On 25 March, with Iwo Jima about to fall, the government formed a "people's volunteer corps"[90] of the entire able population, male and female, from 13 to 60 to work on war production, and later to fight to repel the invaders—using sharpened bamboo spears as weapons. In Tokyo, the non-essential population were under orders to evacuate to the countryside, taking only whatever possessions they could remove. In this total mobilization, the frantic last-minute hope for an atomic bomb to turn the tables played a central part, as it did in Germany. A Japanese chronicle of the war records for July that in "strangely peaceful" Kyoto,

> in the research laboratory headed by Professor Bunsaku Arakatsu, work was going on, both by day and by night, in an attempt to separate U-235 through the ultracentrifugal method.... The Arakatsu project in Kyoto was proceeding under the orders of the navy, which, having lost most of its ships, was now trying to reestablish itself through a new and decisive weapon.[91]

On 21 July 1945—five days after Manhattan's successful test at Alamogordo—a special meeting was held at a hotel on the shores of Lake Biwa, north of Kyoto. Those attending were Arakatsu's main collaborators, several senior naval officers, and a few other scientists. The accounts of the meeting mention four talks: Arakatsu on the Navy project generally; Rear Admiral Nitta—an expert on centrifuges—on the centrifugal separator; Professor Kobayashi on critical mass; and Professor Yukawa on world atomic energy. This last talk, the Japanese source notes,

> was concerned solely with theoretical aspects of nuclear research. It hardly touched on the actual production of an atomic bomb – beyond Yukawa's prediction that, although such a bomb was theoretically possible, it could not be produced in time to be used in the Pacific war. With this the meeting adjourned; the men who had taken part in it continued to discuss in small groups of twos and threes, as summer breezes from the lake fanned their faces, the theoretical aspects of nuclear energy. Then quietly they dispersed....[92]

The account conveys no hint of crisis, devastation, or impending defeat. Was this a Japanese way of addressing the unthinkable without openly stating it, of conveying a research result that remained negative, or of tempering gently the unattainable expectations of the Navy?

Immediately after the Hiroshima bomb, General Groves organized a small, Alsos-type investigative team to send to Japan. Its mission was to assess Japan's progress in nuclear physics and its access to fissionable materials. A third purpose was to find out what Japan could reveal about nuclear physics and uranium sources in the Soviet Union. The team, known as Group Three, was headed by Major Robert Furman, who had been with Alsos in Europe. There were key differences, however, from the original Alsos team: less time for preparation, less need for secrecy, less sense of urgency, and greater reliance on informal scientist-to-scientist relations. As Furman wrote at the start:

> ... we believe that Japanese scientists will be cooperative. It is therefore proposed to study the work in nuclear physics as far as possible by informal and friendly visits to the leading academic workers. A scientific approach will be used, with the authority of the occupation remaining in the background.[93]

Two days after arrival in Tokyo on 7 September, mission members had a first interview with Ryokichi Sagane, a former pupil of Ernest Lawrence, who had helped Nishina to build the first cyclotron in Japan. Sagane, known personally in California by Philip Morrison of the Group Three team, was guardedly cooperative, mentioned Nishina, but played down any notion of "government or military interest . . . in atomic bomb production."[94]

Nishina, interviewed the next day, was more communicative, but more interested in discussing the Hiroshima explosion than in describing his own work. Both Furman and Morrison reported on this interview. Morrison concluded:

> There is no evidence of much work in isotope separation, although they have studied two or three methods . . . . There is no evidence of real government interest in the field before Hiroshima.

Furman's report was very similar:

> Nishina claimed that his staff had been diverted from nuclear physics research to the army and war work, and that very little support had been given his work by the government. The War and Navy Departments, he said, had no research project for atomic energy.[95]

On 14 September, Furman's progress report to headquarters at Tinian was even more direct:

> The following general conclusions have been reached: The government and the military gave no priority to research in the field of nuclear physics and had no program to produce a bomb.[96]

On 15 September, Morrison and Furman interviewed Arakatsu and Yukawa in Kyoto, where Morrison wrote of courtesy, scientific competence, and absence of governmental pressure:

> Our relations were most friendly, and there was great exchange of gifts and courtesies . . . . The general picture confirms the ideas gained earlier in Tokyo . . . . There is again no evidence of government support or unusual pressure for work in nuclear physics.[97]

Furman, although more sceptical about Yukawa, concluded that the nuclear physics they had seen in Kyoto confirmed "the disinterest of the government in this field." After a visit to the Osaka department the next day, they felt similarly reassured.

While these interviews were proceeding, Lieutenant Robert Nininger, a mission member and uranium specialist in Groves's office, was seeking information from geologists. The trail soon led him to Korea, but when he arrived there he found that Soviet troops had already sealed off all access to the northern industrial and mining areas. When he visited the Riken's Korean metals refinery near Seoul, he was told of a stockpile of "three tons of fergusonite . . . [for] atom bomb experimentation."[98] His first reaction was disbelief. Further enquiries produced indications of much activity in distant locations, but little that could be directly checked out. On hearing of these efforts, Furman requested authority to order a full official disclosure, with a five-day deadline, of all uranium stocks, efforts to find uranium, and results of these searches, throughout Japanese-held territory. Such a directive was issued, signed by General MacArthur.

Even as this directive was issued, the Group Three team was suddenly ordered to return home. Furman's 160-page final report is dated 28 September. It was reassuring with respect to nuclear science—"at the level of 1942 investigations in the United States"—and also with respect to geology—the authorities were not extracting "uranium-bearing ores of importance." He cautioned that official returns not yet analysed, or still to be submitted, might modify this picture.

He recommended that, when the required statements were submitted, they should be forwarded to the Manhattan office in Washington "should they reveal anything of importance."[99]

There are signs here of undue haste. Only in the first week of October did Furman meet a representative from the Army, and then another from the Navy.[100] The Army backed up Nishina's story: the Army "definitely had made no program" of its own and "Nishina had no Army project." By contrast, the Navy representative spoke openly of the Navy's decision in the spring of 1944 to build a uranium bomb, and to coordinate its effort with Army research. After a month in Japan, Furman was hearing, probably for the first time, a first-hand account of Navy and Army involvement, but he was now on the point of leaving Japan. He called for "a full report of the Navy war interest and activities in nuclear physics research, including all documents . . . ."[101] When ready, it would have to be forwarded to Washington. On or about 10 October, the Furman mission returned to America, apparently leaving its reports unmodified. One can only regret that this mission to Japan was not conducted by someone of the calibre of Samuel Goudsmit.

This review of Japan's wartime nuclear programmes should also be placed in a wider context. In *Now It Can be Told*, General Groves freely admits his wartime inattention to Japan, and gives four specific reasons for it:

> We did not make any appreciable effort during the war to secure information on atomic developments in Japan. First, and most important, there was not even the remotest possibility that Japan had enough uranium or uranium ore . . . . Also the industrial effort that would be required far exceeded what Japan was capable of. Then, too . . . their qualified people were altogether too few in number . . . . Finally, it would have been extremely difficult for us to secure and to get out of Japan any information of the type we needed. I hoped that if any sizable program was started, we would get wind of it . . . .[102]

Groves's involvement with Japan did not end with the return of his inquiry group. On 24 November 1945, General MacArthur's headquarters began the destruction, on orders from the War Department in Washington, of the five remaining cyclotrons in Japanese university laboratories.[103] When the story broke in the American and international press on 30 November, the destruction was widely condemned as an act of needless barbarism and vandalism. In his memoirs, published in 1962, Groves devotes an entire chapter to the affair of the cyclotrons, admitting that the action had been "stupid" and "a very serious error." He explains that a directive intended "to ensure that they were properly secured . . . but not destroyed" had been misunderstood by an unnamed, new subordinate staff officer, whose faulty text had been sent to Tokyo through a lapse in usual procedures. When the War Department issued an apology for this "mistake," the storm blew over.

Taken in isolation, Groves's explanation might seem reasonable and acceptable, but in a context of other unresolved questions concerning Japan, acceptance may be more difficult. Several of these questions have been noted already: the five missing separation columns ordered by Colonel Suzuki for thermal diffusion; Arakatsu's ultracentrifuge design and its possible construction; the transfers in early 1945 of plants and machinery from the Osaka area to northern Korea; the plants and industries in the Hungnam area, and their fate under Soviet occupation; the persistent quest for uranium, even from low-grade ores and sands in all areas under Japanese control; the shipments of uranium from Germany, including the final cargo on the intercepted submarine U234; the hasty, all-too-trusting reports of the Furman mission. In studying Japan, one senses at once an absence of researched and documented academic studies of the sort that exist adequately for the United States, Britain,

Germany, Canada, or the Soviet Union. On the evidence available to date, the necessary conclusion is that the full story of Japan's wartime nuclear effort has yet to be told.

Vicissitudes of an alliance

In comparison with the German and the Japanese efforts, the Manhattan Project became the model, the shining example, the standard of measurement for other entries in the nuclear race. In personnel, it had the most scientists, many of the best scientists—including many of the ablest German and European exiles—and those most worried by fear of being too late. Manhattan had superior organizational strengths, both before and after Groves's appointment in September 1942. These strengths included: unwavering support from President Roosevelt; successful combination of academic science and large-scale industrial engineering; open-ended finances and willingness to take huge risks on unproven designs; and unrelenting determination to streamline organizational, developmental, and production processes to the greatest extent possible. Combining rare qualities of foresight and capacity for pre–emptive action, Groves assured for Manhattan ample raw materials, industrial building sites, and the highest industrial priority in the American system, during his first few days in office. In research, the huge risks of the project were hedged by backing four different paths to fissile material, and later using four different production methods to obtain uranium 235 and plutonium.

In comparison with work in Germany, Britain, or Japan, Manhattan enjoyed an enviable security from direct enemy action. Indeed a single enemy act, Japan's unannounced attack on Pearl Harbor in December 1941, became a trigger for more rapid mobilization. The black anger kindled at Pearl Harbor became a part of the American decision to use the bomb. On the day of the bomb at Hiroshima, President Truman, in his announcement to the nation, could not refrain from evoking the unforgettable link:

> The Japanese began the war from the air at Pearl Harbor. They have been repaid manyfold. And the end is not yet . . . . Let there be no mistake; we shall completely destroy Japan's power to make war.[104]

American nuclear work was privileged in yet another way. Although serious mobilization began only after Pearl Harbor, the project grew directly from the work of Britain's Maud Committee in 1940 and 1941. The committee's report of July 1941, which led Churchill and the British Chiefs of Staff to decide on moving ahead on 3 September, was also a factor in Roosevelt's authorization to Bush to proceed with research and development at their crucial meeting on 9 October. In 1942, with rapid expansion of research and involvement of the Army's Manhattan Engineering District from June onwards, the American effort rapidly outdistanced its British counterpart, by then reorganized and operating as Tube Alloys. With more limited resources, lower industrial priorities, and a more cautious one-stage-at-a-time approach to risk-taking, Tube Alloys could match neither the scale nor the speed of Manhattan. As the gap widened, a dangerous level of misunderstanding opened up that seriously impeded work in Britain and Canada during the first eight months of 1943.

By persistent effort and tortuous negotiation, this rift was repaired by the first Quebec Agreement, signed by Roosevelt and Churchill on 19 August 1943. It reintroduced into the Manhattan project elements of the cooperation that had inspired pre-Manhattan

British–American exchanges in 1940 and 1941. It took into account that the scale of operations of the projects had become vastly different during 1942. For areas of parallel activity, a limited exchange of information was restored after an almost total blackout for eight months.

The Quebec Agreement also established a Combined Policy Committee, representing the United States, Britain, and Canada, to address issues of common concern and recommend fair allocation of raw materials. This committee in turn led in June 1944 to the creation of the Combined Development Trust, formed to locate and secure relevant raw materials, principally uranium and thorium, on a world scale as a joint American–British endeavour. The final section of the Quebec Agreement provided for "such *ad hoc* arrangements . . . as may appear to be necessary or desirable if the project is to be brought to fruition at the earliest moment."[105] As a result of this added flexibility, a few dozen of the British project's leading scientists were assigned to Los Alamos and other American locations for the last two years of the war. Other experts, mainly from industry, went for shorter periods.

The short-term consequences of the Quebec Agreement for the project in Britain were inevitably negative. With many top scientists absent in North America, the gap between Tube Alloys and Manhattan could only widen. During a long, inconclusive debate over future policy directions for Britain, basic research continued in the universities and in some ICI divisions, but further development was halted at the level of planning and plant design. Among the casualties was the pilot plant at Rhydymwyn in north Wales. As Gowing reports of this period, "work on the planning of a full-scale diffusion plant, and the experiments at Valley, faded into the background."[106]

On balance, the Quebec Agreement of 1943 represented a victory of rationality over narrow national interests. The British loan of scientists to Manhattan was a reminder that the project was founded on joint beginnings and still retained joint features even after the respective industrial efforts became vastly unequal. The loan of personnel would be repaid with a bonus when almost all of these scientists returned to Britain after the war, enriched by a unique experience as pathfinders in the nuclear age.

In evaluating the ups and downs of the British–American alliance, there are human factors that call for careful assessment. Superficially, General Groves, in his extreme concern for security, appears as a consistent opponent of any information exchange beyond what was absolutely necessary for American interests. In the American official history, he is represented as an opponent of the Quebec arrangements: "Certainly he did not go out of his way to expedite interchange. He was carefully and unenthusiastically correct."[107]

Groves's own memoirs reinforce this view. "The gaseous diffusion project as it was built," he explains, "was essentially an all-American effort." At the first British–American discussions on plant design in the spring of 1942, "the views of the British group [Akers, Peierls, Simon] . . . were quite different from the American, and their methods and equipment were dissimilar to ours." During the barrier crisis, when the "strong delegation of scientists and engineers" visited from Britain in the autumn of 1943,

> The barrier material was discussed with them in considerable detail, but their views did not influence the selection of either our first barrier, which was later abandoned, or the second . . . . At this late date, I was

strongly opposed to any major change such as the British suggested, for any one of them would have seriously delayed the completion of the plant.[108]

What Groves does not say in this account is that the "late date" of the 1943–4 mission, which included Franz Simon's third and fourth visits to North America, had been delayed for eight months by the impasse over the Conant rules announced in January 1943.

One could suspect that Groves's repeated denials of British influences on Manhattan decisions may be influenced by postwar Congressional politics and the project's ultimate responsibility to the American taxpayer. A more thorough reading of Groves's account reveals a more balanced appreciation of Britain's contributions. He acknowledges that British suggestions, even if not adopted, were always thoroughly examined, used to recheck American designs, or to solve operational problems in the new plants. Though initially apprehensive about the Quebec Agreements, he became a strong supporter of the Combined Policy Committee, praising its smooth operations and absence of "any serious differences among its members."[109] He also devotes an entire chapter to the establishment and effectiveness of the Combined Development Trust in locating and securing critical raw materials around the world.[110] Beyond the formal institutions, he developed an improbable working relationship of openness, trust, and personal friendship with Professor Chadwick, the leader of the post-Quebec British contingent, whose temperament was so completely different from his own.

Groves adds further balance in his penultimate chapter, when he writes of a new impasse when Attlee and Truman were attempting to replace the Quebec Agreement with an appropriate postwar accord. After explaining how this impasse has arisen, Groves offers a generous two-page summary of the British contributions to Manhattan,[111] which may be compressed further as follows:

1. Encouragement and support at the highest level. "British optimism" as a spur to faster American "laboratory effort" in the early stages.
2. Scientific aid from Chadwick's British contingent and "about a dozen Canadian scientists."
3. Study and laboratory work on separating U^{235}.
4. Studies of nuclear properties of heavy water.
5. Miscellaneous scientific and technical information, such as "seals for the gaseous diffusion plant."
6. Manufacture in Wales by International Nickel of "material needed" to produce diffusion barriers.

Margaret Gowing explains that the 3,000 tons of this special barrier material needed up to June 1945 were "sent meticulously on schedule and . . . paid for by the British Government under reverse Lend-Lease."[112]

Groves then supplements his list with a few personal reflections:

> On the whole, the contribution of the British was helpful but not vital. Their work at Los Alamos was of high quality but their numbers were too small to enable them to play a major role. On the other hand, I cannot escape the feeling that without active and continuing British interest there probably would have been no atomic bomb to drop on Hiroshima. The British realized from the start what the implications of the work would be. . . . Looking back on those war days, I can see that Prime Minister Churchill was probably the best friend that the Manhattan Project ever had. . . . In addition to his other wartime achievements,

Mr. Churchill emerged as our project's most effective and enthusiastic supporter; for that we shall always be in his debt.[113]

This evidence suggests a need for reappraisal of the Manhattan leaders. While Groves, Bush, and Conant were in substantial agreement when the gates were closing on information exchange in December 1942, it was Groves who adjusted most readily to renewed collaboration after Quebec, acting with pragmatism and consideration for all parts of the American–British–Canadian enterprise. Bush and Conant, both old stock New Englanders, still harboured an inner distrust for the British connection, British industry, and British motives generally, a distrust that they could not rationally set aside. Conant, in particular, had a long memory for past conflicts.[114]

The roles of Conant and Bush in the interchange issue have been more closely scrutinized by two American historians, Martin Sherman and James Hershberg.[115] The composite picture that emerges from their studies shows Conant as the "initiator of the policy of restricted interchange" from October 1942 onwards and the one primarily responsible for its defence; the one who "indoctrinated Bush, Stimson, and through them FDR with his reservations about exchanging data"; and the one who crafted the "three options" strategy to win Roosevelt's approval for the American-imposed middle option in December 1942. Bush backed Conant on all these issues. Both anticipated a possible British refusal to accept a unilateral American decision, and both considered any resulting delays to the bomb programme to be of "much less significance" than the main issue underlying both sides of the interchange debate: the postwar international power structure. Where Churchill sought a continuing American–British alliance, Conant envisioned a "new world order" founded on American global hegemony and continuing supremacy in nuclear weaponry.

The Groves volume has one more remarkable passage that invites our attention. For a man who considered himself to have been invariably right, it is unusual for Groves to revisit an earlier decision. Over five pages, he describes Abelson's development of liquid thermal diffusion and the late, hastily improvised insertion of it alongside other Manhattan separation processes at Oppenheimer's suggestion in June 1944:

> Just why no one had thought of it at least a year earlier I cannot explain, but not one of us had. Probably it was because at the time the thermal diffusion process was studied by the MED [i.e. Manhattan] we were thinking of a single process that would produce the final product. No one was considering combining processes . . . .
>
> If I had appreciated the possibilities of thermal diffusion, we would have gone ahead with it much sooner, taken a bit more time on the design of the plant and made it much bigger and better. Its effect on our production of U-235 in June and July, 1945, would have been appreciable. Whether it would have ended the war sooner, I do not know.[116]

On one point this account is incomplete. The obvious reason why "no one had thought of it" is not the one suggested by Groves, but rather because the Army's and Navy's nuclear programmes had been at a sour, non-communicative impasse between October 1943 and June 1944 (see Chapter 6).

Customarily, the world has viewed the Manhattan Project as a standard, a model standing apart, above the others. From a more precise comparative perspective, it did not evolve without flaws. The most obvious were the two extended breakdowns in communication: between the United States and Britain in 1943, and between the American Army and Navy in 1943–4. The time lost in these breakdowns cannot be calculated precisely. The essential point is that the breakdowns were repaired, and work was carried on to ultimate success. The British–American nuclear alliance faltered, closed down for several months, was patiently rebuilt and reshaped under new rules, and functioned again to the benefit of both sides.

Fragile and imperfect as it was, the contrast with the Axis powers was enormous. In nuclear matters, Germany and Japan appear to have cooperated only minimally, with no recorded exchange of restricted scientific information. When Japan asked for badly needed experimental uranium in July 1943, authorities on both sides were content to mask the transaction under a false reason. When Nishina's team chose gaseous thermal diffusion as its preferred method for separating U^{235} in March 1943, it seemingly had no information on the Harteck team's diligent but unsuccessful attempts with a Clusius tube in 1940 and 1941. Only at the very end, with Germany collapsing on both fronts in April 1945, did the German submarine U234 leave Norway on a long voyage to Japan, carrying a variety of weapons, weapon designs, and a reported 560 kilograms of uranium oxide to continue the struggle. Departing too late, U234 would surrender a month later in mid-Atlantic to an American destroyer, to be landed at Portsmouth, New Hampshire, on 19 May.

Looking ahead

If Japan's nuclear efforts are marked by large information gaps, the German effort, thanks to the Farm Hall transcripts, has relatively few. The shocked, intense discussions on the Hiroshima bomb and how it had been achieved faded away after Heisenberg's general lecture to the group on 14 August. Discussion resumed on the detainees' primary concern: their prospects for future employment as working physicists. While this debate has little to do in a narrow sense with the race for the bomb, it casts more revealing light on the aspirations, values, and working relationships of the German participants. The news of the bomb, in the fifth week after their arrival at Farm Hall, required a sudden scaling down of professional expectations. Unlike the missile experts, the nuclear scientists realized instantly that their imagined nuclear superiority had vanished utterly, and that Americans in particular would not be needing German help.

In the weeks following his lecture, Heisenberg continued in his role as leader and spokesman for the group. The older professors, Hahn and von Laue, whom Goudsmit had originally selected for detention as possible leaders or spokesmen for the German postwar scientific community who were untainted by Nazism, usually deferred to the younger man while at Farm Hall. Heisenberg proved to be a shrewd, tenacious negotiator, capable of winning maximum benefit from a situation already tilted in favour of the detainees. It was high policy from the beginning—even before their arrival in England—that the "guests" should be kept in a good frame of mind, and this became a matter of professional pride for the middle-rank custodial officers in charge. With these dynamics, daily life at Farm Hall became a continuous round of bargaining over smaller or larger issues. The detainees, who included six professors of whom two—and soon three—were Nobel laureates, ranked high on a social hierarchy which their British captors recognized and shared.

The main instruments of the guests in this bargaining were anger, psychological pressure, and threats to withdraw their parole. Their anger might be real, or feigned and strategic. Psychological pressures included alternating periods of civility and rudeness, or similar tactics calculated to keep their custodians off balance. For the detainees, confinement without guards rested on an honour system. Each had promised not to run away, undertaking to give advance notice of any withdrawal of this promise. In bargaining, any hint or threat to withdraw was usually sufficient to win concessions. As the transcripts and summaries show, the strategy and tactics to follow were often debated openly by the whole group. It says something for Major Rittner and Captain Brodie that they reacted to these tactics with remarkable patience and a touch of humour, which were acknowledged by the detainees when the hour of their return to Germany approached.

Although the bargaining process was sometimes over trivial privileges, it evolved over time to more important matters: communication with the detainees' families and arranging assistance to families in difficulties; visits from three British scientists and from a group of American officers headed by Major Calvert of Groves's London office; and a meeting of Heisenberg, Hahn, and von Laue with a group of British scientists in London on 2 October. One key visit occurred on 8 September, when Professor Blackett, newly appointed scientific adviser to the incoming Labour Government, arrived for a two-day visit to meet the detainees and initiate talks about their future. Blackett admits and emphasizes his newness to his post: "We were not very well prepared for the war, but we were probably better prepared for war than we were for peace."[117]

Heisenberg was quick to sense the policy vacuum, and to make practical suggestions to fill it. In 11 recorded pages of transcript, this first conversation with Blackett shows Heisenberg's skills in advocacy. On 17 September, he followed up Blackett's visit with a letter discussing future plans, research directions, and a future location for the Kaiser Wilhelm Institute for Physics, of which he was still acting director. The discussion process continued, and, after the London meeting on 2 October, Heisenberg could report to the group on ongoing planning, and add a comment: "I feel that at present we have a fair amount of influence on what will be done."[118] In a report for the week of 19–25 November, Captain Brodie would also comment on the guests' changing perceptions: "Their appreciation of their situation seems somewhat misguided in that they think the big decisions lie with them . . . ."[119]

These discussions looked beyond the future of German science to the future of science in Europe. Heisenberg was adamant that the rebuilding of science in Germany must be significant, and not just "physics on the Roumanian or Bulgarian scale":

> To do modern physics in a small way is of no use at all . . . . I don't want to do petty physics. Either, I want to do proper physics or none at all. If the final decision is that I can't do any proper physics and I go back to GERMANY again, naturally they, too, will realize that I am then going to consider doing physics with the Russians after all.[120]

In his discussions with colleagues, Heisenberg brings out the Russian card on several occasions while at Farm Hall: the Russians will offer high salaries, better equipment, and better working conditions to German scientists working in the Soviet Union, and he, like others, will "consider" working there if conditions in Germany fall short. If this reiterated back-up plan were accepted at face value, and not as a mere bargaining tactic, it would clear Heisenberg of any charges of excessive German nationalism. But the option seems never to have been sufficiently put to the test. Six months after his return to Germany, Heisenberg did receive a letter by courier from the Soviet Union, written by a colleague from the Gottow group, Heinz Pose. The letter praised working conditions in Russia, mentioned Soviet interest in Heisenberg and his work, and invited him to reply if interested. On 29 July 1946, from Göttingen, Heisenberg politely declined, wording his reply more towards British zone authorities than to Pose or his Soviet superiors. He "wanted to stay in Germany for the time being, as long as it was possible to do science under fairly satisfactory conditions and to nourish and bring up his children."[121]

If the transcripts point to Heisenberg's strengths as an influential leader in the German postwar scientific recovery, they also reveal deficiencies. His leadership of the Farm Hall group as a whole is flawed by partiality and favouritism, particularly against the non-academic scientists. In discussions with Professor Blackett on 9 September about the possible futures of detainees after their return to Germany, he gives strong support to his own associates, Weizsäcker and Wirtz, but his praise for the others is distinctly fainter, even condescending.[122] Thus Korsching:

> He is not a brilliant physicist, but he is a very good experimenter and he had a nice idea on separation of isotopes . . . . He is the type of man who has never been abroad. He is German and he has never come out of his German cities.

Bagge:

> His primary quality is great energy. He is a very active man and in so far he has done good work . . . . In some ways he is a proletarian type, he comes from a proletarian family and that is one of the reasons why he went into the Party, but he never was what one would call a fanatical Nazi. In some way I like him quite well.

Blackett asks if Diebner is a physicist:

> He has got his degree in physics, but he is not really a physicist, he is more a kind of *Verwaltungsmann*. He was connected with the *Heeres-Waffenamt*.

The Diebner group's better results with the lattice of uranium cubes go unmentioned.

Another passage from the transcripts illustrates a different side of Heisenberg. This passage, separated in the text from any surrounding context, is a private conversation between Heisenberg and Wirtz on 18 July, and is flagged for its open admission of German atrocities. In a paragraph-long monologue Heisenberg recalls:

> During the war I had five calls for help in cases where people were murdered by our people. One was SOLOMAN(?), HOFFMAN's(?) son-in-law. I could do nothing in his case as he had already been killed when I got the letter. The second one was COUSYNS the Belgian cosmic ray man; he disappeared in a Gestapo Camp and I couldn't even find out through HIMMLER's staff whether he was alive or dead. I presume he is dead too. Then there was the mathematician CAMMAILLE; I tried to do something . . . but it was no good and he was shot. Then from among the Polish professors there was a logistician with a Jewish name – and then with the other Poles . . . SCHOUDER, a mathematician. He had written to me and I had put out feelers . . . . Then SCHERRER wrote me the following ridiculous letter [from Switzerland, giving Schouder's location and assumed name] which was of course opened at the frontier . . . . I have now been told that he was murdered.[123]

In the entire passage (here shortened), there is no hint of regret, of sympathy for the victims, of apology for failure, or of any emotion other than irritation at Paul Scherrer's fatal blunder. Heisenberg is unmoved by this recital, but Wirtz is touched by it. "We have done things that are unique in the world," he says.

One of Heisenberg's biographers analyses this wartime record in some detail, linking it with longer term tendencies in Heisenberg's character:

> Throughout his life, Heisenberg saw himself as primarily responsible only for his own circle of friends, colleagues, and students, and he believed himself incapable of really helping anyone beyond that circle ... when Heisenberg did act, he did so as he had in past situations – offering too little, too late ... his efforts were pitifully weak as responses to the life-and-death situations of those whom he sought to help.[124]

What stands out alongside this failure as a sentient human being is Heisenberg's extraordinary capacity for shutting out the sordid ugliness of his surroundings and concentrating on the scientific work at hand. His course had been set in the first two years of the Nazi regime. On 2 June 1933, in an attempt to save physics at Göttingen, Heisenberg appealed personally to his friend and colleague Max Born to postpone any decision to resign his chair:

> Since only a very few will be affected by the law – you and Franck certainly not, Courant probably not either – the political transformation could take place by itself without any sort of damage to Göttingen physics.... Certainly in the course of time the ugly will separate itself from the beautiful....[125]

Born hesitated, but only briefly. In July he accepted a three-year appointment at Cambridge. In the law dismissing Jewish faculty, there was more at stake than physics at Göttingen.

Two years later, at Leipzig, in the midst of the second, harsher round of dismissals under the Nuremberg laws, Heisenberg joined with a small group of senior colleagues to protest the new dismissals. The protest was brushed aside by the minister, with reprimands for the participants. The incident left Heisenberg bruised, depressed, and more rigorously work-focused than before. In a letter to his mother on 5 October 1935, he wrote of his narrowed horizons:

> ... I must be satisfied to oversee in the small field of science the values that must become important for the future. That is in this general chaos the only clear thing that is left for me to do. The world out there is really ugly, but the work is beautiful.[126]

This self-imposed isolation, however, would prove shortlived and illusory. As Stark's attack in July 1936 and the wider struggle over Deutsche Physik would soon demonstrate, professional survival in a totalitarian society may require unsavoury alliances with a criminal regime.

The story of Manhattan's success is in part the story of Germany's failure. Why did Germany fail to maintain its early lead over its wartime rivals? That failure can be highlighted by a few key errors or omissions at critical decision points by the scientific and administrative leadership of the project that delayed or narrowed the options that lay ahead. These include:

1. The near abandonment by 1942 of isotope separation, after serious attempts by several different methods, as too difficult and probably impossible. This effectively pushed the main research emphasis towards concentration on the chain reaction.

2. The choice of moderator for a reactor was narrowed after Bothe's negative report on graphite in January 1941. Bothe's "error," never challenged by others in the project, left heavy water as the only promising alternative for a moderator.
3. Although Germany was privileged in controlling Norwegian heavy water after 1940, the supply was vulnerable to sabotage and air raids, and was also insufficient in quantity to produce a successful chain reaction. The cardinal error here was failure to develop an alternative or supplementary supply in Germany.
4. By their failure to achieve a chain reaction, the German teams were also barred from following the other route to a nuclear explosion, through transuranic elements (93 and 94) produced in a working reactor.
5. In general, the German project was characterized by inept and inadequate allocations of scarce materials, by scant effort to develop adequate supplies, and by chaotic, changeable leadership.

External factors also played a part. From 1943, Allied bombing drove institutes from major cities, destroyed apparatus being made in factories, and paralysed transport. In the last months of the war, working conditions for research became impossible.

The combined OSRD–Manhattan Project faced these issues and key decisions more successfully. Despite its late start—after Roosevelt's approval in October 1941—the American project had the benefit of British experience of the Maud Committee and Churchill's prior decision to produce nuclear weapons. At the points where Germany had faltered, Manhattan did better:

1. It backed three different methods for isotope separation, discontinued one (centrifuges) in 1942, added another in 1944, and used each of the remaining three in combination to produce fissile U^{235} for a uranium bomb.
2. Because heavy water was unavailable in 1942 in quantities needed, Fermi achieved the first successful chain reaction using graphite as moderator. When heavy water became available in quantity, Manhattan also built the first heavy water reactor in Chicago in 1944.
3. Building on its successful chain reaction, and alongside its success in isotope separation, the Manhattan Project also pioneered the transuranic route to fissile material by isolating element 94 (plutonium) and developing a plutonium bomb.
4. By contrast with the German project, Manhattan was characterized by strict attention to raw material supplies, industrial priorities, administrative clarity and simplicity, and willingness to make huge construction investments before scientific results could be guaranteed.
5. Finally, Manhattan was founded on elements of interallied cooperation which, although far from perfect, delivered sizeable benefits to both major partners before and after the rupture in 1943. By contrast, the image of German and Japanese scientists exchanging basic information internationally, or working side by side in the same laboratories, defies our imagination.

8

SOMETHING REASONABLE AGAIN

> ... we begin to think sometimes ... about the continuation of our peace-time research, and it is really a great relief to think of something reasonable again.

On 17 February 1945, Franz wrote to his California colleague from Berkeley days, Joel Hildebrand. Hildebrand had renewed contact with Franz in 1943–4 during a posting as scientific liaison officer at the American Embassy in London. After his return to America, he sent the Simon family a food parcel. Franz expanded his note of thanks into a report on future plans:

> Dear Joel,
> It was a very nice surprise to get, a few days ago, the box with the wonderful dried fruit from you. As you can imagine, it has not survived very long, but is still a very pleasant memory. Very many thanks from all of us.
> We are now all settled again and have nearly forgotten that part of the family was ever away for such a long time....
> With the war obviously drawing to an end now, I assume you are chiefly busy with post war problems. We have here to a certain extent started to think about the future organisation of physics and have to struggle with quite formidable problems. Industry, which as you know was not so research minded here as it was in the States, has now realized its mistake, and is now asking for enormous numbers of physicists who have first of all to be produced. In addition they want to take some of the prominent people of middle age from the universities to their research laboratories, which means they would not be available for teaching. The personnel question seems to be the greatest problem, while I am glad to say that on the financial side, we will get what we need. Apart from this, we begin to think sometimes in our leisure hours – which of course are still short – about the continuation of our peace-time research, and it is really a great relief to think of something reasonable again.[1]

Futures, however, do not always unfold as one anticipates. In Franz's case, the difference was due largely to increased recognition of his eminence as a scientist. That recognition would take a variety of forms. On 20 June 1945, he was elected to a studentship of Christ Church—that is, to an official fellowship of that college. For the first time, he would be involved in the government of a college. In the same year, the university honoured him by elevating his readership to a personal Chair in Thermodynamics. On the public side, he received a C.B.E. in the New Year's Honours List of 1946. Around the laboratory, he and his colleagues found it an amusing counterpart to his Iron Cross First Class for service in the 1914–18 war.

Further distinctions followed, culminating with a knighthood in 1954.[1a] In 1946 he accepted a vice-presidency of the Atomic Scientists' Association, and was also named by the Royal Society to the board of the National Physical Laboratory or NPL, the British national standards laboratory. He accepted these positions willingly, and others that followed, because he felt that his chair carried a dual responsibility. Primarily, it enabled further pathbreaking work in his field of greatest interest, very low temperatures. Second, but not less important, he saw a more general mandate to educate the public and the government in the proper role of science in the educational system, in the industrial economy, and in society generally. Appointments to public and private bodies provided a vital platform for fulfilling this broader mandate. In time, this broader task would expand to include more associations, public lectures and speaking engagements, scientific journalism, and occasional broadcasts on the BBC.

The autumn of 1945 saw one further initiative that was not made public at the time. It may have remained unknown to Franz himself. Lord Cherwell, as one of a half–dozen British department heads canvassed each year, received the usual letter from the Nobel Committee for Physics in Stockholm, inviting him to nominate one candidate for the Nobel Prize for Physics for 1946. He replied, briefly, on 7 January 1946:

> Gentlemen,
>
> In response to your invitation . . . I beg to submit the name of Professor F.E. Simon, F.R.S.
>
> As you are of course aware, Professor Simon's researches in the field of low temperature are altogether outstanding; indeed I think many would hold that he is the greatest exponent of Low Temperature Physics since Kamerlingh Onnes . . . .
>
> During the war he has been engaged chiefly on problems connected with the separation of Isotopes, especially by the Diffusion Method in which he has made very great progress. No doubt this work will be published in due course, but for the time being it is unfortunately secret.[2]

Given its brevity, and the requirement of secrecy, the nomination was not a compelling one, but it served to place Simon's name and work before the Nobel Committee.

The new and renewed postwar activities would have made for a busy life even by themselves, but Franz also had continuing commitments from the war years that could not easily be broken off. On the nuclear side, the new Labour government, contrary to some expectations, moved quickly to establish an active programme of nuclear development, part of which grew directly from the wartime diffusion project. At a more personal level, Franz had an abiding concern for the future of postwar Germany. The Nazi regime had been defeated in the field, but its legacy still loomed large in German society and German universities. For several years after 1945, British nuclear development and German de–nazification would claim a significant proportion of Franz's waking hours. The combination of new projects with older continuities and obligations would make for an extremely active, highly diversified life.

Approaching absolute zero: experimentation and theory

When Franz wrote to Hildebrand of "something interesting," he was thinking of his prewar field of low-temperature physics. The wartime work on isotope separation was a striking technical achievement, but remote from exploration in basic science. The low-temperature world that he had put aside in 1940 was an unexplored Alice-in-Wonderland world where familiar elements and compounds may not behave as expected. At the Strasbourg conference on magnetism in May 1939, Franz's paper had looked forward to improving methods of magnetic cooling by up to 1,000 times,[3] that is, getting three orders of magnitude closer to absolute zero by utilizing the paramagnetism of atomic nuclei.

After six wartime years, the quest could be resumed in 1945. A first step required equipping the Clarendon appropriately for low-temperature work. The new building, completed in 1939, had been used for defence projects through the war years, including diffusion research continuing up to 1947. To become a leading-edge low-temperature laboratory, it needed equipment for liquid air, liquid hydrogen, and liquid helium, which allowed experiments in the Kelvin temperature range from 20 K to 1 K, as well as high-pressure and magnetic facilities. By January 1947, Born could write to Simon: "Urey told me that he had visited you and seen your new helium plants where you can get liquid helium from taps in every room like water. Well, I have great confidence in you but I regard this description as a little exaggeration."[4] By 1950, Nicholas Kurti reports, this section of the Clarendon "had become one of the largest and most important low-temperature schools in the world."[5]

Simon's experimental work, and its continuities and advances after 1945, have been described by others closer to the scene, notably by Nicholas Kurti and by Simon's Harvard colleague Percy Bridgman.[6] The continuing aim, from Berlin days onwards, was to reach progressively lower temperatures approaching absolute zero (0 K). The means for doing so, using four successive methods of cooling, was through liquid helium, an element interesting both for its own unique properties and as a means for cooling other substances to a similar range. With each new temperature range reached, new territory was opened up for experiments.

For Franz personally, the quest for lower temperatures was shaped and driven by one overriding theoretical question: the validity of Nernst's heat theorem and his claim to have discovered a third law of thermodynamics. The question had been a central one for him since his earliest days as a doctoral student. Nernst himself, building on nineteenth-century discoveries that specific heats for many substances diminish as their temperature goes down, was the first to suggest, in a paper of 1905, that at absolute zero specific heats should reduce to zero. In 1907, the young Einstein, in a paper based on quantum theory, theorized independently that at absolute zero all specific heats should become zero.

With further experimentation, and easier accessibility to lower temperatures, the early 1920s saw the discovery of various anomalies and exceptions that led critics to cast serious doubt on the Nernst theorem. The debate of these years stimulated Franz to examine these anomalous cases carefully. Some proved to be due to faulty extrapolation from known data to temperatures not reachable earlier. Others, such as crystals or glasses, seemed to be clear exceptions. Franz hypothesized that substances of this type were not in thermodynamic equilibrium internally, and therefore were not treatable in thermodynamic terms. As he worked to test and refine the principle, it had become clear to him that the key variable was entropy, the degree of disorder in the system, the difference between total heat and that portion available for conversion to mechanical energy. It is entropy, his revision postulated, that should decline and disappear as temperature reaches absolute zero. By 1927 he proposed a restatement of the third law to make it applicable only to substances in internal equilibrium.[7] Acceptance of this revised version, however, would come slowly, and the decade of the 1930s, as he would later characterize it, would be "a period of utter confusion."[8]

In 1956, after a further decade of experiments, Franz would revisit the central theoretical question one more time. He was invited to deliver the 40th annual Guthrie Lecture of the Physical Society. His lecture, entitled "The third law of thermodynamics: an historical survey," was delivered in London on 13 March, and published in the 1956 *Yearbook* of the society. In 22 pages of printed text, the lecture sets out in an orderly fashion the "very chequered history"[9] of a scientific concept, from Nernst's original paper—published in 1906—through the period of challenges and reformulation to the developments of the 1950s. In this thoroughgoing review Franz was addressing an audience of scientists. His text, complete with numerous formulae and graphs, is challenging for non-scientists. At a non-technical level, however, he makes several points of wider interest.

His account shows how Nernst's original interest was a response to needs of the growing German chemical industry for an improved understanding of gas reactions, in much the same way that the earlier laws of thermodynamics had developed from the technology of steam engines during the earlier industrial revolution. He is careful to show how Nernst, already a renowned physical chemist, remained close to his original ideas about chemical constants when others took up the question from the standpoint of physics and emerging statistical approaches of quantum theory. Even after Franz's reformulations in 1927 and 1930, Nernst resisted the concept of entropy change and preferred simpler language, though the disagreement did not impair a lasting friendship between master and pupil.

A further point is that many of the scientists listed as contributors to the development had been known to Franz personally as teachers, earlier students of Nernst, colleagues, or his own students. For the early period he mentions Einstein, Max Planck, Max Born, Peter Debye, Otto Stern, Charles Lindemann and Frederick Lindemann. After 1920 he includes William Giauque in Berkeley, Kurt Mendelssohn, Martin Ruhemann, G.O. Jones, C.A. Swenson, and Robert Berman. Even Ralph Fowler, whom he quotes as a leading critic of the theorem in the early 1930s, had probably crossed paths with Franz while on scientific postings to Ottawa and Washington during the Second World War. In his conclusion, Franz emphasizes the collective nature of the development: "I hope I have shown you how a great mind tackled a very obscure situation, at first only in order to elucidate some relatively narrow field, and how later, as the result of all the interaction and cross-fertilization with other fields, the Third Law emerged in all its generality, as we know it to-day."[10]

Nuclear legacy

If a long-awaited return to low-temperature research could claim first priority for Franz after hostilities ended, he also had continuing obligations to wartime projects. The line of development runs through his last wartime visit to America, when he attended the Tube Alloys Technical Committee meeting in Washington in May 1944 (see Chapter 6). This meeting had been called specifically to work out future directions for British nuclear policy, and to consider a feasible timetable for them. The committee endorsed Chadwick's insistence on giving highest priority to British participation in the Manhattan programme, but it also backed continuing nuclear research in Britain using whatever scientific manpower remained available. For the path ahead, Gowing notes, ". . . in May 1944, the general consensus of opinion favoured a low separation diffusion plant as Britain's first step towards her own postwar project, and ICI were asked to provide estimates."[11] This "compromise solution,"[12] proposed by Halban to the Washington meeting, was the prudent choice, because at this date none of the three Manhattan routes to the bomb had yet been proved free from problems. The key advantage was flexibility. The output from a low-separation plant could be used either in a reactor to produce plutonium or for higher enrichment of uranium by whichever method should prove most efficient.

In subsequent meetings, the technical and consultative committees continued planning for a postwar nuclear research facility in England. They considered site requirements, equipment needs, staffing, and estimated costs. In late 1944, a group in the Montreal laboratory began to design a low-power, graphite-based reactor for research and for producing radioactive isotopes. In April 1945, Sir John Anderson authorized the research facility in Britain to proceed independently of other projects, and by July John Cockcroft had been recruited from the Montreal laboratory to head it. The site chosen for it was at Harwell, a former airfield close to Oxford. During the same period ICI developed plans and cost estimates for the proposed low-separation diffusion plant. These plans included flexibility to allow for different levels of enrichment. When this phase was completed and the plans were deemed ready for the production phase, they were "put on ice"[13] in April 1945 for the time when they would be wanted. By this point it was clearly advisable to await results from Manhattan before making any large-scale investment.

The new Labour government, in office after 27 July, lost little time in setting a course for nuclear policy. On 28 August, two weeks after the Japanese surrender, Prime Minister Attlee drafted an urgent memorandum for a few senior cabinet colleagues. "A decision on major policy with regard to the atomic bomb is imperative,"[14] he wrote as he tried to assess the implications for Britain of the new international realities. The hard realities included an unbending American monopoly on bomb production, massive Soviet armies stationed across Eastern Europe, and no prospect as yet for international agreement on nuclear weapons control. The new government saw the huge Soviet preponderance in conventional forces as a threat that necessitated a British-based nuclear deterrent, to be built with American assistance if possible, but, if not, by British effort alone. In just over a year, Attlee and his select group would make several key decisions. In December 1945, they approved a reactor for plutonium production; in January 1946, an organizational structure for atomic energy production and an armament facility; in February, a headquarters and development centre for the project at Risley in Lancashire; in March, a uranium-processing plant at Springfields in Lancashire; and in October 1946, preliminary work on a gaseous diffusion plant.

These decisions were made in deep secrecy, the product of a half-dozen senior ministers, a group designated by the unrevealing name of Gen 75.[15] Usually there were delays between the approval dates and the minimum public disclosure needed when plant construction or other consequences became visible to the public. On the armament side in particular, the ultimate purpose of HER,[16] the acronym for High Explosive Research, was not disclosed even to many who worked on that project. Though the general directions were clear as early as 1945, the decision to make British atomic bombs was not made explicit until January 1947, and was disclosed—guardedly—in the House of Commons only on 12 May 1948. On the military side, the Chiefs of Staff appeared almost totally in the dark. When asked to state their atomic bomb requirements late in 1945, their reply, sent on 1 January 1946, showed an obvious incomprehension of the magnitude of the task:

> ... in order to be effective as a deterrent we must have a considerable number of bombs at our disposal. It is not possible now to assess the precise number which we might require but we are convinced we should aim to have as soon as possible a stock in the order of hundreds rather than scores.[17]

Military thinking at this point clearly assumed that the larger the nuclear deterrent, the less likelihood that it would ever be used.

In the light of six decades of later history, these decisions may seem strange. They appear less strange if we place them in an immediate postwar context. Britain had emerged from the war with her Empire and self-governing Dominions apparently intact. She had worldwide responsibilities to fulfil and interests to defend. With a combined land area of about 12 million square miles and a population estimated at about 566 million in the late 1940s, the Empire and Commonwealth together comprised almost four times the land area of the United States and its dependencies, and almost four times its population.[18] The common assumption, virtually unchallenged at the time and occasionally stated explicitly, was that postwar Britain remained a first-class power, and as such had a need and an entitlement to develop atomic bombs. "The discriminative test for a first-class power," wrote one of the project's leading figures in 1951, "is whether it has made an atomic bomb."[19] That Britain had been all but bankrupted by the war was seldom considered, though financial stringency would impose severe limits on the direction and pace of Britain's postwar nuclear development.

Given its early and rapid beginnings, Britain's route to her first bomb test proved long and tortuous. Its official history is chronicled by Margaret Gowing in two further volumes totalling more than 1,000 pages—more than double the length of her volume on the war years—and covering a seven-year period from 1945 to the first test in October 1952. In comparative terms, that period was unduly prolonged. The USSR had detonated its first atomic bomb three years earlier, in August 1949. The British project was held back by continuing financial stringency and economic crises, by repeated but unsuccessful attempts to revive Anglo-American collaboration, and perhaps most of all by an absence of any deep sense of urgency such as had fuelled the Manhattan effort. Though the British effort was nominally headed by an eminent military figure, Lord Portal was in no sense comparable to General Groves. The period also saw another change of government, from Labour back to the Conservatives, in October 1951, with Churchill returning as Prime Minister.

A later book on Britain's bomb effort up to the first test in Australia is titled *Test of Greatness*.[20] The title is apt, but in comparative context the scale and timing of Britain's test in October 1952 serve only to highlight the unchallengeable lead at this point of the two superpowers. From this seven-year effort a triple irony stands out. Britain's vast Empire of 1945, in the name of which the project was first launched, was much reduced by the accession to independence of India, Pakistan, Ceylon, and Burma in the later 1940s. Second, Britain's first test of a plutonium bomb on 3 October 1952 was overshadowed a month later by the first American test explosion for a hydrogen weapon, and nine months after that by the first Soviet hydrogen test explosion. Third, for Britain, the price for a successful weapon test had included a delay in developing nuclear power. When the design engineering team proposed in 1947 to develop the second Windscale reactor as a dual-purpose plant to produce plutonium along with steam for power generation, the proposal was firmly rejected by Lord Portal as untried and likely to delay production of fissile material.[21]

After a successful bomb test in Australia in October 1952, prospects for exploring nuclear power improved. In the spring of 1953 permission was obtained to proceed with a dual-purpose reactor. A site was chosen at the Sellafield complex, and construction began on the reactor later known as Calder Hall. The military priority remained paramount: "Calder Hall was intended primarily as a plutonium-producing factory,"[22] and it was engineered accordingly. Steam for electricity was a secondary goal, but the electricity generated was expected to bring in revenue, and perhaps relieve recurring winter coal crises. The new reactor was opened by the Queen and joined to the national electricity grid in October 1956. Before this date, however, Calder Hall had become the prototype for a bold new plan, announced by the government in February 1955, to build a further dozen nuclear power stations between 1957 and the mid-1960s.

The subsequent history of these stations and their successors is a complex sequence of optimism and disillusionment, activity and reappraisal, technical successes and disappointments, fluctuations in supply and prices of uranium and competing fuels such as coal or gas, a serious reactor accident (at Windscale in October 1957), and mounting public concern over nuclear safety. Simon, however, would not survive long enough to witness most of these developments, his life being cut short suddenly and unexpectedly in 1956, a few days after the opening of Calder Hall. He could confidently predict the age of nuclear power in Britain, but he did not live to see it realized.

Committees, committees

As a scientist increasingly recognized in policy circles and the serious media, Simon faced more and more requests across a growing range of scientific and educational issues. The next two sections will illustrate how these broader activities became a significant part of his life after 1945. An abstract compiled from Franz's monthly calendars lists almost 250 meetings of committees, panels, and other similar groups that he attended in the decade after August 1945. This number is understated, because some calendar entries were illegible, or unidentified, or occasionally purposely disguised. Of this total about 100 can be clearly linked to the nuclear projects, and about 20 more concerned electricity research or other forms of energy. With deceptive simplicity, Franz's curriculum vitae of this period notes membership on the "atomic energy committees."[23] From his monthly calendars and daily diaries one can usually see the committee, time and place of the meeting, and a few names of those attending, but rarely any hint of agenda items.

The most central of the nuclear project's bodies was the Technical Committee. At first the Tube Alloys Technical Committee continued its regular monthly meetings, with Franz attending, from September 1945 into early 1946. When Lord Portal took over as Controller of Atomic Energy in March, a new, larger Technical Committee replaced its predecessor. Franz's calendar shows that on 26 March he dined in Christ Church, and that Portal, who had once been an undergraduate there, was also present. Diary entries fill in more detail. On each of the two days preceding, Cherwell telephoned Franz to set up the meeting. On the 26th, Franz went to Cherwell's rooms in Christ Church at 6.25 p.m., met Portal there and discussed "T.A." until 7.15. After dinner, discussion continued from 9 until 11.25 p.m. The calendar then shows Franz at a meeting of the revamped Technical Committee in London on 24 April, and a further meeting in Oxford of a subcommittee of the Technical Committee on 30 April, which discussed "combined operations," possibly between the Harwell scientists and the engineers' production projects at Risley.

Overall, Franz attended at least 28 meetings of the Technical Committee between September 1945 and May 1956. More frequent were some 35 meetings in his own area of wartime research—membranes and isotope separation by gaseous diffusion. Typically, these were held at Kings Norton in Birmingham, where the T.A. Directorate had acquired a small factory early in 1945 to establish a pilot plant for membrane production. These meetings were variously referred to as the Diffusion Committee, the Membrane Panel, or the Isotope Committee. From February 1952 another series of meetings is listed under the acronym IPDC,[24] which can be identified through Gowing's account as the more secret Isotope Plant Design Committee. The name itself, as she explains, was a disguised name for the giant gaseous diffusion plant being built near Liverpool at Capenhurst in Cheshire. Franz lists nine meetings attended of the IPDC between the first one held at Capenhurst on 1 February 1952 and another there on 21 May 1954, marked "last IPDC." The others include one more at Capenhurst on 9 June 1953, and five others—all in 1953—held in London.

Harwell, which operated as an independent research establishment, had its own projects and committees, wider ranging debates, and freedom to work on longer-run issues. Meetings there were only a few minutes' distance from Oxford. Franz's calendars list recurring meetings of two Harwell committees: one on nuclear power (from May 1946), the other on heavy water (from February 1951). Beyond these, he lists occasional meetings at Harwell on specific problems, such as corrosion (December 1947), fast reactors (May 1948), graphite (February 1950), and a "Full Enrichment Mtg." (September 1953). Gowing's two postwar volumes are interesting for their portrayal of complementarity and mutual support—though with occasional frictions—between two very different types of establishment, an academic-style research institute at Harwell and a deadline-driven, large-scale engineering headquarters at Risley and its associated production plants.

While the industrial plants were under construction, some committee meetings were held as site visits. Franz records a visit, together with Akers, Skinner, and Peierls, on 16 August 1949, to the plutonium plant at Windscale to "see Piles," and a second visit, with members of the Technical Committee, on 29 June 1950. In between, he lists a visit with the Isotope Committee to the uranium metal-refining plant at Springfields in Lancashire on 5–6 January 1950. The Technical Committee returned to Sellafield (i.e. Windscale) for another meeting on 17 August 1951, when the first chemical separation plants for plutonium—urgently needed—were being built and tested amid many difficulties.[25]

Though the nuclear meetings may have had greater immediate impact, Simon's committee work leading in other directions was more extensive and more diversified. One body that met frequently, and usually in Oxford, was a local low-temperature group, for which he records 24 meetings beginning in October 1945. In the same subject area, he also lists numerous conferences and executive links with international associations for low-temperature research and commercial refrigeration. He was singularly active at the Royal Society, serving a term on the Society's council from 1949 to 1951, on its grants committee, its selection committee for new fellows, and several other committees. The diary shows council meetings of the National Physical Laboratory from December 1945 to 1951, and meetings of the Department of Scientific and Industrial Research (DSIR) Scientific Grants Committee from May 1946. For another energy-related committee, the Research Council of the British Electricity Authority, he lists 17 meetings attended between October 1949 and 1953.

The calendars also show a wide variety of shorter-term or ad hoc committees, single meetings on specific problems, and consultations with industry associations or specific firms. There are references to ICI Plastics, the "Tin people," the diamond industry, the "Standard people," a gas turbines meeting, a helium meeting in Oxford, and one with a research committee of Shell. The contexts for some meetings and some committees remain insufficiently identified. Many of these activities cluster in 1950 or 1951, by which time he was becoming more widely known as the Scientific Correspondent of the *Financial Times*.

His connection with the diamond industry was a special case. In 1949 he worked with Sir Ernest Oppenheimer of the De Beers group to establish a collaborative research project between the industry and academic scientists working on the physics or chemistry of natural and synthetic diamonds. It brought industry support for academic research and fostered new industrial applications for synthetic diamonds in high technology. This pioneer model of academic–industrial collaboration grew into an annual Diamond Research Conference. After the 60th Diamond Conference in 2009, the committee secretary wrote that "The impact of nearly 60 years of collaborative research is substantial . . . . We owe a debt of thanks to Professor Sir Francis Simon for his vision in establishing the Diamond Research Programme."[26]

These many committees and connections became a vital, productive part of Simon's life. His biographer, Nancy Arms, at one stage his secretary at the Clarendon, gives first-hand testimony on their place in his life:

> Simon liked committees. He was always on the look out for something new and interesting and they gave him an opportunity for finding out what was going on. He also liked arranging things, even though he was not always very efficient. He liked to make sure that he missed nothing that would further the interests of the Clarendon Laboratory. After a meeting at Harwell to discuss co-operation with the Clarendon someone said: "Do you know what Simon means by co-operation? Sitting at a table and going away with £5000 for the Clarendon."[27]

Most committees, however, did not have money to allocate, though their work was rewarding in other ways. For Simon, committees fulfilled a double personal need: as another outlet for his naturally gregarious temperament, and as a channel for improving and promoting the quality of life for society at large.

Public voice for science

From April 1948 onward, Franz found a larger public forum for his ideas on science in society. He became the regular—but unnamed—Scientific Correspondent of the *Financial Times*, with a mandate to contribute articles from time to time on scientific issues of his choosing. The arrangement was made through Lord Moore, managing editor, and Hargreaves Parkinson, editor of the paper. By November of that year, an additional agreement established a second series of short news items on science in industry, to be selected and approved by Simon from material gathered by student assistants. These brief notes proved popular, and generated a large reader response, much of it seeking further information. In June 1949 these notes were elevated from occasional filler to a weekly column under the heading "Science and Industry."

The longer articles, with a norm of about 1,200 words per issue—some split between two or even three issues—present an overview of Simon's views on science in modern industrial societies. In 1951 a selection of these articles was reprinted in book form as *The Neglect of Science*, and under his own name. This collection, together with other articles written later,[28] argued forcefully for a more rational and efficient science policy for Britain during the years of economic crisis after 1945.

The publication of *The Neglect of Science* stripped away the anonymity of the *Financial Times*'s Scientific Correspondent. Simon had also published under his own name in this paper as early as December 1950. From 1951 onwards his long articles all appeared under his own byline. In November 1952 Parkinson's successor as editor, Gordon Newton, appointed a new Scientific Correspondent without consulting Simon. A few months later he reminded Simon that "our arrangement" will come to an end on 30 June 1953, though further signed contributions from him would still be welcome. After this point, Simon's main articles for the general public are divided between the *Financial Times* and other publications in roughly equal numbers.

The articles fall mainly into three well-defined areas. The first is technological education and the key role of science, technology, and innovation in advanced industrial societies. A second, larger group is concerned with fuels and energy, with emphasis on thermodynamic efficiency. The third concerns emerging scientific rivalries at the international level.

In the first category, the core paper of *The Neglect of Science*, published in June 1948, deplores Britain's failure to develop its own industrial research laboratories, and corresponding technological institutes to produce research scientists on the model of the Swiss Institute of Technology in Zurich, or M.I.T. or Cal Tech in the United States. At a minimum, he writes, "we should at least build two such Institutes of Technology at once."[29] A second essay, titled "Fundamental research," discusses its importance in the rise of twentieth–century physics, and the delicate problems of financing and encouraging it. A third essay, "America revisited," first published in January 1950, introduces comparative elements. It notes the growing supremacy of American industry "at the expense of industry in the weaker countries"[30] and suggests how Britain might counter this trend. Once again, the road to improvement starts with developing high-quality institutes of technology, together with higher salaries and status for scientists in research or management positions in industry.

More *Financial Times* essays, written after publication of *The Neglect of Science*, continue this theme. In "Scientific manpower" (29–30 November 1951), he stresses the need for more rational planning. The grants and plans for university equipment have at this point exceeded the number of young scientists needed to utilize it fully, and interuniversity pooling on larger projects would be helpful, as in the United States. The most serious deficiency is a shortage of technologists, and inadequate technological education. In a second article published in 1955, "The shortage of scientific manpower,"[31] he reviews the overall situation and finds it little changed. The funding and status of science, and especially applied science, are too low in relation to what is at stake: "To-day a country's industrial status is determined by its general level of scientific achievement and the extent to which this permeates both industry and government."

Simon's most succinct statement of the technical educational issue came in the *Sunday Times* of 4 March 1956, entitled "Technologists and technicians."[32] The article arose as a comment on a government White Paper announcing its intention to spend £100 million over five years on technical education, and to build up some 30 technical colleges as colleges of advanced technology. "The distinction between technologists and technicians," he explains, "is the crux of the matter," and the White Paper acknowledges this.

> Technologists are people who are educated at university level in the fundamentals of science and in their application to industry. Technicians, on the other hand, have only an *ad hoc* training in some special field

of technology. Both groups are essential and are needed in much larger numbers. However, no number of technicians can replace a technologist . . . .

In most industrial countries the education of technologists and technicians is quite separate. Technicians are educated in technical colleges; technologists either in university departments or in "Institutes of Technology." This is the pattern in those countries which have overtaken Britain.

Franz welcomes the White Paper and its promised funding, but is apprehensive about some features of the plan: continuing control by local governments, with inputs from local industries; standards to be set by the Ministry of Education, which "has no experience in running university-type institutions . . . . Would it not be better to divert at least part of the funds to creating, say, a couple of really large top-rank Institutes of Technology?"

The more numerous essays on fuels and power are divided between short-term and longer-term possibilities. In the short term, the overriding problem of this period was repeated shortfalls in Britain's coal production. In a core article titled "Why waste coal?"[33] in June 1948, Franz points out that much of the 200 million tons produced annually is consumed very inefficiently. A steam locomotive is about five per cent efficient in energy terms, but with a steam-driven central power plant and electrification of the rail lines "the overall efficiency of a locomotive could be raised to about 20 per cent." The largest, simplest, and fastest savings could be obtained from the 60 million tons of coal used annually for household heating, most of it in open fireplaces. These "contraptions to heat the stars . . . and to soil the earth" are about 15 per cent efficient, whereas even "simple closed stoves" installed in the same fireplaces could be about 40–50 per cent efficient. He estimates possible short-term savings here at 20 million tons of coal per year, or 10 per cent of the country's annual production. "There is an unanswerable case," he writes, "for the elimination of open fires as soon as possible."

Two more essays push the case further. The next, in September 1948, "Economical heat production,"[34] explains the underlying thermodynamic issue: mechanical and electrical energy are mutually convertible from one to the other "with only small losses," and either can be converted efficiently into heat energy. "But one cannot convert heat energy fully into either of the other two forms." Thus existing coal-fired power stations are only converting fossil fuel energy into electricity at about 20 per cent efficiency, and hence, after allowing for transmission loss, using electricity produced from coal for home heating has an "overall efficiency . . . roughly the same as that of the open fireplace."

The third essay, titled "Wanted: a national fuel policy,"[35] argues vigorously for a more decisive energy policy. It appeared in mid-winter of 1950–1. "Once more," it begins, "we have a coal crisis." The remedy lies in a comprehensive national fuel policy to "make the best combined use of coal, gas, and electricity." As before, the largest immediate savings would be realizable by a quick phasing out of open fireplaces and electric heaters for domestic heating. The power thus saved would be needed for Britain's ailing and uncompetitive industries in the growing competition for world markets. This third essay is a double call for an intelligent, science-based energy policy and for bold government action to achieve it. The tone is sharp, earnest, and a little exasperated by the rank absurdities of the situation. It ends with a rebuttal, on humanitarian grounds, to some of his critics. To one who opposes government intervention as an infringement of freedom, he asks:

> Does he realize that his primitive ideas about freedom rob the country of about 100 tons of coal during his lifetime; that he pours many hundredweights of soot and dirt on other people's heads and houses; and that his extravagance condemns one miner to labour quite unnecessarily for about three months?
>
> Does he realize that at present about 100,000 miners are doing nothing else but labouring for the coal needed in excess of what would be sufficient if proper heating methods were applied?
>
> Mining is obviously an unpleasant job which no one is very anxious to do and we have no right to demand that people mine coal which is then only wasted.

Addressing the longer run, Franz briefly discusses both nuclear power and other options. On these topics, the sense of crisis is absent and the time scale is estimated in decades. On nuclear prospects, a first article published in December 1948, "Prospects of atomic power,"[36] is unexpectedly negative, pointing out major obstacles to achieving nuclear power: the cumbersome minimum size of a workable reactor; the unknown availability and cost of uranium or thorium for fuel; the "singularly unpleasant type of plant" required for atomic energy; and the "very tricky problem" of disposal of radioactive fission products. This first short paper is designed to dispel any euphoria that persisted after 1945 regarding nuclear energy as a panacea. "For a long time to come," it concludes, "we will have to rely on coal as our basic source of power."

In a companion piece a month later, "Power sources for the future,"[37] he reviews briefly other energy alternatives. Water power is "highly convenient" but "concentrated in relatively few places." Wind is successful in small installations, but irregular in supply. Tidal power could be developed in a few places, but capital costs would be high compared with coal. By contrast, solar radiation is a huge but neglected source of energy, and one of the most promising for long-term development:

> ... why the general public puts so much hope in atomic energy and 'cold-shoulders' the sun is a mystery – perhaps the sun has not quite the glamour of atomic energy. But after all the sun is nothing else than a nuclear reactor and there is the definite advantage that it is at a safe distance, free for all, and out of reach of politicians.

The tone of this essay is lighthearted and futuristic, with a 50-year time horizon, but it ends with another reminder that in the present, i.e. 1949, the issue is coal:

> By a modest effort we could save enough to make us secure against crippling power shortages. By a determined effort we could save about half of the present consumption; that is, about 100 million tons of coal per year in this country alone! This is the only reasonable and safe way to deal with the power question before new sources of power come into action.

In the five years after publication of *The Neglect of Science*, Simon would write at least eight more articles on fuel and energy. The message, variable in details and presentation, remained essentially simple and direct. Coal production in Britain has effectively peaked, at slightly over 200 million tons a year. More oil imports are costly, requiring unaffordable foreign exchange. Nuclear power will assuredly come in time, but only after a developmental period of 15–20 years before it can be significant. "Reactors," he writes in 1955, "are in about the same state as aeroplanes were when Blériot first crossed the English Channel."[38] During this critical period, Britain will face a serious gap in energy resources compared against her main industrial rivals. In this interval, one available option will be more efficient utilization of existing fuels.

In February 1955, the Ministry of Fuel and Power released a bold White Paper that announced plans to build a further dozen nuclear power stations similar to the one still incomplete at Calder Hall, with a target date for completion of all 12 by 1965. Franz was sceptical of the White Paper's rough estimate that by 1975 nuclear power stations might supply about 17–25 per cent of an expected electricity production of "about 55–60,000 megawatts,"[39] though such a result could be achievable by the "last decade of the century."[40]

As matters turned out, economic and technical factors would delay the nuclear timetable. By 1975, Britain's electricity output was still fuelled 70 per cent by coal, 20 per cent by oil, and 9 per cent by nuclear reactors. By 2004, the percentage for coal had dropped to 33 per cent; in its place was another fossil fuel, natural gas from the North Sea, at 40 per cent. Nuclear power plants supplied 19 per cent of total grid electricity, having peaked at 26 per cent in 1997, before the shutdown of some first-generation Magnox reactors reduced nuclear capacity. Minor sources accounted for the remaining eight per cent. [41]

For the longer run, however, Franz was confident that nuclear power would replace fossil fuels and that at some point hydrogen fusion would replace fission:

> ... the first half of next century will see the ascendancy of nuclear power, whether based on uranium fission or hydrogen fusion. The latter is bound to become predominant sometime because of the abundance of the fuel.... It is impossible to predict when this is going to happen; one may discover a way to tame the thermo-nuclear reaction in ten or perhaps only after 100 years, but come it will.[42]

Half a century after he wrote this, power from fusion, despite prolonged research, seems no closer to daily use than Blériot's fragile monoplane of 1909 was to modern air transport.

The third major theme in these articles is international rivalry in science. In the later 1940s, the overriding question was when the Soviet Union would achieve its own atomic bomb. Related issues included Soviet policy on scientific manpower, the quality of Soviet science, secrecy, security, and the role of espionage. Franz's first article in the *Financial Times* in this area, entitled simply "Atomic energy," appeared on 9 December 1948. In view of his sensitive position as an insider, he proceeds cautiously, and in general terms. After a quick review of nuclear discoveries since the 1930s, he explains how the Smyth Report of 1945 attempted to establish an official line between "fundamental physical facts" that could be divulged—or discovered independently—and "the real secrets," technical matters such as isotope separation, purification of reactor materials, the handling of radioactive fission products, "and scores of similar problems." In practice, this line was hard to achieve:

> The compromises reached in the Smyth Report are often inconsistent. Driven by the understandable desire to show the American public what difficulties had to be overcome and how many different methods had to be followed up, much information has been given away of a kind which must enable any country starting afresh in this field to save two or three years, whereas on minor points information is often obviously and deliberately withheld.... But even with the knowledge gained from the Smyth Report and subsequent publications, the practical release of atomic energy is still bound to take many years for any country which did not take part in the wartime projects.[43]

The highlight of this 1948 article is a four-paragraph passage on the USSR: "The question in everyone's mind.... How long will it take the Russians to produce bombs?" Despite their having "a number of first-class scientists and engineers," two factors will work against them: "An atomic energy project requires technical experience in very many fields," and it also requires freedom from ideologically driven theories, such as those of "the charlatan Lysenko," which led to the purging of "many of Russia's best geneticists."

> On all the evidence it seems a reasonable estimate that Russia cannot have the first bomb in less than, say, another three years and probably not before five years; moreover, one bomb is, of course, of no military value. To produce enough to 'saturate' an enemy would require many more years.[44]

When this article was reprinted in *The Neglect of Science* in 1951, the Soviet bomb had become established fact, and these paragraphs required fixing. Franz allowed his original text to stand intact, but drafted a lengthy footnote that will be examined more closely in the next chapter.

The 1948 article ends with a bleak view of the global nuclear scene. So far, he notes, all attempts at international control have failed. It is not possible to separate weapons production from power production, because "the large reactors which are envisaged for the production of power at the same time produce plutonium . . . ."[45] His conclusion is sobering, but realistic:

> If we weigh the potential benefits of atomic energy in power production . . . and in research, against the potential dangers, it is obvious that with the present standards of international morality we would be happier without it. Some people have suggested a moratorium on the development of nuclear physics . . . . This is, of course, not a realistic proposal, and in any case the damage has already been done. We have to try to make the best of a bad job – as with so many other things.[46]

This gloomy outlook was strangely transformed by 1954, in a paper entitled "The atomic rivals."[47] By this date, with Stalin dead, Franz could sense an improved atmosphere for avoiding war. Paradoxically, the change had come as a result of weapons development—"the *cheap* hydrogen bomb"—and the near assurance of mutual destruction for both sides if it were ever used. Though the United States still probably has a greater stock of bombs,

> . . . there is a *very real danger* that the Russians are going to overtake us in the atomic field by putting a bigger effort into it . . . . Russia's main strength in the nuclear field is now their own excellence in science and technology and the magnitude of their effort.

In the West, the United States is obsessed with secrecy and security issues, but "without an exchange of ideas and the ensuing cross-fertilization, developments are intolerably retarded." Britain in turn is handicapped by misallocation between its large defence budget and too little for scientific education and research. Both countries have lost—though Britain more so—by the postwar breakdown in information exchange.

Two articles of 1955 and 1956 examine the Soviet scientific effort further. In the first, "A Ministry of Science?",[48] Simon emphasizes again the shift in East–West competition from nuclear weaponry to science and technology:

> The world is entering a new phase as the danger of major wars appears to be receding. Not that people have become more reasonable; but they know that in a nuclear war even the "victor" is doomed to perish. Nevertheless at the economic level the struggle will continue . . . . Here, too, the Russians are emerging as very vigorous competitors . . . . Their output of scientists and technologists is now higher than that of the whole non-Communist world taken together. Moreover their expansion of scientific education is gaining momentum all the time . . . . If things remain as they are, Russia will have achieved technological supremacy within a decade or two.

The rest of this article outlines how Britain might respond to this challenge by a massive, all-round expansion of scientific and technical education and by fostering a more scientific culture in the higher civil service.

The second paper, "The Soviet bid for technological Leadership,"[49] originally a BBC broadcast, was printed in *The Listener* of 19 January 1956. It presents a more systematic comparison of a Soviet Union that is rapidly catching up with the West in a non-military form of international rivalry, aimed at "building up a higher-level community of scientists and engineers" larger than those of its counterparts in the West. Further, "their quality is excellent." The means include providing premium salaries and social status for scientists and engineers, and allocating resources sufficient to achieve scientific supremacy. This is done through the Academy of Sciences, or its 100 research institutes, "employing over 10,000 scientists of university standard." This effort is contrasted with the "more or less stationary" output of scientists in the West collectively. In Britain, as previous papers insisted, there is a deplorable lack of science teachers, a 100 to 1 funding imbalance between defence and science education, misallocation within science between atomic energy and less popular fields, and an absence, stemming from divided responsibilities, of any overall scientific policy. Like the 1955 paper, this one ends with a call for a clearly defined scientific policy "and an agency to see it through" in response to the new challenge.

These papers of 1955 and 1956 should be understood in their immediate context. Franz was deeply impressed in August 1955 by the strong Soviet contribution that he encountered at the UN Geneva Conference on Peaceful Uses of Atomic Energy,[50] the first such meeting between Soviet and Western scientists since the 1930s. As David Holloway notes, a "large and serious" Soviet delegation presented 102 papers at this conference, on topics organized and directed by Kurchatov himself. Franz was perhaps too sanguine in generalizing from the physical sciences—where "there is almost no party line to toe"—to Soviet science as a whole. But Simon did not have the inside story. By Holloway's account, "the development of the atomic and hydrogen bombs gave the [physical] scientists a new standing in the regime." Even before Stalin's death, he notes, "the physics community was an island of intellectual autonomy in the totalitarian state."[51] With Stalin gone, scientists attempted to broaden scientific autonomy, but with mixed success. A concerted attempt in 1955 to win greater freedom in biology was angrily blocked by Khrushchev, who continued to support Lysenko.

The message of Franz's three papers on Soviet science and technology would be jarringly driven home two years later. As the two superpowers prepared to mark 1958 as International Geophysical Year, scientists on both sides were working towards the first Earth-orbiting satellites. In October 1957, President Eisenhower announced a well-publicized launching of the American Navy's Vanguard satellite for December. A New York publisher announced a book called "Vanguard! The Story of the First Man-Made Satellite." To the world's astonishment, however, a Soviet satellite called Sputnik pre–empted the American effort on 4 October and orbited successfully for the next three months. As if to emphasize the message, a larger Soviet satellite, with a dog named Laika as passenger, was launched safely on 3 November, and continued orbiting for more than five months.

The American Vanguard launch, watched by the world's press, was set for 6 December. When the engines were ignited, the apparatus rose a few feet off the pad and then fell back in an explosion of flame. If the two successive Sputniks caused serious American concern, the failures of Vanguard spelled national humiliation. After a second failure in February

1958, Vanguard orbited successfully on a third attempt in March, but this was followed by four more successive failures by September. Meanwhile, American honour was partially redeemed by an Army rocket team led by Wernher von Braun, which launched successfully the first American satellite, Explorer 1, on 31 January 1958. Laika was euthanized in space, but commemorated in the Russian monument to space heroes in Kaliningrad.

Some contributions to the *Financial Times* focused on miscellaneous topics. At least two were co-authored with colleagues. An article on "Modern drugs: the role of chemistry"[52] traced the developmental stages of pharmaceuticals from natural remedies derived from plants through chemical synthesis of their active ingredients to development of further drugs from the same chemical family. It portrays the modern industry as a "continuing battle" between new chemotherapeutic products and evolving drug-resistant strains of bacteria or insects. An article titled "Vitamins of the soil,"[53] of 17 March 1949, examines the effects of trace elements in farming soils on the growth of plants or grazing animals. Examples include a boron supplement for improving apple crops, or a cobalt deficiency—requiring an iron supplement—in sheep farming.

These miscellaneous articles could be amusing or serious. *The Neglect of Science* contains examples of both. Its concluding article, called "Some reflections on accuracy,"[54] gives hilarious examples of a human tendency to report large numbers in more detail than is needed or meaningful: among examples, an almanack that reported the 1941 population of India as 388, 977, 955.

On the serious side is an outspoken article in the same collection, "Can the world's population be fed?"[55] It tells of a tiny Pacific island inhabited only by goats, which live free from predators in a condition of Malthusian misery and starvation. Their situation is due to man, for "only man has the ability to upset natural equilibrium." The same phenomenon, Simon notes, is observable in some countries:

> These countries only too obviously resemble Goat Island. Now that the major check, disease, has been largely eliminated, man has done nothing to stop the growth of population . . . . The earth harbours at present a population of about 2,200 million and the number is increasing by about 20 million each year. To feed everyone properly, even at the present time, about twice as much food will have to be produced . . . . We are still a very long way off the ultimate limit to our population. A growing population will, it is true, have to make do with less proteins – particularly animal proteins – but it is probable that a sufficient diet could be provided for a population at least five times as great as the present world population . . . .

With global population estimated at 6.6 billion in 2007, or three times Simon's estimate at mid-twentieth century, and growing by about 75 million per year, this essay repays reading today. Though it ignores new agricultural technology and proposes a politically unreachable goal of a global population policy, it addresses a fundamental issue: that if population remains unchecked, the world will contain at some point more humans than it can adequately feed. A brief footnote reads: "Written in June 1949, but not published," which suggests that some topics were deemed beyond the pale for the editors of the *Financial Times*.

9

SECURITY LAPSES

The Fuchs affair

In 1951, as Simon was reviewing selected essays for his collection *The Neglect of Science*, one article from 1948, "Atomic energy," had already been overtaken by events. Like most Western observers, he had overestimated the time needed for a first Soviet atomic bomb. In 1948 he had predicted at least "another three years and probably not before five years," but the first test explosion in the USSR occurred in August 1949. In revising for the collection, he left the original text unchanged but drafted a long footnote to explain the discrepancy between his forecast and the event. On 8 May 1951, he sent a draft of this note to Michael Perrin for clearance and possible comment.

Perrin replied on 10 May helpful and also diplomatic. "There is no reason why you should not put in the footnote in the way in which you have drafted it," he wrote, "but, as a purely personal comment, I would make the following two suggestions." The first was to give more emphasis to the "chief reason" for the inaccurate earlier estimates, "the efficiency of Russian espionage activities." The second was "not to refer so specifically to Fuchs as the only man who gave information to the Russians." The first suggestion meant changing the order of paragraphs in Simon's draft. Perrin added, "I would myself prefer to see your footnote start something like this:"

> As is well known, the Russians have obtained the atomic bomb a few years earlier than was generally expected. The main reason for this seems to be the underestimate that was made of the effectiveness of Russian espionage activities which has now been disclosed by the Fuchs case and others. It is now clear that very important information was available to the Russians which must have saved them a considerable amount of time and effort.[1]

Perrin's emphasis is easily understood. The nuclear science community and the public at large had been thunderstruck by the arrest and trial of Klaus Fuchs early in 1950. A year later, memories of his betrayal were still fresh and painful. Recognizing Perrin's unrivalled authority on intelligence questions, Simon adopted Perrin's wording verbatim as the first paragraph of the footnote.

In two further paragraphs, he qualified his 1948 remarks concerning ideological pressure on experimental science, and also stressed that Russian scientists had ready access to Western scientific publications:

> Another reason seems to be that the picture of suppressed science given in the article is not entirely correct. Later information seems to indicate that in actual fact the majority of scientists are more or less free to do as they like if only they pay lip service to the official doctrine . . . . For instance, while all official

teaching in genetics now follows strictly the Michurin–Lysenko line, the majority of experiments are planned without any regard for it at all.

It is known that all the more important publications of Western science are very quickly translated and circulated to Russian scientists by an excellent organization. Also, contrary to the impression given by the official Soviet propaganda, Russian scientists have the highest regard for Western science and very quickly assimilate its results into their own work.[2]

As for his earlier view that an atomic project would require "technical experience in very many fields," further comment was unnecessary. The Soviet achievement spoke for itself. What Simon did not know, and probably never knew, were Stalin's decisions to resume nuclear research in 1943 and to go all out for atomic weapons in August 1945 (see Chapter 7). If measured from 1945, Franz's original estimate was accurate enough.

In the aftershock of the Soviet Union's early success with the bomb, followed soon after by the arrest and trial of Fuchs, the question of Soviet espionage became a huge public concern. For how could a scientist who had worked at five sensitive stages of the British and American projects between 1941 and 1949 so successfully avoid detection? What critical information had he disclosed to the Soviet Union? And what would be the consequences of Fuch's betrayal? Because of its widespread implications, the Fuchs case deserves careful scrutiny.

Klaus Fuchs had arrived from Germany as a young student refugee in 1933. He studied physics with Nevill Mott at Bristol, and moved to Max Born's department in Edinburgh in 1937. Highly recommended by Mott, he was soon viewed similarly by Born: ". . . we found him a very nice quiet fellow with sad eyes; he was extremely gifted, on quite a different level from all my other pupils in Edinburgh."[3] Indeed, for Born, Fuchs became a kind of top-of-scale standard against which he calibrated his other students. "X is quite clever," he wrote to Simon, "but compared with Fuchs, simply 0.00."[4] In May 1940, Fuchs was interned as an enemy alien and removed to Canada, but released and returned to Britain by December at British request. Peierls had met Fuchs, knew his written papers, and had invited him to work as an assistant on mathematical problems concerning isotope research. His skills were in desperately short supply. By May 1941 he began to work in Birmingham, with full security clearance. His help, Peierls notes, was "as efficient as I had hoped . . . . Fuchs became a lodger in our house, and he was a pleasant person to have around . . . courteous and even-tempered."[5]

From 1941 to 1947, with growing responsibilities and seniority, Fuchs faced at least six more British security reviews, and each time the security risk was judged less important than the indispensability of his work.[6] He was especially proud of his senior position at Harwell. "I suppose it could be said," he once remarked, "that I *am* Harwell."[7] As Born relates, one of Fuchs's delegated tasks was to escort visiting scientists around the laboratories:

> Cockcroft invited me to give a lecture to his staff on our work in Edinburgh. When I arrived in Harwell, he and Fuchs received me at the station and Fuchs was attached to me as a guide to show me the laboratories. I spent a very pleasant day with him, not only in the cyclotron hall and other scientific places, but also in his tiny and tidy prefabricated house where he gave me a pleasant lunch.[8]

The unmasking of Fuchs's Soviet connections would be slow and tension-ridden, stretching over five months. On the British side, the process began with an invitation by MI5 to Michael Perrin on 5 September 1949 to discuss a secret report received from the British

Embassy in Washington. The FBI had found evidence of a leak from Los Alamos in 1944; they were looking for a "non-American scientist whose sister had attended a university in the United States."[9] Perrin, familiar with the British team at Los Alamos, acknowledged that this evidence pointed towards Fuchs. But it had come from the Soviet wartime code, deciphered only in 1949, and could not be presented in a court without disclosing its source.

This was not the best week for Perrin. Five days after the visit to MI5, the Washington Embassy called him directly to ask him to proceed to the American Embassy in London. There he was given a secure connection to the Pentagon: a radioactive cloud had been detected over the Pacific off Kamchatka.[10] It had crossed the Pacific and the United States, and was heading for Europe. Could the RAF take air samples from it as it passed eastwards? Perrin acted, and the air samples so collected showed evidence of a plutonium detonation.

During these events, Franz was absent on a seven-week tour of North America, due back on 12 October. His diary, normally discreet, gives more hints than usual of events unfolding over the next several months. He had heard—and noted—the first news while visiting the Canadian reactor at Chalk River on 23 September: "News of Russian A-Bomb." On the return voyage, he disembarked from the *Queen Elizabeth* at Southampton on 12 October, and telephoned Perrin from Oxford the same evening. On 18 October, his first day in London since his return, he went to Perrin's office at Shell Mex House for an hour or longer: "disc[uss] new developments, A Bomb, etc." Two days later, the Technical Committee met at Harwell, and discussed "Russ[ian] Bomb. Am[erican] Collaboration."

The Fuchs situation had to be juxtaposed with other events of a term that was busier than usual. The on-and-off question of American–British information exchange was once again under discussion, and an American team was sent over at this point to evaluate British nuclear work. The team was headed by General Kenneth Nichols, who had been General Groves's right-hand officer in the Manhattan Project. Franz's diary records Nichols dining in Christ Church with Cherwell, Cockcroft, and Simon on 30 October, and Franz then discussing the Nichols "story" with Michael Perrin in London on 8 November. With the Fuchs case still under wraps, it was not an ideal time for an American team visit, and Nichols had a past record as "one of the most hostile critics of the British programme." The American team came, "were shown everything," but offered very little comment.[11]

On Sunday 20 November, Niels Bohr was scheduled to visit Harwell and give a lecture. Franz, by arrangement with Cockcroft, spent part of the previous evening arranging for the Bohrs to come to Sunday tea afterwards at Belbroughton Road. The diary for Sunday charts the day hour by hour. At 11, Franz left for Harwell with Maurice Pryce, Born's son-in-law and a theoretical physicist. There they met Bohr, Fuchs in his usual role as guide, and Cherwell. Bohr delivered his lecture, and Pryce and Simon were home in Oxford by 1.20 p.m. Franz had dinner and a short rest before the first guests arrived. Before they did, he drew Dor aside with an odd request: "When they arrive, I want you to show Fuchs around the garden." She had met Fuchs previously when she stayed briefly with the Peierls family in Birmingham while taking a training course with ICI. At first reluctant, the firmness of the request left her no option but to agree. At 3.00 p.m., the Halbans arrived, at 3.20 the Pryces, and then at 3.30 Professor and Mrs Bohr, the Indian physicist Homi Bhabha, and Fuchs. While Fuchs was being shown the garden, the topic of conversation indoors is noted in the diary as "Russian bomb and consequ[ences]." At 4.15, more guests arrived, and the "Tea Party" proceeded more conventionally until 6.30.

Through November, the suspense over Fuchs continued, with no end in sight. One new element was that Klaus's father, Emil Fuchs, accepted an academic post in Leipzig, and moved in mid-October from Frankfurt to the German Democratic Republic. Long active in Quaker circles, the father had made postwar visits to England in 1947 and to the United States in 1948–9. Earlier, his daughter Kristel had escaped Nazi Germany in 1936, through an arrangement made by American Quakers for her to study at Swarthmore College. As Cockcroft and Klaus Fuchs both realized, the father's appointment to Leipzig had implications for the son's security situation at Harwell. Already under suspicion since September from highly sensitive evidence, the move of the father provided sufficient grounds for an interrogation of the son by a leading investigator in MI5, William Skardon. A first meeting with Fuchs, at Harwell on 21 December, produced only strong denials of any leaks. Fuchs was increasingly conflicted, however, between deep ideological convictions and his position as a trusted senior Harwell scientist. Late in January 1950, after five weeks and several further meetings with Skardon, he agreed to make a written statement concerning disclosures to his Soviet contacts. This was done in two parts, a general one to Skardon on 27 January and a second, more technical one on 30 January to Michael Perrin.[12]

Both "confessions" are interesting, and both are reproduced in full by Fuchs's biographer, Robert Williams, but the six-page statement to Perrin is more relevant for present purposes. In a conversation from one scientist to another, Fuchs carefully recalls the information that he passed on to Soviet agents, further queries from Russia on which he had little or no information to give, and topics which the Russians showed little or no interest in pursuing. He describes separately the four locations from which he made disclosures to his Soviet contacts—Birmingham, New York, Los Alamos, and Harwell—and the types of information passed on at each stage. The list is long, revealing, and at points highly specific.[13]

In the first period, working for Peierls at Birmingham on formulae for isotope separation by gaseous diffusion, Fuchs sent mainly copies of his own calculations. He also told the agent "in general terms" of the project in the United Kingdom "and that a small pilot unit . . . was being put up at the Ministry of Supply factory, 'Valley,' in North Wales." He also mentioned "similar work" in the United States and ongoing "collaboration between the two countries." On one occasion he was "very much surprised" to be asked what he knew about "the electro-magnetic method . . . of separating the uranium isotopes," a topic on which he knew nothing at this point.

In the next period at New York, he had a wider view, and the idea of reporting only his own work was abandoned. He knew of the two large production plants for gaseous diffusion and electro-magnetic separation at "Site X" but "did not know where this was." He did know—and reported—the "general scale" and "approximate timing" of the American programme, and "some technical information about the American gaseous diffusion plant . . . . His main contribution was to pass over copies of all the reports prepared in the New York Office of the British Diffusion Mission . . . he handed over, usually, the manuscript of each report after it had been typed for duplication." In the same period, "he paid one short visit to Montreal" and learned of work there on "a small, heavy water pile," but he discounted this work as being of little interest to the Russians.

At Los Alamos after August 1944, his horizons increased exponentially. He realized the importance of plutonium, and learned of the large-scale plans to produce it. At the next meeting with his New York contact, in February 1945 in Boston, he delivered a report of several pages "summarizing the whole problem of making an atomic bomb as he saw it." It discussed the "special difficulties" of making a plutonium bomb—spontaneous fission, detonation by implosion, a smaller critical mass than for U^{235}, and other technical issues not yet settled. At a second meeting with the same agent in late June in Santa Fe, he delivered a written report that "fully described the plutonium bomb which had, by this time, been designed and was to be tested at 'Trinity'." This report contained stunning details: a sketch of the bomb and its components; the important dimensions; the solid plutonium core; the initiator; the tamper; the aluminium shell; the explosive lens system; the two explosives to be used; the expected yield; the date and site of the coming test. This period was by far the most devastating of Fuchs's betrayals.

Fuchs met his contact a third time on 19 September 1945. He said little to Perrin about "several further meetings with him in Santa Fe" or about further information disclosed. His contact, Harry Gold, was arrested in turn, however, and described this third meeting in detail in two statements to the FBI on 22 May and 10 July 1950.[14] By Gold's account, Fuchs was relatively loquacious on this occasion. He spoke of the "tremendous wonderment" of the Trinity explosion at Alamogordo; of the "terrible destruction" of the weapon in Japan; of a "somewhat strained" relationship at Los Alamos between the "British mission and the United States," and tighter security measures "between the two groups"; of his expectations to return to England and continue atomic work there; of efforts by British intelligence to locate his father in Switzerland or in Kiel; of his father's possible visit to England, and the risk that his father might talk of his son's Communist Party activities in 1932 or 1933; and of his deep concern for his father's health and welfare. "I could not give him very much advice," Gold says.

Gold's statements are detailed and interesting, in part because they identify information disclosed orally, of which Gold "made good mental notes" and wrote down "at the first opportunity," information that Fuchs tended to omit or forget. In Gold's account, the September meeting in Santa Fe was the last time he saw Fuchs, though he reports unsuccessful attempts to reach Fuchs through his sister in Cambridge towards Christmas, and again early in 1946. Before parting, they worked out a plan for Fuchs to connect with an agent in London after his return to Britain. They parted, Gold reports, as "two firm friends," a sentiment that he claims Fuchs acknowledged at their last meeting. And this too is interesting, for, in Harry Gold, Fuchs encountered one person—perhaps unique among all his acquaintances—with whom he could put aside his self-declared "controlled schizophrenia"[15] and talk of the double goal that they shared, the advancement of science and Soviet communism.

On occasion, Gold appears as more than a simple carrier of documents. Trained as a laboratory technician and chemist, he had been active in Soviet industrial espionage since the mid-1930s through his own work at Pennsylvania Sugar Company in Philadelphia. Towards the end of his lengthy statement to the FBI in July 1950, he recalls a significant moment from his second meeting with Fuchs in New York in March 1944. In his own words:

> I would like to add that Klaus knew of only two methods for the separation of the isotope from uranium, that is methods as were being pursued here in the United States, and that these methods were. (1) The gaseous diffusion process. (2) The electromagnetic method.
>
> I recall that this last information concerning the methods for the separation of isotopes was given to me on the occasion of our second meeting, when we were walking along 1st Avenue in Manhattan. I also recall that at that time I had mentioned to Klaus the possibility of the use of thermal diffusion as a means of separating isotopes, but that Klaus had brushed this aside.[16]

When the time for confessions arrived, both Fuchs and Gold made extra efforts to ensure that their accounts were correct and complete.

Fuchs's account to Perrin of the Harwell years is a jumble, sharply at variance with his planned, structured reports from New York and Los Alamos. He recounts mainly a long series of requests—some general and some very specific—from the Soviet side. Some of these requests he fulfilled, others he answered partially or with outdated information, or in some areas he withheld information. At this time, he tells Perrin, "he was having increasing doubts on the wisdom of passing information to the Russians." Even so, the disclosures he admitted for this period were serious enough: data on the ill-fated Windscale piles; the plan to build a low-separation plant; more complete details from Los Alamos on the plutonium bomb; questions about the Bikini test (which Fuchs had helped to plan in 1946); and a request in 1947 for "any information he could give about 'the tritium bomb'." Even though "he was very surprised to have the question put in these particular terms," he described what he could remember from Los Alamos "on the design and method of operation of a super bomb, mentioning, in particular, the combination fission bomb, the tritium initiating reaction and the final deuterium one." Fuchs also gave Perrin a detailed list of topics and problem areas about which the Russians had *not* asked him. From these large gaps, and from the specificity of other requests, he was convinced that the Russians had other sources for the areas they did not ask about.

Perrin ended his notes on the interview with a few comments of his own:

> Finally, I discussed with Fuchs the nature of the 'atomic explosion' that had taken place in Russia in the Autumn of 1949. He told me that he would have expected this to be due to a plutonium bomb in the light of all the information he had passed to the Russians . . . . He said that he was, however, extremely surprised that the Russian explosion had taken place so soon . . . . I formed the impression that, throughout the interview, Fuchs was genuinely trying to remember and report all the information that he had given to the Russian agents . . . and that he was not withholding anything. He seemed, on the contrary, to be trying his best to help me to evaluate the present position of atomic energy works in Russia in the light of the information that he had, and had not, passed to them.[17]

On 2 and 3 February 1950, Franz attended meetings at Shell Mex House in London of the Technical Committee and the Isotope Committee respectively. On the 3rd, the diary records his arrival at 10, Isotope Meeting at 10.45, and then "after Mtg. Perrin tells me about Fuchs." Four days after the crucial interview with Fuchs, Perrin's notes must have been vividly fresh as they talked. Over the next fortnight Franz's diary lists about a dozen more discussions of the Fuchs affair, mainly with colleagues close to the scene, including Cherwell (four separate occasions), Cockcroft, Keeley, Halban, Skinner, and Peierls.

At his trial in London on 1 March 1950, Fuchs was sentenced to 14 years imprisonment for violations of the Official Secrets Act, the usual procedure in espionage cases for the USSR. He was released after nine years and allowed to return to East Germany, where he became Deputy Director of the Rossendorf Institute for Nuclear Research near Dresden, and in due course an Academician. His American contact, Harry Gold, arrested in May and tried in a Philadelphia court, was sentenced to 30 years, but released after 15. His first British contact, whom he knew only as "the girl from Banbury,"[18] was more fortunate. This woman, known as Sonja by her Soviet controllers, alarmed by news of Fuchs's arrest and sensational press accounts of his meetings with a "foreign woman with black hair in Banbury," prepared for a hasty departure from England, and flew with her two children from London to Berlin a day or two before Fuchs's trial.[19] Though known to MI5 since 1947 and probably earlier, she travelled without hindrance.

Sonja, born as Ursula Kuczynzki in Berlin in 1907, was a trained Soviet agent with a decade of experience in China, Poland, and Switzerland before her arrival in Britain early in 1941. She came on a British passport, obtained after a second marriage, to Len Beurton, a naturalized British citizen and also a Soviet agent. Her assignment from GRU—Soviet army intelligence—was to recruit other Soviet agents and transmit information to Moscow via short-wave radio or through the Soviet Embassy in London. In Britain, she lived at first in Oxford with her father, the labour economist Robert Kuczynski. In 1942 Beurton joined her, and they rented a cottage in Summertown. While her father and her brother Jurgen—also an economist—were openly Communist and prominent in pro-Soviet causes, Sonja and Len worked under cover. Sonja's meetings with Fuchs took place in the countryside around Banbury, a convenient point midway between North Oxford and Fuchs's work in Birmingham.

Following her flight to Berlin in 1950, Sonja declined further Soviet assignments. After a few years in a government information office, she became a writer, using the pseudonym Ruth Werner. Among her books was a best-selling account of her 20-year career in espionage, *Sonjas Rapport*. An English translation, published as *Sonya's Report* in 1991, contains a somewhat rose-tinted portrait of Klaus Fuchs and a detailed modus operandi of their meetings.

> When I met Klaus for the first time we went for a walk arm-in-arm, according to the old-established principle of illicit meetings. We discussed many things that had nothing to do with the main objective. It was pleasant just to have a conversation with so sensitive and intelligent a comrade and scientist. I noticed that very first time how calm, thoughtful, tactful and cultured he was. We spoke of books, films, and current affairs.
>
> At our subsequent meetings he had questions for Centre which gave me a vague hint as to the nature of his work. I coded them and passed them on over my transmitter, and decoded Centre's replies for him . . . .
>
> Once, it must have been in 1943 when I still lived in Oxford, in Summertown, Klaus gave me a thick book of blueprints, more than a hundred pages long, asking me to forward it quickly. To get in touch with my Soviet contact outside the regular arrangement, I had to travel to London and, at a certain time in a certain place, drop a small piece of chalk and tread on it. [This triggered a meeting with "Sergei" two days later, in the countryside.] . . .
>
> Klaus and I never spent more than half an hour together when we met. Two minutes would have been enough but, apart from the pleasure of the meeting, it would arouse less suspicion if we took a little walk rather than parting immediately. Nobody who did not live in such isolation can guess how precious these meetings with another German comrade were. In this respect he was even worse off than I . . . .
>
> Centre twice acknowledged his messages with 'important' and 'very valuable' . . . .
>
> About two years after we had started working together, Centre asked me to arrange a meeting for Klaus in New York. I had lived there for a few months in 1928, and racked my brains to remember landmarks in the city. I recalled two places, the Prosnit bookshop, uptown, where I had worked, and the Henry Street Settlement at 256 Henry Street . . . . I chose the settlement and found methods of identification and code names. I believe they were used . . . .[20]

Her plan worked. When Harry Gold made his supplementary statement to the FBI on 22 May 1950, he confirmed that his first meeting with Fuchs "took place in late January or very early February 1944 . . . at the Henry Street Settlement on the east side of New York."[21]

The view from the Soviet side

The work of David Holloway on the Soviet bomb makes it possible to round out the full circle of information disclosure in the Fuchs affair by tracing its reception stage by stage on the Soviet side. Igor Kurchatov, the newly appointed scientific director of nuclear research, was a privileged scientist. Having made a good first impression on Molotov, he was given access after Stalingrad to Soviet nuclear intelligence documents, apparently the only academic scientist allowed to see this material. As Molotov later recalled, "Kurchatov sat in my office in the Kremlin for several days studying those materials."[22] Profoundly impressed by these materials, he then wrote two extensive memoranda to Pervukhin, his political superior, describing what he had seen and devising guidelines from it for the future Soviet programme. These memoranda, dated 7 March and 22 March 1943, focused mainly on reports from Britain, for Fuchs would not be in New York until December of that year. When Molotov later asked about the materials, Kurchatov could reply without exaggeration: "Wonderful materials, they fill in just what we are lacking."[23]

In the memorandum of 7 March, Kurchatov wrote of the "huge, inestimable significance for our state and science" of the intelligence material:

> On one hand, the material has shown the seriousness and intensity of research in Britain on the uranium problem; on the other, it has made it possible to obtain very important guidelines for our research, to bypass many very labor-intensive phases of working out the problem and to learn about new scientific and technical ways of solving it.[24]

Systematically, his text grouped what he had read around three problem areas: isotope separation, the chain reaction, and the physics of the fission process. In the first area, he was impressed by the British priority for gaseous diffusion, and by the detailed analysis in the reports of the "process proposed by Simon . . . . It would be possible to reconstruct completely the plans for the machine and the factory on the basis of the material that had been received." He also lists "technical details about Simon's machine that it would be important to know." These included clearances between moving parts, lubricants, and membrane material. Evidently, Fuchs's report on plans for "Valley" had been much appreciated.

The memorandum applied similar analysis to the other two areas. The Halban–Kowarski experiments in Cambridge had suggested a possible chain reaction from uranium and heavy water, but Soviet physicists could not test this further for want of sufficient heavy water. Further results from Halban and Kowarski, and even their present whereabouts—were they in the United States?—were therefore top questions for agents overseas to pursue. Another idea found in the foreign material was that a reactor using uranium 238 might also produce a fissionable isotope of a transuranic element. Kurchatov ended this first memorandum by admitting his changed views on many questions, and proposing three new directions for research: isotope separation by gaseous diffusion; a chain reaction using uranium and heavy water; and closer study of the properties of element 94. In the absence of both raw materials and equipment, this was necessarily a programme for the future, but not so distant a future as some of his older colleagues had hitherto imagined.

Two weeks later, Kurchatov wrote to Pervukhin again. Holloway calls this memorandum of 22 March 1943 a "crucial document" in the Soviet path to the bomb. Kurchatov reports that he has read further on the transuranic elements—93 and 94—and has "attentively looked through" McMillan's work on them in the Berkeley cyclotron, as published in the American

Physical Review in July 1940. He now sees "a *new* direction in the solution to the whole uranium problem." If element 94 has properties like uranium 235, it might provide material for a bomb and thus bypass altogether the difficult problem of separating uranium isotopes. Since the Moscow cyclotron was not expected to be ready until mid-1944, Kurchatov set out four key technical questions about elements 93 and 94 to which answers should be sought, and added a list of American laboratories where such work might be going on. These questions were then forwarded to agents abroad by the Directorate for State Security.

On this basis, the inflow of foreign information continued. Systematic and organized, the espionage project for the bomb had its own code name: Enormoz. In July 1943 Kurchatov wrote another memorandum,[25] this time on 286 reports on various topics concerning the Manhattan Project. These materials, he noted, were not very detailed, and less complete than earlier material from Britain. Two areas of special importance for seeking more technical detail were the uranium–graphite pile and further research on element 94 by Seaborg and Segrè at Berkeley. The July memorandum also mentions one more topic not yet addressed in the Soviet Union but deserving serious attention: the heavy water–uranium pile. Since Fuchs was still in Britain at this point, his role in these American reports is unknown, but Kurchatov's priorities for further material to be sought out would be on the agenda for his first meetings in New York with Harry Gold, early in 1944.

From 1943 onward, Kurchatov began to build a team for bomb design. Headed by a Leningrad colleague, Julii Khariton, this group lacked at first even basic data on explosive fissile materials, but their work was transformed after Fuchs was sent to Los Alamos in August 1944. Research that summer had revealed the tendency to predetonation in plutonium, and the need for an alternative method for detonating a plutonium bomb. Fuchs had joined the Los Alamos group working on implosion, and reported on this work to Harry Gold at their meeting in Boston in February 1945. On 7 April Kurchatov, who knew something about implosion from an earlier source, wrote of the "great value" of the newer material—from Fuchs—on implosion, and the "exceptionally important" data on spontaneous fission. As Holloway concludes, "Kurchatov's memoranda confirm Fuchs's importance as a spy."[26]

Fuchs made his fullest and most damaging disclosures when he met Harry Gold at Santa Fe on a Saturday late in June 1945 and again in the same city on 19 September.[27] Gold reports that, on the first occasion, Fuchs "gave me a sizable packet of information," and, on the second, he was given another "packet of information relating to atomic energy." By a fixed procedure, handovers came just at the moment of parting. For these two packets there is no record of a memorandum from Kurchatov, but on 18 October Beria received from the head of the government security service "a report describing in detail the design of the plutonium bomb. This report, which was evidently based upon the information that Fuchs had supplied, described the components of the bomb, the materials from which they were made, and their dimensions."[28] Julii Khariton, interviewed in 1992, remembered this report as "detailed enough to enable a competent engineer to produce a blueprint for the bomb."

Kurchatov and Khariton studied the report and made a rational decision. They were under orders from Stalin to produce an atomic bomb in the shortest possible time. This was a proven design, twice tested at Trinity and Nagasaki. It seemed by all odds the quickest, surest, and safest route to a nuclear bomb. In the circumstances, there could be no stigma in copying the Manhattan design, and industrial espionage as a route to faster technological development was an established strategy of Leninism. The plutonium bomb, on the

Manhattan model, did prove to be the fastest route to the Soviet atomic bomb. Its test detonation in August 1949 astonished Western scientists, but this was only the first signal from a fast-growing, large-scale nuclear industry that would produce more powerful fission bombs by 1951, the hydrogen bomb by 1953, and the Obninsk power station by 1954.

Whatever the consequences may have been for wider East–West relations, the Fuchs affair also had tangible consequences closer to home. When news of Fuchs's arrest and betrayal reached Washington, any faint hope that remained of reviving Anglo-American cooperation in nuclear weaponry was "killed stone dead."[29] Personal friendships, however, survived. When Harold Urey wrote to Franz on 10 May 1950 of a planned visit to England as part of a Forthcoming global circuit, he touched on the American reaction:

> We are awfully sorry about the Fuchs case, for it has had a most harmful effect on all public relations in the United States. The psychology of people of that kind is unbelievable. I more or less have come to the conclusion that anybody who has ever played with the idea of Communism is a poor security risk.[30]

Franz replied five days later, inviting the Ureys to Oxford, and added:

> Yes, the Fuchs case was an awful affair from many points of view, it also of course had a very bad effect on the morale at Harwell. Fuchs was a very intimate friend of Skinner and Peierls, in both of whose houses he had stayed for many years and nobody ever had the slightest suspicion. He never mentioned politics at all and people who knew about his Communist past were absolutely convinced that he had long foresworn it.[31]

The chain of consequences extended further. Under closer security checks, Bruno Pontecorvo, a leading theoretical physicist who had worked since 1941 at the Montreal laboratory, Chalk River, and Harwell, was suddenly deemed a security risk because of Communist connections of his family in Italy. Pressed to leave Harwell, he reluctantly accepted an alternative, non–sensitive appointment at the University of Liverpool. Over the summer, he chose instead to work in the Soviet Union, travelling with his wife and children from a holiday in Rome via Sweden and Finland to the USSR in early September 1950. With no credible evidence of improper disclosures before his departure, Pontecorvo had a long and distinguished career in nuclear physics at the Joint Institute for Nuclear Research in Dubna, near Moscow, dying in Dubna in 1993.

After the scientists, the security focus shifted to public servants. The first thunderbolt struck in 1951, when two middle-level diplomats from the Foreign Office, Donald Maclean and Guy Burgess, boarded a cross-channel ferry to France on the weekend of 25–27 May and disappeared, to surface a few years later in Moscow. Both had been posted to the British Embassy in Washington, Maclean from May 1944 to September 1948, Burgess from August 1950 for nine months until he was recalled in disgrace, for "dissolute and insulting behaviour,"[32] in May 1951. Maclean, as First Secretary, became Joint Secretary in 1947 to the ultra-sensitive Combined Policy Committee for the nuclear projects. At the time of his flight, he was in London as head of the American Department of the Foreign Office.

Maclean and Burgess were forewarned and aided in their flight by a third member of the so-called "Cambridge Five," Kim Philby. Philby, a Soviet agent since 1934, had been recruited—through Burgess—into MI6 in 1940, and appointed its representative in Washington in September 1949. Two weeks after the defections, Philby was recalled to London in June 1951 and interrogated, but inconclusively. After more than a decade of denials and incomplete evidence, he too defected to Moscow from Beirut in January 1963 when confronted with irrefutable evidence of his work for the Soviets.

The three prime defectors, who lived out their lives and died in Moscow, are not the entire story. One other member of the Cambridge Five—the so-called "Fifth Man" in some accounts—played a pivotal role in unauthorized disclosures a full decade before the defection of Maclean and Burgess in 1951. Klaus Fuchs believed that the USSR had other nuclear informants than himself, and this is amply confirmed by Soviet archives. On 25 September 1941, the intelligence resident at the Soviet Embassy in London informed his government of a secret meeting held nine days earlier.[33] The participants, he reported, had discussed the possibility of a uranium bomb and a time estimate for making it; a fusing mechanism for such a weapon; a contract let to Metropolitan Vickers for "a 20-stage machine"; a contract to ICI for uranium hexafluoride. This discussion can be identified easily enough as a meeting of the scientific panel set up to assess the Maud Committee reports. Eight days later, the embassy resident described this panel's report to the War Cabinet; he had received a copy of this report himself.

The source of these reports, as Holloway shows, was "almost certainly John Cairncross, the 'Fifth Man' of the 'Cambridge Five'," who became a Soviet agent while still an undergraduate. He was recruited to the Foreign Office in 1936, but deemed unsuitable for a foreign posting because of his prickly temperament and broad Scottish accent. In 1941 he was private secretary to Lord Hankey, who at this time was chair of the Cabinet's Scientific Advisory Committee and also of the panel reviewing the Maud Reports. As Hankey's private secretary, Cairncross had access to almost every Foreign Office document. When he moved in March 1942 to the code-breaking centre at Bletchley Park, his opportunities for disclosures continued. Overall, Soviet intelligence archives record that 5,832 documents were sent to the USSR by Cairncross between 1941 and 1945.[34] His work for the Soviets was first uncovered in 1951, through documents found in Burgess's flat after the latter's disappearance. After an investigation, Cairncross was required to resign from the civil service in 1952, but he was allowed to leave Britain and work on the Continent. He faced further interrogations after Philby's defection, but on each occasion he defended himself with such vigor against growing evidence against him that formal charges of espionage were never laid.

Civil servants could pass on early and continuing policy decisions, but participant scientists such as Fuchs could provide detailed plans for factories and weapons. Both categories were systematically sought out by Soviet authorities as a means to shorten the interval to the first Soviet bombs.

The years from June 1941 to 1945 were unusual. Robert Williams's study of Fuchs describes a time of confusion, suspected double agents, and crossings of normal barriers as intelligence agencies of three countries made common cause in a deadly struggle against Hitler. The Kuczynskis, brother and sister, both German Communists, served the same cause in different ways: Ursula as an undercover agent of the GRU, rising through the military ranks to become a colonel in the Red Army, while Jurgen worked in London for the American OSS (Office of Strategic Services) and as an analyst for the US Strategic Bombing Survey, holding the rank of colonel in the US Army Air Force.[35]

10

GERMANY IN THE BALANCE

Signals of survival

As soon as hostilities in Europe ceased, Franz began to renew contacts with colleagues and friends unreachable during the war years. His first enquiries sought answers to very elemental questions: who had survived the war, and with what consequences for the individuals addressed, their families, and their colleagues. Beyond these basics of physical survival were two further considerations. Franz was now in a position to offer help of various kinds to colleagues and friends, from short-term emergency aid to plans for expanded laboratory facilities in Oxford, but he also had a corresponding desire for information from trusted informants on the Continent. His experience in post-1919 Germany and in exile had left him acutely aware of how much was at stake in the government of Germany under Allied occupation. His correspondence of these years recognizes this rebuilding of contacts as a process of reciprocal benefits, and an essential part of a continuing struggle against the Nazi legacy.

A high priority was to contact colleagues in countries freed from German occupation. On 11 June 1945, he wrote to Professor Michels in Amsterdam:

> My dear Michels,
>
> I just see that one can write to Holland again and hasten to get in touch with you, and find out how you have come through these hard times. How are you all – your family and your colleagues . . . .
>
> I wonder whether we can do anything for you or whether you have any urgent needs which we can fulfil. We would only be too glad to do this.[1]

Among specifics, he asks about possibilities of a visit, or "any plans for sending children to this country." He describes the planned switch-over of the Clarendon to low-temperature work. At the new Clarendon, he will soon be in a position to reciprocate for his earlier research visits to Amsterdam.

Over the next few years, similar contacts would follow with many others, some initiated by Franz himself, others arriving unexpectedly from former colleagues and friends who had read or heard of his wartime achievements. A representative example was a closely typed, undated three-page account of the war years received in December 1946 from Clara von Simson, a former student and research collaborator from Berlin days. Before talking of herself, she begins, she must write something "from the soul."[2] She is one of the last people to have had regular contact with Miss Gerstel's mother until she was taken away to Theresienstadt, and "I have seldom met so splendid a person as this old woman who remained so young inwardly." She asks if the daughter now living in Oxford knows of her mother's fate.[2a]

Simson then writes of her own wartime experiences: of a brother, his wife, and one daughter killed by a bomb; of taking over the upbringing of his two surviving children; of another nephew wounded in action; of a complex series of moves around Berlin and as far away as Lübeck; and of housing crowded by bombed-out and displaced families in the last phase of the war. Now back in Berlin, she has the two children with her and a job for herself at the technical university. She describes daily life there, the Wednesday physics colloquium reinstated in "our dear old auditorium in Bunsenstrasse," coming choral events, the children's schooling choices. She asks also for news about the former colleagues in exile and sends warm greetings generally. Do the Simons, she asks, still live in the *Rosenhaus* in Belbroughton Road?

Franz's reply was delayed by illness that had started just as Simson's letter arrived. On 23 December he expressed the family's delight "to have at last some direct news from you."[3] He wrote a two-page "first instalment" of basic information, with promise of more to follow. The house "is still the same house 'of roses,'" though after seven wartime years with no repairs possible it is "looking a bit shabby." He summarizes the past several years for his family: the escape from Germany of all "near relatives," but losses of uncles, aunts, and "many more of the younger generation"; a four-year separation when the family was evacuated to Canada; the loss of his sister Mimi's house in the London blitz, and the move of her family to Belbroughton Road; the deaths earlier in 1946 of his sister Ebeth and his mother; and studies of Kay in Oxford and Dor in Edinburgh.

He then gives brief reports about "your other friends": Kurti (just married, "a great surprise"); Mendelssohns (have five children); Ruhemanns (he is working in industry, separated from his wife); Wohls (in the USA since 1941); Heinz London (remarried); "Old" London (at Duke University); Gerstel (lives at 13 Park Town, Oxford, and knows already of her mother's fate); Paneth (at University of Durham); James Franck (married "some months ago"); Otto Stern (retired and in Berkeley—"He just came through here on his way to collect the Nobel Prize..."). For one colleague, Marckwald, who went to Brazil, Franz has no recent information.

Franz then outlines plans for natural science in Oxford: "a beautiful big Laboratory," modern and well equipped, coupled with strength in theoretical physics and a long Oxford tradition of excellence in chemistry. For low-temperature work, a new high-volume hydrogen liquefier "works beautifully" and liquid helium is expected to be "flowing in January." He sends cordial best wishes, but adds a little request: "Please remember me to those of my former colleagues to which one wants to be remembered. You will be in a position to make the proper choice!"

Another early contact was with Clemens Schaefer, a long-time colleague who had taught theoretical physics at the University of Breslau from 1926 until the collapse in 1945. Schaefer signalled his survival on 5 November 1946,[4] in response to a query from Franz dated 25 October. He had lost everything in the expulsion from Breslau, but had been able to teach again at the University of Cologne in his native city, where life was also "very difficult." Of his four sons, one had been killed and two others reported missing in the USSR 18 months previously. Simon's reply, on 2 January 1947, apologized for the delay (he had been ill at home "for a full month"), mentioned his own lost relatives, and raised the issue of renewed scientific relations.[5] Schaefer replied with a long letter on 10 February emphasizing the obstacles. Among these, he reports that almost the entire Breslau medical faculty had belonged to the SS, and "many unseemly and unmentionable things"[6] had been done there.

Franz replied on 7 March, suggesting a different approach. It would be better to know which colleagues have "behaved correctly in every respect."[7] Schaefer's reply, dated 10 April, contains a list of 17 physicists whose behaviour has been "beyond reproach (*tadellos*)."[8] Somewhat unexpectedly, the list contains five of the Farm Hall detainees. While Hahn and von Laue were beyond doubt, it also included Heisenberg, Gerlach, and Wirtz. While some on Schaefer's list might raise questions, Franz expanded the idea by seeking similar reports from other scientists.

A year later, Schaefer wrote on 16 March 1948 to report that after three years he had received official notice of his missing sons' deaths. Both had been killed in the closing days of the war, in March 1945. When Franz replied on 12 April, he suggested that Schaefer might consider visiting Britain, as Bonhoeffer and Heisenberg had already done. "We want your help and advice on the situation in Germany,"[9] he writes. Schaefer did eventually come to Oxford, but only after several postponements, in June 1953, by which date he was 75 years old.

In the years after 1945, these contacts would multiply. Some wrote to Simon with anxious—even urgent—requests. In a letter dated 31 December 1946, a former student from Breslau days, Hugo Gutsche, wrote from Sweden to report that, as a non-Jewish, German refugee, he was facing deportation from that country if he could not provide proof of his anti-Nazi beliefs, "and I am afraid you are the only person alive that can do it."[10] Franz wrote a strong letter of support, mentioning the Breslau years, post-1933 meetings in Paris and Oxford, and attempts to find work for Gutsche outside Germany. He ends with a firm declaration: "I can vouch absolutely for Dr. Gutsche's character and in particular confirm that he has always been a convinced and uncompromising enemy of the Nazis."[11]

A more conventional request for certification arose when Franz received a long letter from a distinguished colleague of long standing in low-temperature research, Walther Meissner. Delayed in delivery, it is dated from Munich on 15 December 1947, but was not received in Oxford until early March 1948. In a long letter, Meissner describes renewed de-nazification proceedings and requests a testimonial from Simon for his own forthcoming re-hearing. He explains how he managed during the war to continue some low-temperature research on the pretence of its expected military value, thereby saving "a dozen men" from service at the front:

> I am writing you this in order to conceal not a thing. I can almost say I did sabotage by my works. But perhaps it won't be of much use to me in the wind that is blowing now (perhaps, but I do hope for the contrary). Anyhow in my conscience I am absolutely pure. I was an enraged opponent of Nazism from its very beginning and in my words and my doings an opponent of Antisemitism.[12]

He also explains that he had helped to rescue two detained scientists from the camps, one of them Franz Pollitzer, freed from the Gestapo in 1938. As a chemist and specialist on gases for the Linde refrigeration firm, Pollitzer had worked closely with Simon since the early 1920s.

Franz reacted quickly on 10 March, sending his letter through an American diplomatic channel to avoid further delays. His letter testifies to Meissner's courageous, unwavering anti-Nazi behaviour that Simon had known in the prewar years, and to reports of his constancy during the war years:

> I know personally very well that not only have you never been a nazi but that you resisted them actively as much as it was within your power.... You have always kept faith with your friends whether they be aryan or non-aryan. When I saw Pollitzer in Paris in 1939 he told me that he owed his life largely to your intervention....[13]

Franz was sufficiently disturbed by the Meissner case and the delayed letter that he wrote to von Laue about it the next day, asking if von Laue "could let me know more concrete data"[14] about the affair. It is clear nonetheless that Meissner's career continued. He had become President of the Bavarian Academy of Sciences in 1946, and also served as the first President of the Bavarian section of the German Physical Society, now decentralized by occupation orders.

One other request from this period is backed by no written document. It survives only in the memories of Kay and Dor. Freda, the housekeeper and children's nanny who wore her hair in a long, circular tress over her head, and who had resigned abruptly in 1933, contacted Simon after the war. She requested a letter of reference to get to Argentina with her two sons. She was asked to show that she had had no Nazi attachment, but investigation showed otherwise, and that she had been one of the highest ranking female party members in Breslau.

Now and then came appeals for help on a larger scale, requiring collective effort. One day a letter addressed to Simon arrived from Paul Harteck in Hamburg.[15] Dated 29 November 1946, it concerned Karl-Friedrich Bonhoeffer. Bonhoeffer and Simon, both students of Nernst, had been near contemporaries in student days in Berlin, and both achieved the rank of *extraordinarius* in 1927. Bonhoeffer had worked with Harteck, and they became co-discoverers of orthohydrogen and parahydrogen in 1929.

The letter announced a health crisis. Bonhoeffer was suffering from severe angina pectoris, a combined result of family war losses and current overwork, to the point that only a period of rest and relaxation, removed from current pressures and anxieties, could offer any hope of recovery. The family's war losses were a matter of public record: his two brothers and two brothers-in-law had been executed by the Nazis. The surviving brother was left to provide for his own four children aged 7–13 and the "numerous children of his murdered relatives." His current work pressures included a newly established deanship in Leipzig University and a call to the Chair in Physical Chemistry at Berlin University. Harteck's letter to foreign colleagues was primarily about the possibilities for a rest period outside of Germany, but also about food parcels and family support to ease the strain and anxiety.

On 11 December, Franz responded quickly and positively, though he was ill at home himself for most of the month: "I will try to do what I can for him."[16] He approves the suggestion of "a few months in another country – say Switzerland" and asks who else Harteck has already approached. On 20 December he wrote to Otto Stern, then visiting Zurich after receiving his Nobel Prize in Stockholm:

> ... I think one should do everything one can to help him. The trouble is that Bonhoeffer lives in the Russian Zone. Perhaps an invitation to Switzerland might be the least objectionable to the Russians and I wondered whether you could discuss the matter with the Swiss scientists. I have also written to Urey... asking him to help in this matter.[17]

Urey, among others, made a contribution in cash.

Harteck wrote to Simon again on 23 January 1947 to thank him for his help and to report on further developments. Michael Polanyi was inviting the Bonhoeffers to visit Manchester, and Hugh Taylor, at Princeton, was trying to set up "an extended stay in Sweden by Bonhoeffer and family,"[18] with backing from the Rockefeller Foundation.

How this health crisis was resolved is not spelled out in this correspondence, but Bonhoeffer did make a visit to England. Franz's calendar and diary for 1947 show Bonhoeffer in Oxford for the week of 20–25 May and again briefly on 10–11 June. The visit is mentioned in a letter of 24 June from Franz to another former Berlin colleague: "Bonhoeffer's visit was most interesting to us and he definitely enjoyed it – in addition he was putting on weight to the tune of 1 pound per day!!"[19] Bonhoeffer took up the Berlin Chair in the same year, moved to Göttingen in 1949, and by the early 1950s was in regular correspondence with Franz. The remarkable thing about this case is that scientists in several countries—including Harteck, Simon, and Urey—cooperated without hesitation to assist a respected colleague.

The case of Wilhelm Jost, another friend and colleague of the Berlin years, had similarities to Bonhoeffer's, but is also more complex. Jost sent a "survival" letter to Oxford in September 1946. He now has a Chair in Marburg, and he also sends regards to Miss Gerstel. The exchange of letters continued, but in early March of 1947 Franz received a letter reporting that Jost had tuberculosis, and needed sanatorium care. Franz replied on 15 March: he was "profoundly shocked" by news of the illness and promised to help. A new round of letters followed, as in the Bonhoeffer case, but this time Franz was the principal coordinator. By 22 April, he assured Jost that help was on the way: both food parcels and funds for a stay in a Swiss sanatorium. His frequent letters of support to Jost continued, and in these we can trace Franz's developing picture of the wider German situation. When Jost sent letters of thanks in May and early June, Simon declined credit for the results:

> By the way, you now attribute too much to my influence. It is true I have started some of the actions, but as we cannot send money from England the funds have been collected in America and Switzerland. Urey, Kuhn and Halban are responsible for them. Of course it was quite right to acknowledge the parcels directly! The Paneth who handed on our parcels is the daughter of the chemist – she is working with the Quakers in Germany.[20]

The direct delivery of parcels was proving necessary to ensure arrival. A similar food parcel sent from Britain to Schaefer had been looted in transit and delivered empty in Germany.

The effort to help Jost ran into serious delays. Though he had an invitation to a Swiss sanatorium by June 1947, his exit permit was issued and then blocked in the bureaucracy because scientists of possible interest to the American authorities were not permitted to leave the American zone. Despite lobbying from American scientists, approval from Washington was delayed until early 1948. In the meantime, Jost lived in a sanatorium in the Black Forest. By January 1948, still unable to enter Switzerland, he wrote to Franz in bitterness and despair:

> I was a fool that I did not join the Nazi Party. If I had, I would not have had to suffer during the war, would have been properly fed and not contracted tuberculosis, from which I am now suffering. In addition I would have been de-nazified and would certainly not have been treated worse by the Allies than I am now.[21]

On 30 January, however, Franz writes to express delight at the arrival of the exit permit and hope for Jost's rapid recovery in Davos.

Jost's troubles were not ended by a year in the Swiss sanatorium. Though his health was restored, he faced a difficult situation in returning to Marburg. After he left for Davos, the ministry reassigned "a considerable part" of his team's laboratory space to a man called Schenck, an elderly ex-Nazi metallurgist, who was barred from teaching students. With Jost's approval, Simon wrote on his colleague's behalf to the Education Minister of Hesse, but received a reply that he felt showed ministry bias in favour of Schenck. A more tangible consequence of Franz's intervention was that Jost was sent a severe ministerial reprimand for "discussing these matters with a foreigner," a letter so threatening that Jost's wife did not dare to forward it for fear of jeopardizing her husband's recovery. From this point, the efforts of Franz and his American colleagues were focused on finding a suitable position elsewhere for Jost, most probably in the United States. In the short run, however, he faced a return to Marburg, described by his wife as a place "full of hate, intrigues, and no possibility to do something positive."[22]

Still acutely concerned for Jost, Franz wrote on 22 February 1949 to the education branch of the US military government, outlining the case in full and describing the social pressures and social ostracism being felt by "every non-nationalist and people like Jost" in a setting where "the nationalist movement . . . has grown by leaps and bounds." Could the authorities, Franz asks, "take an interest in this matter?"[23] On 25 March, he received a timely and positive response from a top education official in the military government of Hesse. The official had reviewed the relevant ministry files. He discounted Simon's perception of pro-Nazi bias, but undertook to ask the US Resident University Officer in Marburg to contact Jost after his return and "to take any steps in his power to protect Professor Jost from recriminations." Further, if a position elsewhere were found for Jost, "I can assure you of the wholehearted cooperation of this office."[24]

After years of tribulation, Jost's career took a turn for the better. In 1951 he moved to the Technische Hochschule in Darmstadt, and two years later he was invited to a prestigious Chair in Physical Chemistry at Göttingen. With it went the directorship of the university's physicochemical institute, a position he would hold for 18 years until his retirement in 1971. During the transition, the hostile climate at Marburg was offset by at least two visits to England. Jost appears in Oxford in Franz's calendar for a few days in January 1950 and again in June 1951. Through bad times and good, their correspondence continued, and Jost became one of Franz's most trusted informants on postwar Germany. Their correspondence occupies three complete file folders in the Simon Papers.

The family connections with Jost and Schaefer would continue even after Franz's death. When academic memorial sessions for Simon were scheduled in divided Berlin in 1959 and again in 1960, Lotte asked for advice from these two trusted friends on whether or not she should attend.

Although the Jost case became for a time a medical emergency, its more enduring issue was protection of an outspoken anti-Nazi in a setting where such views were rare. For Franz, Jost was not the first such case. In April 1946 he had become active in a similar case at the Brunswick Technical University in the British zone. A former professor there, R.S. Hilpert, had carried resistance to Nazism a step further by leaving Germany for France when war loomed. After the French surrender in 1940, Hilpert was handed over to the Germans and sentenced to death. Since chemists were needed, however, his sentence was reduced to imprisonment. After the war, Hilpert sought exoneration and academic reinstatement, but encountered "only deaf ears not only with the Germans but also with the Occupation Forces."[25]

Following consultations in Oxford to find a suitable channel, Franz wrote on Hilpert's behalf to a Colonel Blount at Control Commission headquarters in Minden. The problem centred on the rector, of whom Franz could speak from direct personal knowledge:

> He tells me that Professor Hartmann is still in office in Brunswick and has the confidence of the British. If that is true – which I have to assume – then this is an incredible case. I know Hartmann quite well from the time when I was a Professor in Breslau and he was a Privat Dozent at the Chemistry Laboratory. Scientifically he was always absolutely zero and had very little chance of getting anywhere. When nazism broke out, he switched over to the Nazis at once and behaved in the most vile way. He made life unbearable for some of the more decent fellows ... and owing to his typical nazi behaviour he was one of the most hated men in Breslau. If people like Hartmann are supported by the British occupying forces, then we certainly cannot expect a better new generation of Germans to grow up.
>
> I do hope you will be able to do something about this.[26]

The matter moved with studied slackness. After seven months, Franz received a reply from an education branch official dated 9 December to report that the case of Helmut Hartmann

> ... is still under review. The facts mentioned in your letter have been investigated but there appears to be as much evidence for him on his present behaviour as there is against him. The case has been hitherto characterised by a marked disinclination on the part of those who knew him in the past to make any statement that could be taken in evidence either for or against him.... We are therefore appealing to you, sir, to help us towards a decision.... Mention of any specific examples of his having behaved "in a most vile way" would be most welcome.[27]

The clear subtext of this letter is that early conversion to Nazism, and a pattern of thuggish behaviour towards colleagues thereafter, are insufficient grounds for the education branch to do anything.

On 3 January 1947, Franz wrote a more detailed report on Hartmann's record at Breslau, as seen from his own vantage point as dean of the faculty in 1932–3. When this produced no reply for three months, he took the matter to the Master of Balliol, Lord Lindsay, whose connections with the government proved effective. In a letter of 9 July, Hilpert reported receiving a grudging formal exoneration. The process had taken 15 months since Hilpert had contacted Simon, and, as Hilpert acknowledged, it would have failed altogether without Lindsay's intervention. By this point Hilpert wanted to leave Brunswick, but Franz urged caution, since jobs outside Germany were scarce and Hilpert, born in 1883, was close to retirement.

The Hilpert connection was mutually beneficial, for Hilpert sent Simon two reports on German universities: a brief general overview—to be discussed in the next section—and a longer case-by-case evaluation of 31 of Hilpert's current and former senior colleagues. Of these, he considers about half to have been party members or close sympathizers, the remainder to have been without known connections, or recognizable anti-Nazis. Each group, however, has shadings. The Nazis are ranked from "very dangerous" to two who are "absolutely decent," or "absolutely objective," beyond reproach as academics. Of the non-Nazis, about half are passive, deferring to an energetic, capable, ex-Nazi Rector, Gassner, who is both feared and admired. This report also reviews the Hartmann case and suggests that the negative evidence may have been suppressed or ignored.

Among all of Franz's postwar contacts with Germany, one stands apart. While the others connect in some fashion with his academic career, Leonhard Günthermann was a former army comrade from Franz's Bavarian regiment. Later, he owned a toy factory in Nuremberg. In 1933, just as Franz was preparing to move to England, he received a letter, dated 14 June, from Günthermann. He and his wife wanted to say to the Simons that "our feelings of friendship remain the same as they have been hitherto." This relationship had been forged during long service together on the battlefield, and on his side he assured Simon that it could not be undermined "as long as you want it so. I condemn the generalization of the present government's position against the Jewish race."[28]

The Günthermanns have read much in the press about the disturbances in Breslau, but he hopes that Simon, as an old, tested front-line officer, is keeping everything under control (*walten und schalten*). In any case, he concludes, please let us know how things are going. Would they like to visit Nuremberg with their family? "We would be glad to welcome them at our place." The letter was much appreciated by Franz, and all the more so because support from non-Jewish academic colleagues in Breslau had been conspicuously absent. He stayed in touch with Günthermann after moving to Oxford.

In May 1949, Simon sent Günthermann a brief note, to which the latter replied on 7 June. This "first sign of life after the catastrophe" had brought great joy for the family, and would be passed on to old comrades from the regiment who have had no information in recent years. Günthermann summarizes his family's situation—health, business conditions, daughters' careers—during the war and postwar years. Despite their difficulties, "it is forgotten how many thousands are much worse off."[29] He enquires about any possibility of a Simon visit to Germany, but asks in any case for more complete news of the family's experiences in these years.

Franz replied with a similar overview on 11 June, mentioning losses of relatives, wartime evacuation to Canada, his own trans-Atlantic visits and work on the bomb project, his daughters' current work in London, food rationing, and holidays on the Continent. On the question of visiting Germany, "I have had various invitations, but so far have always declined."[30] The reason, he explains at some length, is his perception of the current situation in Germany: a general public refusal to accept any responsibility for all that has happened—"for example, the murder of 6 million Jews"—and the return of reactionaries and Nazis in government and the universities. In the circumstances, he prefers to meet his old friends from Germany in England when this is possible. Would conditions in the next few years enable a business trip—if possible with his wife—to visit the Simons there?

Günthermann replies on 23 June. His family are happy to hear the good news about the Simon family. On the matter of a future visit to England, he is pessimistic. A foreign travel visa is required, his firm cannot do business deals in England, and possession of foreign currency is forbidden. He does not believe this will change in the next few years. Though Günthermann and Simon both intend to keep in touch, the momentum for a personal meeting has clearly diminished, and no evidence of a meeting has been found.

The Pakenham letters

As early as June 1947, the situation in German universities generally was very much in Franz's mind. From Bonhoeffer's visits to Oxford in May and June he had learned a good deal, and he also had received two written reports, of several pages each, one from Jost and another from Hilpert. On 24 June he writes to Jost to thank him for two recent letters and "your long report." He asks permission to have it translated, show it to friends, and "also make some official use of it."[31] By this point, Franz had a clear plan for action. His college, Christ Church, had scheduled a Gaudy for 25 June. At the festivities, Franz made a point of talking with his colleague and lecturer in politics, Lord Pakenham, on the question of Germany. Pakenham, who had run for Parliament unsuccessfully in the 1945 election, was given a peerage by Prime Minister Attlee. He was at first made a lord in waiting, responsible for guiding various government bills through the House of Lords. In April 1947 he was promoted to be minister for the British sectors of Germany and Austria, a non-cabinet post.

On 4 August, Franz followed up with a carefully drafted three-page letter, which reminds Pakenham of their discussion at the Gaudy and the reports he had promised to send the new minister. On the main point, his letter does not mince words:

> From letters and particularly from discussions with German colleagues I formed the impression that the Universities in Germany are practically completely in the hands of reactionaries many of whom had been quite openly active nazis. I know it is unavoidable that there are only a few professors at the German Universities who have the right mentality to educate the next German academic generation. What however does not seem to me unavoidable is that these few people have been given so little influence.
>
> The situation at the present moment is that people who were actively anti-nazi and who tried to sabotage the German war effort or at least to do as little as possible to help, are regarded by the Germans as traitors and I have reluctantly come to the conclusion that many officials of the occupying forces share this view. I admit that it is very difficult for an Englishman to understand the position, as in this country it is unthinkable that any government could be as criminal as the Nazi government was.[32]

In the pages following, he refers to the Bonhoeffer executions and the Hilpert case as typical illustrations, and vouches for the authors of the two enclosed reports.

Both authors, he explains, are well-known physicists. The first document, sent on to Simon by Hilpert, is by Professor Hartmut Kallmann, currently head of the Haber Institute of Physical Chemistry at Berlin-Dahlem: "he is half jewish and married to an 'aryan' and thus he managed to survive in Germany." The second is by Professor Jost, "an excellent physical chemist, pure Aryan . . . one of the very few people who can be trusted."

> While Kallmann's report is in a way on negative lines as he believes that as a non-aryan, he cannot stay any longer in the country, Jost's report is more positive and suggests a few remedies for the present situation.

Franz's own letter has little to suggest by way of specific remedies. His stated purpose is simply to report that the few former anti-Nazis, weak, marginalized, and vastly outnumbered by a close-knit network carried over from the previous regime, have also been unrecognized and ignored by the occupying powers.

Pakenham replied with a two-page letter on 8 September, most of it devoted to an unneeded rehearsal of the magnitude of the problem. He admits that the "biggest problem" concerning the universities is the professors, "highly conservative and, indeed, reactionary . . . very strongly entrenched and in a position to dominate the younger members of the staff . . . [and] they are in a very strong position to nominate successors with ideas like their own . . . ." However, "I do not believe that an effective form of denazification is the answer." British policy, which his educational advisor, Robert Birley, is developing, will be "somehow to bring outside influences to bear on the universities" to encourage reforms. "This is a big task and it is one which we are firmly convinced the Germans must perform themselves . . . . It will take a long time." As a "first step" in this long process, he continues, there are hopes for creating advisory commissions to the ministers of education of each *Land*, to advise on university matters and discuss reforms "if suitable people can be found."[33] The sole practical suggestion in this letter is that Simon should meet Birley for more discussion.

Franz, back from a holiday on the Continent, replied on 30 September. He would be glad to meet Birley. While in Switzerland, he reports, he had met three former colleagues from Germany—two of them now professors at Zurich, the third in Switzerland for a tuberculosis cure—and all three "of the decent type." Discussions with them had reinforced his previous information and given it a new perspective:

> They emphasised one point which seems to be of very great importance and I believe they are right. It is the fact that more often than not our representatives in Germany have picked the wrong people and they mentioned particularly how discouraging it is to the very few decent ones to see how we fall for those who are most adaptable and able to flatter us in a clever way, while those who are really decent have little or no influence. You will remember that I previously emphasised this point to you but from what I have heard recently I believe it is even more important than I realised at the time.[34]

A "good illustration" of this, he adds, was the notorious Justi case, which had arisen while Franz was away, and which will be explained shortly.

By this date, Franz had a specific remedy to suggest for the faulty decisions: "I have often wondered why our representatives in Germany do not attach to their offices a German of the right type as adviser . . . . I personally can give you the name of one." He names Paul Rosbaud, and emphasizes his qualifications for such a post: wide experience with a full range of German scientists through working for Springer publications; an active anti-Nazi through the war years; a main interest in rebuilding Germany "properly." Finally, "as he is a 'pure aryan,' he would not be confronted with the many difficulties that refugees – particularly jewish refugees – would have to encounter." Franz urges strongly a meeting with Rosbaud, and gives the latter's address, telephone, and several possible referees.

There followed several weeks of related meetings and discussions. Simon met with Birley on 6 November. Birley met Rosbaud on 10 November, and they talked of education and university questions at some length. Rosbaud's first impressions, reported to Simon the same day, were favourable, though he had been unable to give information on some of the humanists and historians asked about by Birley. The Simon–Pakenham contact was resumed in late January. They met in Christ Church on 24 January, and Pakenham was given a revealing article by "a young and able German scientist, Miss Martius," published in the November 1947 issue of *Deutsche Rundschau*. Simon followed up with another three-page letter on 28 January, containing new evidence, new enclosures, and practical suggestions for better decision-making.

His prime evidence was the Martius article itself, the sharp reaction surrounding its publication, and extracts from a covering letter sent by Martius to Simon. Ursula Martius, an assistant to Professor Kallmann, had attended the 1947 Physical Society meetings in Göttingen in September, and the article in *Deutsche Rundschau* begins with a description of her shock and surprise at what she saw there: "People who are still appearing to me in my nightmares were sitting there, living and unchanged, in the first rows. Unchanged, that is, unless one considers sleek blue suits in place of uniforms and absence of party insignia as 'change'."[35] The article goes on to name five former party members who attended, among them the notorious General Schumann (see Chapter 5). For each one, she lists past and present posts, together with one or more instances of past misdeeds.

In the covering letter to Simon, Martius explains that circulation of an early draft had led to an attempt to have publication cancelled, not for factual inaccuracy but for its harmful effect on "national honour." The result was a modified draft "as mild as one can make it without falsification." The most critical thing about the present situation, she writes, is a tendency of "many scientists" to shelter tainted colleagues under a cloak of distorted patriotism, out of a sense of fulfilling a national duty. "Such behaviour . . . makes a cleansing almost impossible."[36]

Whether mild or not, the Martius article evoked an immense reaction in the German Physical Society, the British zone section of which had sponsored the Göttingen meeting. The article, its context, and the society's reaction became the subject of a lengthy article by Gerhard Rammer in 2007 in a collection of studies on the society during the Hitler period. Rammer terms the Martius article an "absolutely exceptional case,"[37] an unprecedented breach of a tight, unwritten code of behaviour that required academics not to criticize the political behaviour of their colleagues. In response, von Laue, the President for the British zone of the postwar decentralized society, initiated a widespread consultation and debate, the outcome of which, Rammer says, was a clearer policy for the society. The "code of silence" or non-criticism was upheld for academics who had successfully completed a denazification procedure, based on the prevailing principle that "research and politics are two separate spheres (*Bereiche*)."[38] Exceptionally, however, serious past offences, such as denouncing or damaging another colleague, were recognized as unforgivable.

In this consultation and debate, the society's overriding concern for the five individuals named by Martius obscured the more important part of her argument. The last three pages of her four-page article make a carefully reasoned case for a new concept of civic responsibility in universities. In the post-Hitler years, professors should be not only scientists but also teachers (*Lehrer*), concerned for the values of a student generation that has known only the brutal authority structure of the Nazis. "This generation is not beyond rescue, but it is in very many cases unstable, lacking advice, and in search of new ideals." At this juncture, "university teachers are no longer just individual personalities, but are charged as strictly as any member of parliament"[39] to set a moral example for their students.

Martius, who became Ursula Franklin, was living in Toronto while this chapter was being written and graciously agreed to an interview. On 4 June 2008, in her office at Massey College at the University of Toronto, she explained in detail the background to her 1947 article. The setting was Berlin, where four-power control was less constraining than in the four zones of occupation each run by a single occupier. In the first postwar winter of 1945–6, a group of young scholars—activists, returned exiles, and survivors from labour camps—began to meet in a single, heated room to discuss issues of higher education and university reform. Drawn from different disciplines and ideologies, they developed a sense of community born from the shared hardships of restoring a city in ruins. They were inspired by a few outstanding scientists—Otto Warburg, Iwan Stransky, and Kallmann were mentioned by name. As scientists, anti-Nazis, and reformers, their aim became to act as a catalyst for change in a previously closed system: "We were not prepared to clone the previous situation."[40]

When physicists who shared these tendencies attended the 1947 meeting in Göttingen, they saw and heard things that astonished them. Franklin recounted how one German exile working in the Netherlands was baffled by a round of audience applause on his arrival, until he realized that the applause was for a former *Gauleiter* entering just behind him. Her letter to Franz describes how a professor at Brunswick Technical University, Eduard Justi, "went so far as to take the chair one day as 'deputy (*Stellvertreter*) for the *Reichsmarschall* for atomic physics'." Such examples, she wrote to Simon, were "only a part of the completely horrible political and social impressions that we experienced there." Worse still, "not one of those attending raised any question or took offence at these things."[41]

In the interview, Franklin emphasized the "watershed" quality of the 1947 Göttingen meeting. Anti-Nazis, who had viewed this meeting as an opportunity for change, were suddenly confronted with a brutal reality: "We all realized that there was no place for us. It was absolutely clear."[42] The realization diverted them from dreams for rebuilding to plans for leaving Germany. Kallmann left Berlin in 1948 for a US Army research laboratory near New York; Martius in 1949 on a postdoctoral grant and research position in Toronto. In North America, both would have distinguished academic careers. They were not alone in leaving. Rammer cites other examples, but notes that this postwar emigration has been "hitherto insufficiently explored."[43]

The second enclosure in Franz's letter of 28 January to Pakenham was part of a letter from Clara von Simson. The extract corroborates from other sources what Martius had reported about the Göttingen meeting; in particular, its key message:

> That it will be easier, especially in the British zone, to obtain a call to a chair if one was a PG [i.e. *Parteigenosse*, or former Nazi party member], and that such a professor takes care to select like-minded assistants. It is also increasingly reported that attempts to block this run into difficulties from the occupying power.... How this should be possible is totally incomprehensible for us here ... but the more harmful, very sad consequence is that the people of good will gradually become discouraged, and do nothing.[44]

Franz sums up the message of the two enclosures in a few blunt words: "The real power at the universities is again in the hands of the supporters of the Nazis."

His letter goes on to discuss other topics, among them, a 12-day lecture tour in Belgium and Holland earlier in the month. In the Low Countries, he found scientific colleagues "extremely worried" about Allied policy for German universities. Having been occupied in wartime, they understand more fully "what is threatened if the wrong people get to the top again."

In the rest of the letter Franz makes three specific suggestions, all focused on the critical need for advice from people "who do understand something of the situation and have really first-hand knowledge." As a new proposal, he suggests "a small committee, say of the four or five German scientists who are Fellows of the Royal Society, with people from other fields, such as Jacobsthal [archaeology] and ask them for their opinions." Second, that "someone with real knowledge should be in a permanent position in the British Administration to advise on matters of scientific policy and personnel." Franz urges once again the "unique knowledge" of Paul Rosbaud for such a position. "You mentioned to me a few difficulties against his appointment," Franz notes, but the benefit of such an appointment would outweigh "these difficulties." Third, Franz reminds Pakenham of the latter's invitation to visit London and "discuss some more details." He "would like to do that very much," offers to bring along Max Born—currently lecturing in Oxford—and proposes three possible dates for a meeting in mid-February.

Pakenham acknowledged this letter by a handwritten note on 3 February, promising a longer reply after a visit the following week to four universities: Göttingen, Bonn, Cologne, and Münster. His three-paragraph reply, unsigned and undated, but mailed by his secretary on 25 February, shows him moving more firmly towards deliberate inaction.

> I deplore as much as you do any difficulties which are put in the way of anti-Nazi professors by their colleagues, but there are obvious limitations to the extent to which the Control Commission can interfere. They could hardly, for example, intervene in support of the appointment of particular professors, since if they did so the position of these among their colleagues and students would be quite impossible. Nor can they intervene directly to remove individual professors unless they have committed some definite offence. Indeed, I am strongly of the opinion that the time for witch-hunting is over and that everything possible must be done to encourage the development of stability in Germany.[45]

Pakenham was not the first to use the term "witch-hunting" as a code word for de-nazification. By this point it was becoming an overworked cliché.

He was equally dismissive of Simon's suggestion of an advisory committee of "refugee German scientists in this country." He had considered the suggestion "carefully," but "I do not feel that this would fill a real need." He would, however, always be glad to receive "information or representations" from individual scientists. He is also, he adds, "still looking into" the Rosbaud question, and hopes to write about this "in the near future." The curt dismissal of Simon's committee proposal pointedly avoids any mention of the Royal Society, or its long-established role in dispensing scientific advice to government, or that most of its Fellows of German origin were by this date British citizens with stellar academic careers in both countries.

The only faint bright spot in this letter was a middle paragraph in which Pakenham mentions a proposal just announced by the military governor to set up a zonal advisory commission of "Germans from different spheres of life . . . supplemented by one or two people from outside Germany," to survey the whole field of universities and recommend needed changes. "It is through broad actions such as this that we can hope to influence the universities rather than by individual intervention in particular instances."

It says something for Franz's self-control that he replied to this letter on 11 March in remarkably civil language. He focuses on the sole positive point, and then gently fends off the charge of witch-hunting:

> I was very interested to hear about the commission you are going to set up and I hope you will get the right people. I must say I am a bit disappointed that you think I have been advocating a "witch-hunt", which was certainly not my intention. The only thing I wanted to ensure was that decent people should get preference before the Nazis.[46]

In the rest of this letter he reports on his recent encounter with Heisenberg, who had been in Oxford earlier in the week. He also refers Pakenham to Goudsmit's recent book on Germany's atomic energy project, *Alsos*, and offers to lend his own copy "if you have not got one."

By this date, Franz had invested much time and effort over eight months in his attempt to link with those shaping policy for occupied Germany, but with little or no tangible effect. He had a longer-term connection through Lord Lindsay and Balliol. Lindsay's interventions often produced rapid results but usually dealt with individual cases of injustice. With Pakenham, Simon aimed at a more enlightened policy on a broader front, particularly in the crucial universities sector. This plan received a setback, however, when Pakenham was removed from his portfolio, after just over a year in office, on 31 May, and made Minister of Civil Aviation. The removal was unrelated to any action or inaction on the universities front. His major misstep was an outspoken campaign to halt the dismantling of the Ruhr steel industry, a question of high inter-Allied policy. By doing so, he was also crossing swords with his own senior minister, Foreign Secretary Ernest Bevin, a hard-bitten trade unionist who had scant sympathy for any of the Christian Socialist ministers in the Attlee government.

For Franz, the problem was that Pakenham was untried and unpredictable at this point, still in the early stages of shaping what his wife would charitably describe as a "halo for eccentricity."[47] In his later years, he wore this reputation with pride, and developed a formidable capacity for stirring public outrage. In the German portfolio there were no previous indicators. As a Christian Socialist by philosophy and Catholic convert by religion, he was impervious to advice from mere human sources. As a member of the House of Lords, he was immune from electoral consequences. For most matters, his own conscience, unencumbered by situational complexity or inter-Allied agreements, was a sufficient guide. When he exchanged the German portfolio for Civil Aviation, it was the French ambassador to London who made the appropriate comment: "Now Lord Pakenham will be able to have his head *always* in the clouds."[48]

Surprisingly, Simon's letters to Pakenham continued even after the latter's change of portfolios. Presumably Franz felt that even an unpromising channel to government might be better open than closed. Even after the transfer, Pakenham continued speaking out on German issues, though his voice, as his wife concedes, "had lost some of its authority."[49] He continued as a member of the government, and even spent a few enchanted months as First Lord of the Admiralty, though still not in the Cabinet, from May 1951 until the Labour government's electoral defeat in October 1951.

With Labour out of power, Pakenham returned to Christ Church to teach politics again, and doubtless saw Simon regularly at meetings of the college's governing body. The three known letters of this period from Simon to Pakenham—two in 1949, one in 1952—are worth noting for two reasons: they add more evidence on events in Germany, and they show Franz's commitment to a deeply felt cause. Though he often told German colleagues that he had put Germany behind him forever after the betrayal of 1933, his inner logic reminded him that no such simple separation was possible. In these years he became convinced that the ideological battle for democracy in Germany was crucial beyond Germany, crucial for the future of Europe, and crucial for global peace.

The first of these three later letters was the result of a discussion in college. Dated 7 March 1949, it is brief and largely self-explanatory:

Dear Pakenham,
    When we met last Friday and discussed our eternal problem subject, the position in Germany, I also gave you a few examples of the recently growing anti-Semitism in Germany. Your reply was: "What do you want; look at the Jews in Berlin who protested against the showing of the film 'Oliver Twist'."
    I have not seen this film, but now that I have got some information about it, I must say that I think that the protest was quite justified. There is, of course, no objection against showing such a film here, but to show it in Berlin, after all that has happened in Germany in the last fifteen years and where the minds of the people cannot yet have returned to anything nearing normality, I think is unwise, to say the least . . . .

> I do not know who was responsible for releasing this film in Germany. If it was the British administration, they certainly showed a regrettable lack of common sense; if it was a German official, he might possibly have done it out of downright mischievousness.[50]

As contemporary press accounts show, the opening of the 1948 film version of *Oliver Twist* in Berlin was a colossal and predictable blunder. When completed, the film had been blocked from distribution in the United States by protests from Jewish groups, and was eventually allowed in—after cuts of several minutes—only in 1951. It was banned in Israel as antisemitic. When it opened in February 1949 at a cinema in Charlottenburg in the British zone of Berlin, the Jewish *Gemeinde* likewise protested, and Berlin mayor Ernst Reuter and some councillors asked the British to ban the film. The point of contention was the role of Fagin, overplayed by Alec Guinness as a stereotypical Eastern Jew in a manner all too reminiscent of the caricatures in Streicher's *Die Stürmer*.

The British military authorities refused a ban, but posted a guard of 30 Berlin police—50 by some accounts—around the cinema to keep order. However, the cinema, the Kurbel, was in Giesebrechtstrasse, in a neighbourhood "heavily populated by Polish Jewish displaced persons," and a local crowd of 300 rioters overran the police line and attacked the cinema marquee. While the British dithered, the German manager of the Kurbel hastily switched programmes to something less provocative, "a dull B picture about copper mining in Ireland."[51]

The second letter of 1949 from Simon to Pakenham, dated 4 November, returns to the earlier style, a letter rich in new evidence reinforcing Franz's previous warnings of Nazi resurgence. With it Franz encloses a translation of a most remarkable letter, "written to a former German physicist, who has now a high position in Canada, by one of his former colleagues." The letter is a starkly frank private letter, written from friend to friend, concerning a forthcoming official letter to sound out the recipient about his interest in returning to a chair and an institute in Germany. In the translated copy, Franz has omitted the names of the correspondents and the university in question, as protection against possible damage from the carelessness of "our representatives in Germany." In the original German text, dated 28 February 1948, the addressee and the writer are clearly identified: Gerhard Herzberg and Walter Werfl.

The writer of this letter, as a dean and an obvious antinationalist, feels obligated to disclose the situation fully:

> Now as much as I would like to have you here, I have to tell you honestly how circumstances are here and probably at all universities of the West .... Very grave doubts I have however because of the spiritual development of our colleges in general and I must say that sometimes I am close to despair and would rather like to depart for good from my native country. One fights a hopeless battle against the stupid nationalism you know so well from before, only that it has become much worse than it ever was. 95% of all colleagues are governed by the following things: 1) A blind hatred against the occupation power .... 2) An actually passionate sympathy and preference for the "poor P.G.s" for whom one tries to procure all possible advantages; conversely a profound distrust and an insurmountable aversion against all those who suffered through the Nazis .... 3) The definite conviction that in a few years Germany will again have its nationalistic regime, and the frantic fear not to be able at that time to prove one's "national reliability." The Nazi scare was never as great as the fear which the average German has already now of the Nazis of the future ....
>
> I have always felt that we could not escape that sort of mental attitude unless we could get back again part of the colleagues gone abroad. But on the other hand one cannot expect any man to commit spiritual

or perhaps even physical suicide by returning to Germany. So far none has returned to . . . yet. It is true that not many have been called back, because in general one does not want them here, but would much rather throw out the few adversaries of Nationalism.[52]

Herzberg did receive an official enquiry from the Education Minister of North Rhine-Westphalia regarding this chair, but it was dated 29 August 1948. In the summer of 1947, after exploring several possibilities, he had accepted a senior research post at the National Research Council in Ottawa, to begin in 1948. The German offer was declined.[53]

The two-page letter of 4 November from Simon to Pakenham contains other data. Franz had spent some six weeks in North America in September and early October, and a further week in Paris later in October to attend a UNESCO conference. Both trips yielded new evidence on Germany. In the United States he had met "several decent Germans, who have recently escaped from Germany," whose own reports were similar to those of the enclosed letter. Just how Franz received a copy of the letter to Herzberg is not spelled out to Pakenham, but his calendar shows that he gave a lecture in Ottawa on 24 September and then spent the evening with the Herzbergs themselves.

At the UNESCO conference, Simon met J.W.R. Thompson, "a Canadian who is in charge of German affairs in UNESCO," and showed him the recently acquired letter to Herzberg:

> He told me that this letter might have given a good picture of the situation at the time when it was written (February 1948) but that the position is now very much worse and that he was in complete despair over the whole situation. He told me as a further example that when one of the "Länder" recently instituted their own government, they dismissed 1800 of the employees of their Ministries in order to replace them by the former Nazi occupants of these positions and that 800 of them belonged to the Ministry of Education!

Thompson worked for UNESCO from September 1947 to 1954 on re-education in Germany, and was influential in preparing the way for Germany's entry to this body.

In three further paragraphs, the letter to Pakenham revisits familiar territory: "This situation could have been avoided if the proper people had been picked in Germany in the first place"; all of Simon's efforts to arrange consultation with experts on Germany "have been in vain"; and time for effective intervention may be running out:

> The "Herrenvolk" variety is on top again, while the decent people have been let down by the Allies. Now of course it is much too late to try and change this state of affairs, but I think it is our duty to rescue the few decent people, who are now being trampled down by the Nazis.

But if intervention is henceforth impossible, are any policy options still open? At least, Franz urges, alert the public to the new realities:

> Apart from [a rescue operation] I think it is extremely important to enlighten public opinion about the matter. Once before the people of this country ran blindly into a catastrophe which could have been foreseen. Now as then there were actually people who welcomed the Nazis as a bulwark against the Bolshevists . . . .
>
> Is there anything being done by the Government to enlighten public opinion about the state of affairs in Germany? I know that although you are personally interested in these matters you are no longer in direct charge. I would therefore be very grateful if you could hand on this letter to the official who deals with German affairs. I also think it would be a good idea if he could ask Mr. Thompson of UNESCO . . . to come and see him.

By this point, Simon's link to Germany had become tenuous. He apparently did not know who had taken over Pakenham's former functions, but he could still call for an informed public as the best hope for avoiding disasters like those of the 1930s.

By 1949, the German political environment was changing rapidly. The year brought adoption of a new federal constitution, the Bonn Basic Law, in May, a first election for the Bundestag in August, and formation of the Adenauer government in September. A new Occupation Statute, in force from 21 September, placed control over senior public appointments in German hands, and Adenauer, reversing initial Allied practice, "usually"[54] preferred ex-Nazis of proven efficiency in past posts over democrats with lesser experience.

After a three-year gap, Franz wrote the last of his known letters to Pakenham on 18 October 1952. By then he had been to Germany to receive a medal, as will be explained shortly. The letter is at once an apology for an argument, an explanation, and a hint at closing the file:

> Dear Pakenham,
>
> I am afraid our discussion last Saturday got a bit heated – doubtless my fault. Of course I have a somewhat different outlook on the events in Germany than you have. Of my family nearly forty were 'killed' – to use this rather merciful word for what really happened, – from a ninety-two year old aunt of mine to the two year old son of a cousin. Therefore, if I see that just those very people responsible for such acts are on the way back to power, I naturally feel very strongly about it.
>
> I have looked through our old correspondence and I think that what I told you all the time (and what you thought was highly exaggerated) exactly corresponded to the facts. It is my firm conviction that the Allies could have avoided this and brought to power the real democratic elements in Germany if they had heeded advice from people who had first hand experience.[55]

He goes on to underline the tension he had seen in Germany between a "small class of highly undesirable people who live in luxury and the greater part of the population, in particular the refugees from the East."

The more outspoken tone in this letter is traceable to a changed political setting in Britain. The link to the Labour government, in which Franz invested so much effort and achieved so little satisfaction, is gone. Pakenham was out of Germany, and Labour was out of power. After the election of October 1951, Churchill was again Prime Minister, and Cherwell was back in Cabinet. The earlier, better link was restored.

What Simon omits in this last known letter to Pakenham in October 1952 is any mention of a law on the public service introduced in the Bundestag on 11 May 1951. This law, and its far-reaching consequences for the public service, have been examined carefully in a landmark study by Curt Garner.[56] By Article 131 of the Bonn Basic Law of 1949, the Bundestag was obligated to "regulate the legal status" of those federal public servants who had been employed by the Nazi regime as of May 1945 but who were subsequently unemployed or working in positions inferior to their former rank. The groups covered by the law included persons whose agency had been dissolved (e.g. the military), expellees and refugees from the lost territories or from the Soviet zone, and persons who had worked in the three Western zones but had lost their positions through de-nazification procedures.

The main thrust of the law was to reintegrate these groups into the public service establishment by establishing a "placement obligation" requiring administrations at each level of government—the federation, the Länder, and local governments—to set aside 20 per cent of their public service positions, and 20 per cent of their budgets for salaries and pension rights, to employ these excluded groups and support their families. Stiff financial penalties were set for any administration failing to reach the specified targets. Counting employees, pensioners, and their dependants, the beneficiaries of this law totalled more than a million persons, most of them of voting age. The measure enjoyed a smooth passage through the Bundestag, with all-party concurrence and no member voting against it. The proportion of employees politically compromised through de-nazification was officially estimated at 23 per cent, but revised by Garner to 33 per cent. A solution to this sensitive question was "discussed and agreed upon behind the scenes during parliamentary deliberations." Further, the law "was rendered more palatable to many on the Left by coupling it with a restitution act for all former public employees who had been victims of Nazi persecution."[57]

Garner's article also shows increasing levels of compliance with the placement obligation during its first four years of operation.[58] In July 1951 the federal public service was already compliant at 24 per cent, and remained so in March 1955. Länder administrations increased from 14 to 19 per cent between those dates, and local governments from 8 to 15 per cent. Though the pressure of numbers eased in the later 1950s, the legal burden of the placement obligation continued for ten years, until August 1961.

It seems clear beyond doubt that Simon correctly detected the trend towards re-nazification in the public service after 1949, but he apparently lacked specifics about its constitutional origins, its statutory base, and its sweeping legal extension to all three levels of government. Had he known more about these specifics, he would undoubtedly have denounced them in even stronger language in his last letter to Pakenham.

Who should be invited to Britain?

In the late spring of 1947, and prior to his first meeting with Pakenham, Franz received a letter dated 12 May from Göttingen:

Dear Simon,
 I think it is time that we met and talked about things in Germany. I have had this in mind for many months now, but thought I should like to have really full information and background, gained on the spot, before looking to you for such help and advice as you may be able, and I am sure are willing to give.[59]

The letter is signed by Ronald Fraser as scientific adviser to the research branch of military government in the British zone. He proposed a meeting during a forthcoming visit to London in early June. Simon's reply is missing, but his calendar shows that they met in Oxford and dined in Christ Church on the evening of Sunday 8 June. Prior to this meeting, Simon wrote to Fraser—who was also a physicist—with a small request: that he "bring along any published material"[60] on Heisenberg's ongoing work on superconductivity, for a conference on this topic to be held in Oxford on 27 and 28 June.

Franz's diary gives brief details of the meeting on 8 June: at 4.30 p.m., "fetch Fraser" followed by talk; dinner at Christ Church, with Cherwell and others present; after dinner, 9.45–10.45, more talk with Fraser. As usual, the diary gives no details on the discussion, but over several months evidence would accumulate that this meeting had not gone well. In retrospect, a first clue to the difficulties may be seen in Fraser's original letter. His self-described immersion in the Göttingen scene may have led him to mistake the local situation—which he ruled as a personal fiefdom—for the wider universe. And is there a hint of condescension in his request for "such help and advice as you may be able . . . to give"? As later developments would show, Franz would be deeply troubled by the outcome of this meeting. Seventeen days later, he would discuss the German situation seriously for the first time with newly appointed Minister Pakenham.

Three months later, Franz received an equally troubling letter dated 15 September from his Bristol colleague Nevill Mott:

> Dear Simon,
>
> Dr. R.G. Fraser (C.C.G. Göttingen) suggests bringing over Heisenberg and Justi to Bristol for a short visit to discuss their work on low temperatures with English physicists. What would your reaction be to this proposal? Would you be willing to come if the meeting was some time in November?
>
> Heisenberg, as you probably know, has recently been staying with Bohr who, according to Fraser, is convinced that this work is important and sound.[61]

Several points here invite comment. The letter is dated just eight days after the end of the 1947 Göttingen Physikertagung. It is unclear whether the proposal originates from Fraser, or Heisenberg, or from wider discussion at the meeting, where Fraser had welcomed the delegates and spoken in favour of increased international contacts. Mott himself had attended the smaller Physikertagung of October 1946 but is not listed among the foreign guests at the 1947 meeting.[62] What the letter does not mention is that Fraser had accompanied Heisenberg on his visit to Bohr in Denmark. Michael Frayn, in a "postscript" to his play *Copenhagen*, refers jocularly to Fraser as "Heisenberg's British minder,"[63] assigned to guarantee his return to the British zone.

Franz replied, swiftly and negatively, the next day:

> Many thanks for your letter of the 15th concerning the proposed invitation to Heisenberg and Justi. You know that I am very much in favour of inviting German scientists who have behaved decently during the Nazi time. I am afraid however that Justi certainly does not belong to this category. He was one of the first Nazis and a rather vile one as I know from personal experience. He remained so to the end as I know from Bonhoeffer and also from Clusius whom I saw recently in Zurich. In addition he is, in our opinion at least, not a very good scientist – many of his experiments are sloppy and wrongly interpreted. The main point however is that as a character he is one of the most despicable people I know and I certainly would not attend any meeting to which he was invited. I know that Fraser likes him – "he is so amusing" he said to me a short time ago. But I do not think that Fraser always picks the right person!
>
> One also has to consider very seriously the impression which it would make on the few decent German scientists if one of the first to be invited to England was well known to have been a Nazi, is probably still one at heart, and in addition has no particular scientific merits. Heisenberg is of course another question . . . .[64]

Franz lists both pros and cons regarding Heisenberg: on the negative side, his attitude to the Nazis, and his "very cleverly distorted picture" of German nuclear activities during the war; more positively, "everyone would be pleased to discuss scientific matters with Heisenberg," and relations with him will soon be imperative. "However, I think there are many more important things to discuss with him than his theory of superconductivity."

Fraser's letter raised an important issue of public policy. Since Britain was a free country, any group in Britain was entitled to invite anyone of their choice as a visitor. Thus Rosbaud notes in 1948 that von Weizsäcker, the first apostle of the Farm Hall myth (see Chapter 7), was being invited to Britain by "some Christian associations."[65] Where the visit was official and public money was involved, however, there was a double obligation, in Simon's view, to choose outstanding scientists whose values were compatible with the political objectives of the occupying powers. Justi fell short on both criteria.

The proposal to invite Justi troubled Simon considerably. His diary for that week reports special events at Harwell, starting on 17 September and involving several invited scientists for a small conference on nuclear physics. One of those attending was Paul Scherrer from Zurich. On the morning of Saturday 20 September, the last day of the Harwell events, the diary records Franz showing Scherrer and Sir Charles Darwin around the Clarendon laboratory, but with an added notation "disc[uss] German scientists." They went on to a laboratory at Harwell, apparently to meet or hear Sir John Anderson. Franz then lunched with Scherrer and Mott, and once again the diary notes "(German scientists)." These extra notations suggest a discussion of serious dimensions.

Fortunately for posterity, first Mott and then Simon were moved to put their positions on record by an exchange of letters. On 6 October, Mott wrote:

Dear Simon,
I would like to try to explain my feelings about the Justi business.

What I want to do – so far as I have any influence in this matter, which is not much – is to re-establish scientific contacts between German science and that of the rest of Europe as soon as is humanly possible. I do not honestly believe that any other course does any good at all, though I have the utmost sympathy, and I hope some understanding, of the feelings of people who have had your experience, and of people who have lived under German occupation. In particular, I am very sceptical about the wisdom or possibility of distinguishing now between those who were decent and those who were not. This is because I do not believe that it will help any class of German to be picked out and favoured by the occupying power.

This does not mean that I want to have Justi! If Fraser is bringing over two men to talk about low temperature work, that is obviously a subject about which you know much more than I do; and, as your opinion of Justi's scientific and personal qualifications is not high, I am writing to Fraser to say that he is not the man to bring. I have suggested Laue, which would obviously be better in every way.

With very many thanks for your help and advice,

Yours ever,[66]

Simon's reply is dated two days later, 8 October:

Dear Mott,
Thank you very much for your letter of October 6th. I am very glad about your decision and there is of course no question that Laue is in all respects the right person. I absolutely agree with you that one should re-establish scientific contacts with Germany as soon as possible and actually I think we should already have started this last year. My only point was that the choice of one of the persons was wrong and that his invitation was not in the interests of the cause in which we are both interested.

After our discussion a few days ago I looked up a few relevant data concerning Justi . . . one can sum up by saying that his whole career proves that he is the worst type of opportunist who collaborates with anyone who just happens to be in power. I would not object if he came along in say five years together with other people when we are back to more or less normal conditions but I strongly object to his being one of the first to be invited particularly owing to the effect it must have on the decent scientists. Here I am afraid I disagree with you; I definitely think one has to distinguish between the decent and the non-decent people. What the world needs now are not so much good scientists – there are plenty of them – as honest people.

I was very glad to have had an opportunity of discussing this with you and I believe I have to apologise for being somewhat emotional during our discussions!

With kind regards,

Yours ever,[67]

After a month, Simon received a note from Mott dated 12 November to say that Heisenberg was expected in Bristol on 1 December, after a visit to Cambridge, "but I have heard nothing definite from Fraser as yet . . . . If he [Heisenberg] comes, he is coming alone."[68] The same note mentions that Mott is proposing one of the refugee physicists at Bristol, Walter Heitler, for fellowship in the Royal Society. Would Simon co-sign? With their differences clearly stated and mutually respected, their professional collaboration and personal friendship continued undiminished.

There is no mention of a Heisenberg visit to Bristol in Franz's calendar or diary on or near the date mentioned by Mott, but Max Born confirms two separate visits to England. In a letter to Einstein dated 31 March 1948, Born describes a visit from Heisenberg to Edinburgh in December 1947, finding him "as pleasant and intelligent as ever, but noticeably 'Nazified'."[69] In the next sentence, Born adds that he has "recently talked to him again in Oxford." This visit of Heisenberg to Oxford will be examined in the next section.

After Simon's conversation with Darwin and Scherrer in the Clarendon on 20 September, his concerns began to be heard in official channels. Darwin wrote to Simon that he had taken up the matter up with Sir Henry Dale and told him that Simon would also write. Finding that Dale was absent in America, Darwin also raised Simon's concerns with I.G. Worsfold, the secretary to the Foreign Office committee headed by Dale that dealt with such questions. Simon himself drafted for Dale a lengthy, masterly summary of the entire issue: the choice of Justi as a political mistake and scientific disaster; the importance of setting a good example in Germany, both morally and scientifically; and the need for procedures to avoid such mistakes in the future. This draft is dated 15 October.

According to Simon's summary, the problem arose from making a junior scientific officer responsible for a senior scientist of high standing:

> It seems to me what has happened in this particular case in Göttingen is that Heisenberg has Fraser completely under his influence. Heisenberg is of course one of the greatest living scientists, but on the other hand he is also a very shrewd politician and was always a nationalist of the first water. He has never been a real Nazi but he was quite prepared to share the spoils – one might call him a "fellow-traveller". Fraser feels extremely flattered that such a world-famous man, with whom he would never in the ordinary way have had the slightest contact in scientific matters, has taken him into his confidence . . . . He has now made up his mind that the picture which Heisenberg and his friends have given him about the situation in Germany is the right one and he refuses to listen to any other point of view. That at least was the impression I got when I saw him here and this has been strengthened by his later actions, particularly in the case of Justi.[70]

Among other things, this letter reveals more of what had gone wrong at Simon's meeting with Fraser on 8 June. Even with Dale, however, Simon withheld the point that had shocked him the most at that meeting. It would emerge, almost incidentally, three months later in one of his letters to Pakenham:

> I remember a discussion with Fraser, our representative in Gottingen, who told me that he has a higher respect for a scientist who had worked for the German army than for one who tried to sabotage the

German war effort. I do not think I succeeded in persuading him, in our rather long discussion, that "my country, right or wrong" cannot be applied in the case of such a criminal government as the Nazi government was.[71]

For Simon, who had worked closely with Germans and other Europeans who had risked their lives to oppose the Nazi regime, Fraser's position crossed a line. As events would show, however, Fraser was not alone in holding this view.

Dale's committee was scheduled to meet on 12 November. As the date approached, Worsfold wrote to Darwin on 7 November to inform him that Simon's expected letter had not yet been received. On the 8th, Darwin wrote to Simon enclosing Worsfold's letter, adding that "a word direct from you, rather than indirectly from me, would be desirable."[72] On the 10th, Simon wrote to Darwin to explain a change in the plan:

> Actually I had composed a letter to Dale but have not sent it off, the reason being that when I discussed it with Cherwell he said that although he agreed with every word I had written, he thought it better if matters of such a delicate nature are discussed verbally and not committed to writing . . . .
>
> As your meeting is the day after to-morrow, I am suggesting a compromise. I am enclosing the original letter for your own private information and you can then put my point of view to your Committee. I should be very grateful to you if you could do this. Should the Committee be interested in further details, I think it would probably be best if I go and see Mr. Worsfold.[73]

He had not relished writing the letter to Dale, and he deemed it prudent to follow Cherwell's advice. Even before the meeting took place, however, Worsfold's letter of 7 November to Darwin showed that Simon's message had been heard and and that change was coming:

> We are very sensible here of the importance of encouraging the right German scientists and not doing anything which would give an impression such as I fear Professor Simon and his friends have received. It is of course very difficult for our people in Germany to know all the background of the many German scientists whom they come across, but I think that Fraser has made a mistake in the past through not consulting people who might be able to give him valuable advice. Quite apart from this particular instance about which Simon complains, we have decided to give instructions to the Research Branch that no arrangements for visits of German scientists to this country should be undertaken in future without reference to this office first of all. I am afraid this looks like putting another cog in the already complicated machinery but it is the only way I am afraid.[74]

The outcome of Franz's lengthy effort gave him considerable satisfaction. Henceforth, the choice of official visitors would shift back to London, and it could reflect priorities determined in Britain, rather than in Germany. Ten months later, Simon himself would propose a German visitor to Oxford. On 3 September 1948, he wrote to the British Council asking for travel support for Professor Klaus Clusius, inventor of the Clusius–Dickel tube, to come to Oxford to give a lecture on isotope separation. Simon's choice, and his reasoning for it, are interesting. Clusius had been an active member of the German Uranium Club, but his wartime work on uranium and heavy water was of high interest for Harwell and Clarendon physicists. Further, "Clusius was perhaps not one of the really active anti-Nazis, but he has always behaved decently, so that from that point of view there would also be no objection."[75]

The council gave a quick approval, and Simon wrote a letter of invitation on 12 September. The visit took place in the spring. Franz's calendar and diary show Clusius in Oxford from 5 to 10 March 1949.

Heisenberg's visit to Oxford

The timetable of Heisenberg's two-day visit to Oxford in 1948 can be traced with some precision. Franz's calendar lists an afternoon meeting with Heisenberg on Monday 8 March, followed by dinner in Christ Church that evening. On 9 March, it shows an afternoon meeting, followed by an evening with Heisenberg and Cockcroft. The diary gives more detail, including a general label for the topic discussed in each session: on Monday afternoon, from 3.20 to 5 in the laboratory, "Heisenberg, partly with Cherwell, (disc[ussion] on Germany and science dis[cussion?])," and later with Mendelssohn and Heisenberg again; at 7, dinner in Christ Church with Heisenberg and Born, followed by "long talk with Heisenberg" until 10.45, then home via Magdalen College, where Born had been staying while giving the Waynflete Lectures.

On Tuesday, the diary reads "9.30 fetch Heisenberg from Ch[rist] Ch[urch], short walk. Lab." Until 11, the discussion is on "He[lium] & superconductivity," with Cherwell arriving—as usual—before 11. Simon then had another meeting and a luncheon commitment, returning at 2.30 to find others talking with Heisenberg. At 3.30 Simon gave Heisenberg a tour of the laboratory, after which Cherwell, Mendelssohn, and others "all talk on He etc."

At this point the diary, even if read with the calendar, becomes ambiguous. The evening entry continues: "6.15 home with Hsbg. Dinner, talk on Germany. 8.45 Cockcroft. talk – 10.15." What is imprecise here is the time of Cockcroft's arrival and the time of Heisenberg's departure. Were they both at dinner? If not, did they overlap? This imprecision is regrettable, for in time this evening at Belbroughton Road would give rise to legends.

Heisenberg does not appear in person in the diary again on this visit. It is noteworthy that he was not offered a public platform in Oxford. His departure time, mode of transport, and next destination are not recorded in the diary, which is unusual. However, the discussion generated by his visit continued in the Clarendon over the next two days. On Wednesday 10 March, Franz is in the laboratory by 9, and notes "Talks to Kurti, K. M[endelssohn], Bijl about Heisenberg. Write." This is the sole diary entry until Franz goes home at 12.30 for lunch. The afternoon shows a lengthy meeting of the Governing Body of Christ Church.

On Thursday 11 March, the diary lists seven talks in the morning with laboratory colleagues, but more time is found to "write about Heisenberg." He then saw Arthur Cooke and Born and notes: "With Born home for lunch. Talk on Heisenberg and Schrödinger." He was back by 2.30 for a busy afternoon. The diary for these two days shows that, when Franz wrote about this visit to friends and colleagues, his views on Heisenberg had been shaped after much discussion.

His first report is a three-page letter to Perrin, dated 11 March. It is marked <u>CONFIDENTIAL</u>, and it appears intended for intelligence records. Simon notes that Heisenberg "wanted to talk to me about Goudsmits' [sic] book, which obviously infuriated him very much." *Alsos* had been published in 1947. Perhaps owing to the controversy with Goudsmit, the Farm Hall myth had advanced well beyond its first formulation in 1945, even beyond previous Heisenberg versions. Franz summarizes the essentials in a single paragraph:

> Heisenberg claims that German scientists had no other wish than to prevent Hitler from getting the bomb. They knew about everything . . . He said that if he had gone to Hitler at the beginning of the war and told him what he knew, then he was quite sure that Germany could have developed the atomic bomb just like the Allies![76]

Franz had been in a difficult position in the dialogue with Heisenberg. He could not speak of some sources "from which we know for certain that his story does not correspond with the facts." Instead, Simon pointed out that the "great majority of German scientists" seem to have behaved otherwise. "At the end of our lengthy discussion," he reports, Heisenberg conceded this: "the German scientists had not behaved very well except for a few, for instance Hahn and Laue, and more or less himself."

Franz then describes what he perceives to be Heisenberg's peculiar mental condition. Whatever the motives and precipitating conditions for it, two conclusions seem clear. First, Heisenberg is prone to inventing falsehoods that do not correspond to verifiable facts. He is even capable of statements conflicting with earlier statements of his own, as for example attributing avoidance of a German bomb development to distrust of Hitler in 1948, but to Germany's insufficient industrial capacity in his article in *Die Naturwissenschaften* of December 1946.[77] The second peculiarity of Heisenberg's case is his "complete sincerity and conviction" in telling these stories.

The final paragraphs of the letter to Perrin point to some practical consequences. It is difficult for an outsider who knows neither the facts nor "the workings of this corner of the German mind – or should one say soul – to get a true picture of the situation," and it is almost impossible for Fraser to do so. Simon notes that Heisenberg has "the highest praise for Fraser," and Fraser "has accepted Heisenberg's point of view." These matters, he emphasizes, are important. Heisenberg is influential in policy matters, and likely to become more so as his older contemporaries retire. The importance of understanding his peculiarities will therefore increase. This report to Perrin is rich in detail that will not be examined here, but the full text is reproduced in the Appendix.

On the same day that he wrote to Perrin—11 March—Franz wrote also to Lord Pakenham, enclosing a copy of his just completed report to Perrin. It would be the last letter from Simon to Pakenham before the latter's removal from the German file. He refers to Heisenberg as

> the foremost German scientist and probably the most outstanding theoretical physicist of the world. During the war he was in charge of the German Atomic Energy effort and, as you can imagine, our Intelligence people know a great deal about him. I enclose a copy of a letter which I have written to Mr. Perrin, the Deputy Controller of Atomic Energy, as I believe it may prove of interest to you.[78]

Simon's letters to friends and colleagues add details that sharpen the picture. To Jost, by now installed in a tuberculosis sanatorium at Davos, he offers support and cheerfulness, including a lighthearted view of the latest Heisenberg spin:

> I had a long talk with Heisenberg a few days ago. He now pretends that the German scientists knew everything about the atomic bomb all the time but they did not want to let Hitler have this weapon!!! We know, however, that this is not the true picture. Have you read "Alsos" by Goudsmits [sic]? You will probably be able to get hold of it in Switzerland. It is definitely worth while reading.[79]

To Max Born, he would eventually reveal the darker side of Heisenberg's visit to Oxford, but only in a different context, and more than 3 years after the event.

Heisenberg wrote his own evaluation of his 1947 and 1948 visits to Britain in two letters of 1948 to Arnold Sommerfeld, dated 5 January and 31 March, respectively. They are briefly summarized in a combined note by Mark Walker: "Heisenberg thought that the trip to England had been a success. The time he spent there had been pleasing in every way. Born had been as 'friendly and nice' as in the old days. Simon and Peierls had also been very hospitable, but Heisenberg believed that they found it difficult to free themselves from the injustice they had suffered."[80] Walker's summary is important, because the second of the two Heisenberg letters on which it is based is currently unobtainable.

In the first letter, available in full text, he reports a very nice reception from colleagues but notices "an undertone of conscious anti-German feeling especially among the Jewish colleagues." The Borns in Edinburgh were an exception, "totally, charmingly hospitable, just like in the old times."[81] For the second letter, the evidence is partial, piecemeal.[82] An abstract of it remains in the Arnold Sommerfeld Nachlass of the Max Planck Institut für Physik in Munich. The abstract mentions Peierls and Simon by name, but for specifics one must be guided by Walker's summary. The summary seems to hint that Heisenberg finds some deficiency in Peierls and Simon for their inability to forget and move on.[83] He was slow to realize that for many refugee scientists his wartime work for the Nazi regime made a friendly return to the "old days" impossible.

Paul Rosbaud, living in London and well connected, was better positioned to assess Heisenberg's 1948 visit. Writing to Goudsmit on 25 April 1948, he emphasized its divisive effect in Britain:

> Heisenberg has been in this country and stayed several weeks . . . He was a great success and had an enthusiastic reception in Cambridge . . . Lord Pakenham was apparently deeply impressed by him. He also had talks with Simon and Peierls, who were not so much impressed. I have, of course, not seen him. Heisenberg's position is very strong and, I think, people over here look at him as at some German apostle . . . he will always be arrogant. People will look at you or Simon or Peierls as being biased, at me as an illoyal [sic] and unreliable citizen and at Heisenberg as a truly patriotic and reliable German of the best type. And thousands of Germans who suffered under the Nazis . . . will soon be forgotten.[84]

The innermost details of Heisenberg's 1948 visit to Oxford would emerge almost accidentally. Franz met Heisenberg again in 1951 in Copenhagen, and their further discussion on Germany once again evoked an outrageous prediction from Heisenberg. In relating this to Born, Franz was reminded of the earlier occasion at Belbroughton Road in 1948. Simon's letter of 11 August 1951 describes twinned conferences in Denmark and then recounts both provocations together:

> I went to the Union de Physique in Copenhagen and Lotte came along too. We went a few days earlier and looked in at the Bohr Conference. It was a very pleasant occasion and we saw quite a lot of Bohr and Mrs Bohr. How friendly and unpretentious the Danes are, particularly in comparison with most of the Germans – Heisenberg for instance looks and acts more and more like an actor . . . .
>
> I had some discussion with Heisenberg in Copenhagen about the whole matter [of trends in Germany]. He said quite shamelessly that of course in say 10 years time Germany will again be ruled by the Nazis, perhaps under another name; after all then the generation which was educated under the Nazis would be of the right age. And you can be quite sure that he will "mitmachen" with the greatest pleasure. I believe I have told you that a few years ago he told us here in our house: "If one would have left the Nazis for another 50 years, they would have become quite reasonable."[85]

Because of their unbounded arrogance, some pronouncements of Heisenberg from this period took on a life of their own in the oral tradition, but this is the only example found that is traceable to Franz directly. Other reports[86] describing this same occasion are quoted by Paul Rose, who cites a similar version in English from Mott and Peierls and a German "original" from Nicholas Kurti. The interval of "50 years" was fleshed out a little for lecture purposes during Heisenberg's visit to the United States in 1950. As Goudsmit explains in a letter to Rosbaud, Heisenberg has a theory of revolution,[87] similar to the views he expressed before the war, and the time normally required for "reasonable people" to replace "low-grade people" might be "fifty or one hundred years." Through this reasoning, Goudsmit reports, "he still goes on defending all the evil things in Germany as being the normal by-products of any social revolution."

Another much-quoted Heisenberg saying reveals his utopian postwar fixation on what might have been: "How fine would it have been if we had won this war."[88] Unveiled prematurely at a private dinner on his visit to Zurich in December 1944, it was reported back to Germany and picked up by the Gestapo. Although the war was clearly lost by this date, it was forbidden to say so. With Gerlach's intervention, the investigation was resolved by a reprimand. After 1945 Heisenberg continued to repeat it, and it is likely that Lotte heard it from Heisenberg directly when he dined at Belbroughton Road in 1948. It became a phrase imprinted in her memory of that evening for the rest of her life—*Schade, es wäre so schön*.... There was nothing new in these sayings of Heisenberg. What was noteworthy was his compulsion to repeat them in Britain and North America, and especially among refugee colleagues who were guaranteed to be infuriated or appalled.

Following the two conferences in Copenhagen and the meeting with Heisenberg there in July 1951, Franz appears to have had little or no further contact with him. As scientists, they had agreed to exchange scientific papers, and on 8 June Simon's secretary sent Heisenberg a marked list of those that he wanted. As Simon predicted, however, Heisenberg would go on to play a leading part in the postwar rebuilding of German science. To a degree, he had been groomed for the role by British education branch officials, Colonel Blount and Ronald Fraser, who secured a large Göttingen mansion for him and his family, along with the directorship of the revived Kaiser Wilhelm Institute of Physics. Broadly, British policy worked for concentration of key academic physicists in the British zone and favoured centralized administration of science policy.

The longstanding conflict between Bund and Länder over control of science policy reappeared in a new institutional form in 1949. The revived Notgemeinschaft, a decentralized association of Länder, cultural ministers, and rectors, was challenged by a re-established, centralizing Deutscher Forschungsrat, located in Göttingen and headed by Heisenberg, which claimed to speak as "sole representative of all German science." For the next two years a complex struggle ranged centralizers against decentralizers, academic scientists against industrial scientists, and federal officials against Länder authorities in competing for research funding. David Cassidy gives a capsule overview of this struggle, and Heisenberg's role in it.[89] In the struggle, Heisenberg's undiluted nationalism stands out clearly. As head of the Forschungsrat, he sought its recognition as the sole advisory group to the Chancellor and as the only way to achieve "our wishes for the centralized direction of research." When the conflict was eventually reduced by a negotiated fusion of the two bodies in August 1951, Cassidy notes, a "reluctant" Heisenberg yielded to pressure but "regarded the whole affair as a defeat both for himself and for German democracy."

Throughout this period, Heisenberg retained a coterie of admirers in Britain. Among the scientists, these were influential enough to secure his election to the Royal Society as a foreign fellow in 1955. Laue, more respected for his record as an anti-Nazi, had been honoured in 1949. From the admirers, however, there would be one notable defection. In time, Ronald Fraser would outgrow the role of "slavish admirer" that Franz had described in 1947. A letter from Simon to Born of 18 October 1948 comments on the early signs of change:

> Fraser seems to have changed quite a lot during the last year. He is no longer such a slavish admirer of Heisenberg. I had a short talk with him in Birmingham but Rosbaud told me more details of our discussions with him. I understand that he is now going to leave as he does not get on with his former colleague and present superiour Blount.[90]

Born replied two days later, explaining that Fraser is leaving because "his post is disestablished" and he hopes to join the European Branch of UNESCO, a plan that Born approves:

> I do not think that [Fraser] was ever a 'slavish admirer' of Heisenberg . . . but Heisenberg is without doubt, one of the most gifted men in theoretical physics, and nobody can resist his cleverness and charm. In regard to politics, Fraser is quite clear about Heisenberg's ideas, as my wife and I are, too. The problem is whether to mix these things up with scientific questions. I think that you and I also differ in this respect, as I have seen in your attitude to Schroedinger.[91]

Born's comments reopen a fundamental issue. For Simon, neither scientific giftedness nor charm can compensate for moral inadequacy.

In Fraser's later years, his disillusionment with Heisenberg became complete. In 1966 he received a letter from the historian David Irving, who was seeking confirmation of Heisenberg's claim that his 1947 visit to Bohr had been an effort to repair the rift created by his fabled wartime visit to Bohr in 1941. Fraser, living in antipodean retirement in New Zealand, had nothing to lose by speaking out. In a reply to Irving dated 27 August 1966, he dismisses Heisenberg's versions as typical fabrication:

> Your version . . . and even more your marginal source are way off the beam. The whole story of "a kind of confrontation" in the matter of his 1941 natter with Bohr in the Tivoli gardens is a typical Heisenberg fabrication – maybe a bit brighter than a thousand others, but like them all a product of his *Blut und Boden* guilt complex, which he rationalizes that quickly that the stories become for him the truth, the whole truth and nothing but the truth. Pitiful, in a man of his mental stature . . . .[92]

Fraser's letter is also significant for his own crisp recollection of the 1947 visit. He had arranged it, and he states clearly its threefold intended purpose: (a) to discuss Heisenberg's recent work on superconductivity; (b) "I used (a) as a lever with Bohr to persuade him to receive his prodigal son . . . this was but one of my countless moves to open the windows of the German scientists' Nazi prison to the west";[93] (c) to arrange via Bohr extra Danish food for Heisenberg's large family. The language of (b) is interesting for the light it casts on Fraser's view of his role in Germany.

Fraser was an active participant in a larger plan, launched by Colonel Blount, to build up the strongest possible concentration of German academic scientists in Göttingen, and in the British zone generally, beginning from the nucleus of former Farm Hall detainees. For Britin, such a policy might bring important benefits. German science and scientists were valuable trophies of war in all zones, and all four occupying powers placed their own national interest first in dealing with them. But in the British zone, keeping the academic scientists contented also had a high cost: a willingness of the occupiers to overlook wartime records and keep silent on de-nazification.

If we understand the strategy behind this bargain with the German scientists in the British zone, many of the events and outcomes in this chapter become more understandable: the "deaf ears" faced by Hilpert; the foot-dragging in investigating Hartmann; the refusal of Pakenham to consult Rosbaud, or any of the Jewish refugee scientists; the favours to Heisenberg and Justi; British acquiescence in preferring ex-Nazis over anti-Nazis; the blanket refusal of Pakenham to intervene in specific cases; or Fraser's declaration of his greater respect for a scientist who "worked for the German army" than for one who "tried to sabotage" the war effort.

Rosbaud widens this view beyond Fraser. In a letter dated 11 June 1947—two weeks after Simon's first meeting with Fraser—Rosbaud writes to Goudsmit: "It is a very great danger, if high officers . . . declare that they cannot blame German scientists, who have done such brilliant war work, for their loyalty to Hitler Germany and if they say that they like these scientists much better than the Germans who have been disloyal to their country."[94] On this evidence, British compliance with the German scientists' collective decision to ignore the past was more than the work of any single individual; it was systemic, deliberate, and it had a clear purpose. It also left little room for principles of restorative or remedial justice that were fundamental for Simon and his anti-Nazi colleagues.

Would Simon visit Germany?

If the issue of who should be invited first to Britain was important for Simon and for anti-Nazis generally, the other side of renewed scientific contacts was whether Franz himself—and others in his position—would visit Germany. As the next few pages will show, the question had more complex dimensions for Franz than might appear at first glance. His correspondence with friends and officials addresses the question directly and forcefully.

An example from 1947 is a letter to Jost, who is by now in a sanatorium in Würtemberg under treatment for tuberculosis. Franz, in regular contact and arranging "care" parcels through his Simmon cousins in New York, writes on 5 August:

> You asked me whether I would come and visit Germany. I have been asked before to visit Berlin and Göttingen but so far I have refused for various reasons. One of them is that I cannot see how I could simply talk to the people there as if nothing had happened and on the other hand I cannot very well go and tell them how badly most of them have behaved. I do not know whether you can understand it but emotionally I am not very much concerned with the future of Germany. After what has happened I have made a complete break; the only thing I am interested in is to help my friends there who I know had the right spirit.[95]

Franz, just leaving for a month in Switzerland, regrets that they will not be able to meet there, because Jost's permission to enter a Swiss sanatorium has been delayed by the military.

Two years later, Simon received an invitation of a different order: to visit and give a talk at the Technical University in Berlin-Charlottenburg. The invitation arrived through formal diplomatic channels. It was accompanied by a covering letter dated 9 August 1949 from the German section of the Foreign Office, which urged acceptance "in view of the great importance we attach to contacts between Universities in Germany, and in particular in Berlin, and those in this country."[96] The invitation, signed by the Rector, Professor Apel, on 18 July, leaves details to be worked out with the Faculty Dean, Professor Stranski.

The invitation arrived at an inconvenient time. Franz was about to leave on a 7-week tour of lectures and a conference in North America, his first transatlantic visit since the war. On the eve of his departure, he wrote two letters, a holding letter to the Rector in Berlin, and a covering letter to the Foreign Office. To the Rector he explained the time pressures and proposed to get in touch again after the American tour. In the letter to the Foreign Office, which is marked <u>Confidential</u>, he raised an additional question, security concerns:

> I have now had a talk with the security people of the Department of Atomic Energy of the Ministry of Supply about my suggested visit to Berlin, and there are two different opinions concerning the advisability of such a visit. One of them thinks that it would be a good idea, while the majority think it would be too risky. Meanwhile they are going to make enquiries with their people on the spot and also wait for the return of their senior colleague.[97]

He encloses the Berlin letter for transmission, with copy for the Foreign Office file.

On 10 February 1950, Franz writes again to the Foreign Office about the Berlin invitation and the security issue:

> In the meantime I have heard from the Department of Atomic Energy that in their opinion it would be too dangerous for me to go to Berlin, although they would not forbid me to do so. Under these circumstances, of course, I must refuse the invitation.[98]

However, he raises the further question of how to reply: "Should I explain the true position, or would it be wiser to invent some excuse?" The Foreign Office replies on 18 February that "rather than explain the true position you should merely send a non-commital [sic] note saying that you are afraid that you cannot fit this engagement into your programme."[99] Franz writes to Berlin on 25 February, mentioning "accumulation of work" during his absence and listing commitments for the weeks ahead. He apologizes for "this negative reply," and thanks the Rector for the Technical University's "kind invitation."[100]

Three weeks after his letter declining the invitation to Berlin, Franz writes to Bonhoeffer, who had also enquired about a visit to Germany. It was time for frank talk between close friends on the two real reasons for hesitating:

> The question of my visits to Germany is a rather complicated one. Certainly the money question is quite immaterial . . . . There are however other reasons which perhaps you can guess, which make a visit from me to Berlin under the present circumstances highly inadvisable; I have recently written to the Rektor saying that at the present moment I am afraid I will not be able to come.
>
> I certainly would enjoy seeing you again and many other of my former friends; on the other hand I must tell you quite frankly that I am afraid I would not get on very well with some of your colleagues, as politically they do not seem to have learned anything at all. Thus the position is that although in many respects I would like to come, I certainly do not want to spend my time over there in political discussions which would be unavoidable – and with people who seem to be beyond repair.[101]

In thinking of the risk factor, one may recall that the Berlin airlift had ended less than six months earlier, and occasional kidnappings of scientists were still occurring in Berlin. The Soviets, having built their first atomic bomb with borrowed technology and tested it in September 1949, were also pressing to complete other nuclear projects of their own design.

They would have found Simon, as wartime designer of the Valley plant and a known insider of current British nuclear projects, a tempting target for Soviet attention.

The issue surfaced again a year later, when Bonhoeffer sent Simon a renewed invitation to attend the meetings of the Bunsen Gesellschaft, the German association for physical chemistry. On this occasion the security issue is not mentioned, but Franz makes his condemnation of the scientists more specific. They have never made any collective expression of regret for what happened:

> In my opinion German scientists as a group lost their honour in 1933 and have done nothing to get it back. I admit that one can say that it is not everybody's concern to risk one's position or one's life, but after the war that was no longer necessary. The least you could expect after all that happened was that German scientists collectively, or through their scientific associations, would state publicly and clearly that they regretted what had happened. I did not notice anything of this kind. If I am wrong, please correct me.[102]

The situation changes in 1952, and some detail is necessary to trace the reasons behind the change. On 7 April Franz replies to a letter from Jost, who has plans to visit America in the coming summer:

> It would be nice to see you again this year and indeed I will be on the Continent and am actually going for a few days in Germany!! The reason is that the Low Temperature Society is going to give me the Linde Medal and after careful consideration of the pros and cons I have decided to accept it. Unfortunately the time coincides more or less with your journey to America, as the presentation will take place at the General Assembly of the Society in Stuttgart on the 25th 27th September . . . . I have been asked to keep the Linde Medal business strictly confidential, although I do not pretend to know why.[103]

In early July Franz received a letter from Laue, who had heard of the forthcoming visit to Stuttgart. He invites Simon to visit Berlin also to attend the Physical Society meeting. Franz's reply on 5 July explains that security concerns still apply:

> Yes, I shall go to Stuttgart for a few days at the end of September to receive the "Linde" medal of the Kälteverein. It would be very tempting indeed to go on to Berlin and I am sure that I would benefit a lot in various ways . . . . The trouble however is that I belong to the inner circle of the Atomic Energy Project and we have been asked – rightly or wrongly – not to expose ourselves to the, admittedly slight risk connected with a visit to Berlin (this is of course private information for you alone).[104]

He thus cannot accept, but he thanks Laue for his "very friendly invitation," and would like to attend the meetings another time, when the society "meets in Western Germany."

Franz's calendar and diary outline the highlights of the meeting. He drove from Zurich with Lotte, arriving in Stuttgart around 4 p.m. on 24 September. On the 25th, he was up by 7: "prepare my lecture." At 10.30 the meeting started, the Linde Medal was presented, and Simon responded with the prepared lecture. The diary continues: "Lunch with Fues, Rector & president Rector Conference. disc[uss] his intention to reply to my Pollitzer remarks." In thanking the society for the medal, Franz had spoken of Franz Pollitzer, a long-time Linde chemist and friend of Simon since the early 1920s, as a more suitable and deserving candidate, given the medal's original industrial connections. But Pollitzer had fled Germany, and worked in France until he was rounded up in 1940 and killed in Auschwitz in 1942.

On 26 September, the diary lists another brief private meeting with Fues just before the morning session opened. After the session, Franz and Lotte attended a lunch at the Rector's house, with a notation: "Discuss. Germany." The conference ended that evening with a formal dinner, for which the only programme item noted is "Cabaret!" It seems likely that Fues's response to Franz's remarks on Pollitzer was given just before the morning session's scientific papers, and this is supported by a cryptic "Before Fues, Pollitzer etc." at that point in the diary.

No texts or speaking notes on Simon's remarks on Pollitzer or on Fues's response the next day have been found. The most complete account of Fues's statement is a letter that Simon wrote to Bonhoeffer on 6 October after his return to England:

> I was particularly impressed by Fues, now President of the "Rektoren Konferenz". After a lecture I gave at the meeting of the Kälteverein when I received the "Linde Medal" and in which I alluded to Pollitzer's death at Auschwitz, he got up and made a very decent statement about the responsibility of the Germans for what has happened and that the scientists will try to make amends, as far as is possible, for the harm that has been done. At last somebody has said something on these lines openly and it is only a pity that it was not done before. As Fues spoke definitely in his official capacity as President of the Rektoren Conference, I hope that this statement will be given some publicity.[105]

Erwin Fues, a theoretical physicist of Franz's generation, and pupil of Arnold Sommerfeld, had responded to Simon's tribute to Pollitzer with sympathy and understanding. His statement, the first that Franz had heard from any senior academic, was an act of reconciliation, one that Franz had been waiting to hear since his departure from Germany 19 years earlier.

Franz seems to have written no similar summary of his own remarks on Pollitzer, but for these there was an additional witness. In the transcript of Lotte's BBC interview, a page and a half are devoted to the Stuttgart visit. Speaking in her early eighties, her memory was incomplete on background and proper names, but the seminal moments are recalled with clarity:

> We did not go back until [1952] . . . when he was given the Linde Medal. When they offered it to him his first reaction was, thank you very much, no . . . . But with this offer of the Linde Medal, came a personal letter of Bonhoeffer, who said, "I know exactly what your reaction will be. But please think of us. This is the first time we are active to make a gesture to say we regret what has happened. If you refuse now they'll always say, 'Well, there you are. They don't even want it.' Please come."
>
> So for that reason alone he accepted. And we went to Stuttgart. And it was a very uneasy atmosphere from our side. It was a huge meeting . . . hundreds of people.[106]

She then briefly explains the Pollitzer case, but omits—or possibly never knew—its earlier background: that Franz had worked for Pollitzer's release when he was first arrested and sent to Dachau concentration camp in November 1938; and that by further enquiry after the war Simon had learned in March 1947 of Pollitzer's deportation from France to Auschwitz on 9 September 1942.[107] Lotte continues:

> So when my husband was supposed to say thank you . . . He told me beforehand, "They'll be surprised if they hear the thank you I'm giving them." So when he got up he said, "I had been thinking who ought to get this medal," which had only been given once ten years before and . . . it was for low temperature work mainly in industry . . . . And he said, "I could only think of one man who really deserved it." And we just heard that this man, after having fled to France, was caught and killed in Auschwitz. And then he gave his name.

And I watched the faces of these hundreds of people. And they did not like what they heard at all. But we both felt we couldn't just pretend as if nothing had happened.

In forming general impressions of their visit, Franz and Lotte were conditioned by years of austerity and rationing in Britain and repelled by the level of inequality they observed in Germany. In his letter of 6 October to Bonhoeffer, Franz writes:

In the meantime, I have been in Germany myself for a week. We have seen a lot of interesting things, but as a whole I must say I found it very depressing to see the luxury on the one side and the poverty on the other. It seems to me that this cannot end well.

To Jost on 27 October he is more circumspect: ". . . our impressions of the situation there are somewhat mixed. Perhaps we can discuss that when we meet." Apart from a shared revulsion against inequality, Franz and Lotte retained rather different memories from their days in Stuttgart. For Franz there were positive memories of encounters with old friends, and a sense of accomplishment after his interchanges with Fues. For Lotte the memories that lasted were of unpleasant encounters and conversations that were "not exactly friendly or agreeable."[108] In all likelihood she had missed hearing Fues's statement on the second day.

The conference in Stuttgart was followed by a drive down the Rhine through Bonn and Cologne to the Dutch frontier. It was the last stage of a typical six-week, multi-country Simon summer holiday. These holidays will be described further in the next chapter. What is not mentioned in Franz's letters or Lotte's interview is that the Stuttgart visit had been preceded, in the first week of the same holiday, by a similar five-day trial drive across southern Germany from Trier to Salzburg, and stopping two days in Munich to visit the Meissners.

The calendar and diary show Franz and Lotte visiting Germany briefly again, without fanfare, during their summer holiday of 1953, with an oddly similar travel agenda. This too was a double visit, placed at both ends of a longer holiday in Switzerland. In late August, en route from Luxembourg to Zurich, they drove southward via Trier, the Palatinate and the Black Forest to the Swiss frontier at Waldshut. Three weeks later, after a fortnight at Soglio in the Engadine and a conference in Zurich, they returned to Germany for another two-day conference on low temperatures, this one a meeting of the IIF, the Institut International du Froid, at Baden-Baden. After the conference, they drove down the Rhine valley again, with brief stops to visit the Jost family in Darmstadt and the Schaefers in Cologne. Visits to Germany were no longer an issue, but they would occur only for adequate reasons.

Any after-effects of these visits are difficult to assess. In April 1954, after a request from Bonhoeffer, Franz agreed to be listed among the collaborators of the revived *Zeitschrift für physikalische Chemie*, the journal of which he had earlier been a co-editor: ". . . with regard to my old connections with the *Zeitschrift*, I shall be pleased to accept your offer if you believe that I can be of any use."[109] Later that year, he drew the line more stringently. He declined an invitation to stay with the Brickweddes during a conference in Washington, in order to avoid meeting their other conference guest, an official from Germany's standards office:

I believe between old friends I can be quite frank . . . . I really do not want to risk having to make polite conversation with a man who may have been instrumental either actively or passively in the deaths of a few dozen of my relatives who, starting with an old aunt of 92 and ending with a young boy of 2, ended in the gas chambers. I hope you will forgive my frank statement but of course people like ourselves have a different attitude than those who have not our experience.[110]

Franz's memories of his dead relatives would stay with him—and within him—to the end of his life. He could speak openly of this burden with his contemporaries, but with us of the next generation they were never, ever discussed.

His calendars show no visit to Germany after 1953. He preferred to meet German colleagues and friends in Oxford or London, or on the Continent, or by visits to the refugee diaspora across North America. There were also new countries to explore, notably Israel and Greece. These patterns of travel and discovery will be traced in the next chapter.

11

A ROUNDED LIFE

Franz Simon's life was more than a simple series of peak efforts and achievements. He lived a rounded, integrated life, marked by good manners, a sense of humour in the darkest times, fortitude, a strong moral sense, and inexhaustible curiosity, which focused on the laws of nature and also on the failings and foibles of the various human societies around him. As his calendars and diaries show, he kept meticulous records of the hours and days of his own busy life. He enjoyed good food, and believed in long holidays. He cherished foreign travel and visits to professional colleagues, particularly in North America. He won recognition for his achievements in his own lifetime, but he was also vulnerable, especially in his later years, to occasional serious illness. In the end, he would be struck down, unexpectedly, at the summit of his career. This chapter traces a few areas of his life not yet explored: holidays, other travels, and health.

Summer holidays

For the 10 years after 1945, Franz's calendars and diary chart each day of holiday time as faithfully as they chart his working days. For what we can label "normal" years, the holiday period is about six weeks, arranged as a single block of time. One exceptional year, 1947, shows a total of 60 days, in two blocks, but at this point he was convalescing, after a month of enforced rest for heart trouble in December 1946. Four other years—1948, 1949, 1953, 1954—fell a week or two below the six-week norm, but each of these years saw travel for professional reasons offsetting the shorter holiday period. Three of these four years included visits to North America. Franz believed in longer holidays, for himself and for others. The point is emphasized and repeated in his articles in the *Financial Times*. Basically, he saw holiday time as an indispensable feature of civilized living, and for scientific researchers in industry he argued that longer holidays might lead to more effective research careers.

Most of the 10 summer holidays that began in 1946 had certain features in common. Normally, Franz and Lotte travelled by car. This offered flexibility, and freedom from elaborate prearrangements. The only fixed points were the reservations needed for Channel crossings at the outset and on return, usually by ferry, once or twice by Channel air freighter. Most of these holidays were divided between itinerant days, moving from place to place a day or two at a time, and stops of 1–2 weeks in one location, which became hubs for further exploration of that region. Increasingly, as holiday tourism grew, they sought out locations off the beaten track, until it became a challenge in microgeography to pinpoint some locations. Further, Franz was prepared to spend a few holiday days attending scientific congresses, or visiting continental colleagues, or meeting friends.

The destinations of these holidays, taken year by year, reveal a good deal about Franz's and Lotte's cultural values. In 1946, the first destination was various villages in the Bernese Oberland, with a move to Ticino when the diary records "very bad weather" north of the Alps. In 1947, the year of two "vacations," the main summer destination was a three-week stay in and around Champéry in the lower Valais. Franz also took a shorter spring break in Ticino that year. He was accompanied by Kay and they travelled by train, but this break was part of his convalescence from heart problems in December 1946.

In 1948 and 1949, the summer holidays would be spent in France. In both years, the holiday period would be shortened because of working lecture tours, to the Low Countries in 1948 and to North America in 1949. In 1948 the holiday destination was southern France, mainly in the Rhone estuary and along the coast eastward to Menton. In 1949 they returned to their preferred destination of mountains. After brief stops in Switzerland they crossed into France and settled at Sixt as a hub in the mountains of the Haute Savoie.

In 1950, the itinerary emphasized their preferences for the Swiss Alps, for places off the beaten track, and for less conventional travel routes to get there. They began by a Channel crossing from Southampton to Cherbourg, drove southward to Angers, then up the Loire valley to Tours and Chenonceaux, continuing eastwards across France towards Geneva. On the fourth day from Cherbourg, they drove through Geneva and Interlaken and on to Hohfluh. This small village, in a valley above Meiringen, became their base for 11 days. After a visit to colleagues in Zurich, they found another base in Vaud for a further 12 days on Mont Pelerin above Vevey. In the final week, they spent a few days in the Netherlands, where Franz was presented with the Kamerlingh Onnes Medal at Leiden University.

The 1951 holiday was different: it was mainly devoted to Italy. Franz and Lotte drove directly from the Dover–Ostend ferry through Belgium, France, and Switzerland to the Austrian Tyrol, entering Italy through the Dolomites and down to Venice. There they met Franz's cousins on the Mendelssohn side, the Forsellinis, and divided the second week between Venice, still humid from the summer heat, and the Cortina region in the Dolomites. A third week was spent mainly in Florence, and a fourth in Rome, with minor stops and excursions along the way. The last 10 days of the 38-day journey took them from Rome via Assisi, Arezzo, Volterra, Pisa, and along the Italian Riviera to France and return north to Dieppe.

Exceptionally, Lotte kept a travel diary of this holiday. Its early entries are routine and short, but from the first hours in Venice she was clearly in her element, recording her appreciation of art and architecture, textures, patterns, landscapes, the play of light on buildings, interiors and exteriors of churches, madonnas, and more. Favourite places from earlier visits were seen again and reassessed. In Venice, the Piazza San Marco was "more beautiful and unreal than remembered." They went for an evening gondola tour with the Forsellini cousins: "very romantic in moonshine along the Canale Grande . . . nice views of lit up rooms with old ceilings and chandeliers." In Rome, the Pantheon was "most beautiful, even more so than remembered." Not all of the reassessments were positive. At Siena, the Duomo was "even less interesting than remembered, small, stripy, sugar façade." At St Peter's, the "new approach [is] no improvement," for aesthetic reasons that she explains at some length. This was urban Italy, the art and architecture she had admired as a young traveller in 1921.

Her diary also describes an interesting thermodynamic event for Franz. On 5 October they visited Larderello, near Volterra, to meet the director of a unique power plant at the site, which at that date was the world's only industrial-scale producer of geothermal electricity.

After a tour of the plant, Lotte's diary notes: "Most interesting the 'Fumaroli,' steam escaping with deafening noise at 200° C under pressure."

The holiday timetables for 1952–5 are more irregular, owing in part to various conferences in which Franz participated. The 1952 and 1953 holidays, each with its double visit to Germany, have been described already in Chapter 10. The year 1954 was a year of considerable extra travel, and a holiday reduced accordingly to four weeks. Even these weeks were fitted around a preparatory meeting in Padua and a conference in Grenoble of the Institut International du Froid (IIF). The result was a month of frequent movement between France, Switzerland, and Italy, and no lengthy pauses.

The 1955 summer holiday, six and a half weeks long, was different in two ways. Leaving their car at home, Franz and Lotte travelled by air, trains, and taxis. This holiday also included two conferences important for Franz. The first was the United Nations Conference on Peaceful Uses of Atomic Energy, held in Geneva in the middle weeks of August. During the fortnight of meetings, the Simons—and some other delegates—fitted in a few days in Zermatt. After Geneva, they moved to another favourite holiday spot, Mont Pelerin above Vevey, for another eight days. They then flew to Paris for the nine-day congress and associated meetings of the IIF. After Paris, they flew to Majorca for a further 10 days.

Viewed collectively, this ten-year sequence of holidays reveals much about the Simons' holiday preferences: for mountains over seashore, for mobility over ties to a single location, for smaller villages over areas of mass tourism, and for new regions and new routes over ones already seen. Holiday time was often happily combined with professional interests, reunions with colleagues, attendance at selected conferences, and visits to laboratories or industrial plants.

Amid a flood of minor data on weather conditions, hotel and restaurant ratings, and physical well-being in Franz's diary, there are occasional reminders that travel conditions were not always ideal. Days of "rain" or "heavy rain" are duly reported, and these sometimes led to a change of location. More serious were days of excessive heat or humidity, such as that experienced at Munster in the Vosges region in 1949. In an age when cars lacked air conditioning, neither Lotte nor Franz could adjust well to hot weather, and for Franz's heart condition such days could be acutely uncomfortable. On rare occasions, such as on the "very close and hot" west-to-east drive across France in August 1950, the diary notation "Lo[tte] drives" is a signal of Franz's discomfort. Symptoms recorded in the diary, however, were private, and rarely disclosed to colleagues, friends, or even to family members other than Lotte.

Professional travels

Franz's holiday travels, even with their occasional working digressions, were quite distinct from his rather strenuous tours for professional purposes. The most obvious examples of the latter are three extended tours in North America in 1949, 1953, and 1954, each one built around a scientific conference. The professional purpose of these tours was information exchange, and the benefits ran both ways. The lectures Franz gave at various universities and research laboratories enabled him to publicize the low-temperature work at the Clarendon Laboratory in Oxford, and in turn kept him informed on developments in institutions he visited and in North America generally. Honoraria from lectures were important for financing extended tours, because Britain's financial situation imposed strict limits on foreign currency for travel abroad. Beyond professional reasons, there was a human factor. These travels enabled Franz to renew contact with others who had shared and survived the trauma of the Nazi regime: scientists of the refugee diaspora, surviving relatives, and friends. Their experience had created ties of community of a depth that outsiders could never share.

For the 1949 visit, Franz travelled by sea, on the *Caronia* on 23 August from Southampton to New York, and back to Southampton on the *Queen Elizabeth* by 12 October. The focal point was an international low-temperature physics conference in Cambridge, Massachusetts, from 6 to 10 September. Franz's diary notes, among other things, the presence of colleagues from Oxford, Meissner from Germany, a dinner with the "IIF people" (low-temperature work), and "Oak Ridge people" (nuclear projects).

When the conference ended, he began a round of travel, spending a few days in Chicago, a few in New York, and then a dozen other places for one or two days each. Most of these stops were for reunions with refugee colleagues, but he also stopped briefly in Toronto, Chalk River, and Ottawa. In Toronto, there were reunions with friends who had helped the family there during the war years. Also, the Physics Department of the University of Toronto was seeking a new head, and was interested in Nicholas Kurti for the post. At Chalk River, Franz toured the British–Canadian heavy water reactor, and while there on 23 September he heard first news of the Soviet atomic bomb. A week later, in New York, he lunched with Manson Benedict and Percival Keith, colleagues of Manhattan days. The diary notes "long talk on bomb, England, etc."

In reunions with colleagues of the refugee generation, the topic most frequently mentioned was the alarming situation in Germany. The diary notes "talk on Germany" or a similar phrase at meetings with the Oldenbergs at their summer residence before the conference began; with James Franck in Chicago on 16 September, and again with Meissner also present three days later; with the Herzbergs in Ottawa on the 24th; with the Brickweddes in Washington on the 27th; with the Ladenburgs at Princeton on the 28th; and with the Herreys in New York—Mrs Herrey was his former student in Berlin—on 2 October. A high point was a visit with Einstein in Princeton on 29 September, at which they discussed Germany and other questions. Soon after Franz's return to England on 12 October, he wrote the second last of his letters to Lord Pakenham on 4 November, the letter in which he reported having met "several decent Germans . . . recently escaped from Germany" on his North American tour (see chapter 10). One of these recent arrivals was Ursula Martius. He met her at the University of Toronto physics laboratory on 20 September and again on the 21st, when his diary again notes "Germany" as the topic discussed.

In 1953, the plan for Franz's next tour was similar to the one for 1949: a scientific meeting in Boston, extended by visits to colleagues and relatives, with the extension paid for by lectures at universities he visited. There were, however, a few differences. The meeting in this case was a bicentenary celebration honouring the pioneer American scientist Benjamin Thompson, Count Rumford, born in Woburn, Massachusetts, on 26 March, 1753. The post-meeting travel plan was more ambitious: it included a visit to former colleagues in California. Further, the intellectual climate in the United States had changed appreciably in four years. In the interval had come the Fuchs case, defections of British diplomats to the Soviet Union, and heightened security fears in the United States. By 1953 many American academics were living in fear of irresponsible, damaging denunciations unleashed by Senator Joseph McCarthy.

An added incentive for this trip was that Dor and I were living in Cambridge that year. As Franz explained to a Harvard colleague, the first three days would be "mainly ... with my daughter whom I have not seen for almost three years."[1] He flew directly to Boston on 12 March, spent three days in Cambridge visiting us and contacting a few colleagues. He then went on to New York and Washington for a first round of lectures and visits, returning to Cambridge on 24 March for the Rumford celebration. Dor and I were living that year in a cramped one-room flat in a student rooming house behind the Harvard cyclotron. By chance, another room became vacant shortly before Franz's arrival, and Dor rented it so that he could stay with us while in the area. Though nothing was said, Franz was visibly appalled by our year of marginal existence.

The bicentennial celebration centred around the Rumford medals and medallists. Rumford himself, a wealthy man from two successive marriages to well-to-do widows, had provided in 1796 equivalent sums of £1,000 to the Royal Society in London and $5,000 to the American Academy of Arts and Science (or AAAS) in Boston to establish a biennial medal in his name for scientific discoveries involving heat or light, his own major areas of research. For the bicentenary, living medallists from both academies were invited to attend, and the conference guest list includes 19 previous medallists from the AAAS, four from the Royal Society, plus three new AAAS medallists to mark the special occasion. One of these was Enrico Fermi. The scientific portion of the meeting consisted of three half-day symposia, open to the public, reviewing recent developments in thermodynamics, atomic spectroscopy, and nuclear physics, respectively. For the session on thermodynamics, all four presentations were by Rumford medallists: Percy Bridgman (AAAS, 1917), Sir Alfred Egerton (RS, 1946), Lars Onsager (AAAS, 1953) and Franz Simon (RS, 1948).

On the convivial side, the celebration began with a birthday banquet, "an opulent dinner"[2] on 26 March. It featured dishes either invented by Rumford or improved in their preparation by his inventions: Rumford soup (a nutritious recipe first developed to feed the poor), meats cooked in a closed roaster, Boston-style Indian pudding, and coffee, "one of the most valuable ... luxuries of the table unknown to our forefathers." The banquet speaker was Sanford Brown, a physicist at the Massachusetts Institute of Technology (MIT) who became the leading authority on Rumford's chaotic scientific and administrative career.

On the next evening, the three new medallists were presented, with appreciations of their work, to receive Rumford medals from AAAS President Edwin Land. Enrico Fermi was presented by Peter Debye, himself a Royal Society Rumford medallist of 1930. Was it by chance or design that the scientist who had first produced the self-sustaining chain reaction should be presented by the scientist who had headed the Berlin institute requisitioned by the German army for its own unsuccessful attempt to do the same thing? Debye's three-page text reviews Fermi's work in the 1920s, and also in the 1930s, but passes over the period after 1939 in a single sentence, making no mention of the experiments in the Chicago football stadium, or Los Alamos, or the reasons that led the Fermi family to leave Rome for New York in 1939. For the audience, this was a moment of deep irony.

A report on this meeting for the AAAS *Proceedings* was written by the astronomer Harlow Shapley, who had chaired the organizing committee. He noted that "the academicians did well by themselves ... both in science and in wassailing"[3] at the celebration. Franz Simon wrote a note for *Nature* after his return, thanking the host society for a "quietly efficient" symposium, "lively discussions," and outstandingly successful meetings.[4] As Nancy

Arms testifies, "Simon was a great admirer of Rumford and was well acquainted with all his writings."[5] Yet the two men could scarcely have been more different in temperament or in political convictions. Rumford, an irascible, eighteenth-century authoritarian from Massachusets, had a world view far removed from Simon's, but they shared a remarkable interest in applying scientific discovery to improvements in everyday living. For Simon, Rumford's flaws of temperament and character were more than offset by his contributions to fireplace design, chimney design, household heating, kitchen ranges, more efficient cooking pots, the roaster oven, improved oil lamps, and more fundamental experiments to demonstrate the nature of heat. And a scientist who could spend years on methods for making the finest and tastiest coffee could hardly fail to win Franz's approval.[6]

After the Rumford meeting, Franz left Boston to continue his tour, heading first for California. Before he left, however, he had returned one evening to our flat with an unexpected, troubling discovery. He was deeply upset, sufficiently to speak out on things never mentioned before nor ever again within my hearing. It was Conant, he explained, who had been the architect of the sudden clampdown of security in the Manhattan Project in January 1943, and the subsequent breakdown of Anglo-American nuclear cooperation for eight months of that year. Franz had been talking with several former Manhattan associates and other Harvard and MIT colleagues at these meetings, and doubtless he heard this part of the story—or confirmed it—from one or more of them. When the official history of Manhattan was published in 1962, Conant's role in precipitating the rift would be documented in full, but in 1953 this was apparently known only to insiders.

There was every reason for me to remember vividly Franz's discovery and his distress. This was the first and only time that he ever spoke to me on any aspect of the atomic projects. His discovery that day forced him to revise fundamentally his view of the project. For 10 years he had attributed the security clampdown to the American military. This was understandable, even bearable. What was deeply hurtful was to discover that the source of the rift had been a trusted scientific colleague.

Franz's dejection was put aside during the trip to California. He was accompanied by Dor, and was in a particularly good mood as he renewed old friendships from his visiting term there in 1932. After eight days in Berkeley and Stanford, they flew back to Chicago. From there, Franz continued his tour for two more weeks, and Dor returned to Cambridge.

The third North American visit, in October–November 1954, was focused around a two-day symposium in Washington DC on temperatures. The tour was shorter—it fell in the middle of term—and Franz was accompanied by Lotte. As previously, the main stops were in New York, Washington, and Boston, but enough time was found for visits and lectures at Houston, Columbus, and Lafayette, places not seen previously. Towards the end, they had two days in Princeton. On 10 November they arrived in time for a lunch party at the Oppenheimers, an afternoon talk at the institute for Advanced Study, and dinner with the Oppenheimers and the mathematician Hermann Weyl, who came to the institute in 1933 after resigning from Göttingen. Over dinner, the diary notes "general talk, political, etc."

The next day, Franz went early in "wonderful weather" to the physics laboratory for general discussion and a laboratory visit. At 11 he talked with Niels Bohr, and at 11.30 had a further talk with Einstein, later joined by Lotte and Einstein's niece. For both meetings, the diary notes "disc[uss] general situation." Over lunch, however, Franz talked again with Hermann Weyl, and the diary notes "disc[uss] Germany, Weyl's reaction." By this date, the

German issue had receded but was not forgotten. After lunch, Franz and Lotte took a train to Newark Airport and a plane to Boston for the final days of their tour. Their time in Princeton had been memorable. Einstein and Weyl would both die in 1955, Einstein in April, Weyl in December.

Overall, 1954 was also a year of some new travel. Their American visit followed a summer holiday of 27 days and a month-long spring visit to Israel and Greece. The invitation to visit Israel came from the Weizmann Institute, Israel's centre for scientific research at Rehovoth near Tel Aviv. The first discussions of a visit had occurred during Franz's North American tour in April 1953. A formal invitation, dated 15 May, followed a month later from the academic secretary in Rehovoth: "We wish to start work on developing a Department of Physics and should very much like to have the benefit of your advice on planning, choice of subjects, etc."[7] The institute offered travel expenses for Simon and his wife, accommodation while in Israel, flexible dates for the visit, and an undertaking "to make your stay as pleasant and interesting as possible. I hope that we shall have an opportunity also to show you some of the country."

Franz was immediately interested, but because of other commitments the dates were eventually arranged for the interterm break in March 1954. In the meantime, with preparations continuing, he received a letter dated 14 December 1953 to announce that he had been elected to the institute's board of governors at its latest annual general meeting. "We all appreciate the great interest you are taking in our institute and we hope that our collaboration will be strengthened after your visit to us."[8] After this point, the preparations take on a more formal tone. The special arrangements began on their arrival at Heathrow Airport on 14 March. An official appeared, who boarded them first onto the plane. After stops in Zurich, Rome, and Athens, they landed at Israel's Lydda Airport at noon the next day, where they were met by Professor Isaac Berenblum and a colleague, and taken to Rehovoth. Berenblum had worked in Oxford for 14 years on cancer research before joining the institute at Rehovoth in 1950 as head of experimental biology.

The next morning, they would be driven to nearby Tel Aviv, where Franz called at the British Embassy, met the ambassador, and also visited the British Council office. In the afternoon, back at Rehovoth, discussions began. Over the next two weeks, he would meet department heads at the institute and other scientists, both collectively and individually, to discuss previous development and future plans for physics at the institute and elsewhere in the country. The institute had been founded to commemorate the 70th birthday in 1944 of Chaim Weizmann, who would become Israel's first president. By 1954, it had departments for specialized research in several areas of chemistry, physics, biology, and applied mathematics. Weizmann himself, a distinguished chemist and naturalized British citizen, had gone to Palestine and founded a predecessor institute for chemistry at Rehovoth in 1934.

In the following week the discussions widened, with a visit to the Technion at Haifa, followed by a drive further north, then east towards Galilee and Nazareth, south along the Jordan valley and return to Rehovoth. Two days later, they were driven southward to Beersheba, site of the government's Institute for Arid Zone Research, and then further into the Negev desert to visit a new kibbutz where Ben Gurion was living. The practical reason for these travels was to show Simon the country's economic and scientific potential. One resource was beyond reach. In the first week, Franz's diary noted: "No visit to Dead Sea owing to Arab attitudes."

A highlight of their first week was a private lunch on 19 March with Mrs Vera Weizmann on the Weizmann estate adjacent to the institute. Her husband, Israel's first president, had died in office 16 months previously, and the estate had been left by his will to the new state. Lotte's travel journal describes their hostess admiringly—"very good looking, slim, upright and elegant." After lunch they were shown the house, with its modernist design by Erich Mendelsohn, "elegant and tasteful furniture . . . Chinese pottery, sculptures, pictures . . . portraits and busts of Weizmann. . . . Afterwards walked through extensive, beautifully kept gardens, where Franz took photographs of flowering trees and flowers."

In the evenings there was informal socializing with institute faculty and visitors from other countries. There was also time to meet Franz's relatives living in Israel. Franz's diary records that on the morning after their arrival, after meeting the British ambassador in Tel Aviv, their next priority was to see "Dr. Mendelssohn's Zoo!" Heinrich Mendelssohn was a first cousin on Franz's mother's side. One day later, in the evening after his first public lecture, the diary notes "Fritz Kochmann turns up!" Kochmann, a first cousin on the Simon side, found them at the institute's guest house. The next evening, they went to the Kochmanns, who lived nearby in a "little cottage," for a longer visit. They would meet both cousins on Saturday 27 March, the Sabbath preceding their departure. In the morning, they went with the Kochmanns to the gardens of the institute. In the late afternoon, Heinrich Mendelssohn came from Tel Aviv, apparently alone, for supper and some serious discussion ("country's future, general conditions etc.").

Although both of these cousins had arrived in Palestine by the mid-1930s, their lives had been very different. Heinrich Mendelssohn, a Zionist in Berlin from the age of 12, had emigrated to Jerusalem in 1933 after completing a first degree in zoology and medicine at Berlin University. He obtained a doctorate in Jerusalem and became an eminent wildlife biologist, department head, and Dean at Tel Aviv University, bringing his parents from Germany in 1935.

On the Simon side, the family tree shows that Franz's aunt, Clara Simon, married Fritz Kochmann from Kattowitz in Upper Silesia in 1898, and that their family stayed on in Kattowitz after the area became part of Poland in 1919. The only son (named Heinz in the tree but Fritz in Franz's diary) reached Palestine by 1935, as evidenced by the birth of his son, Gad or Gadi, in March of that year in Gedera. Neither Franz's diary nor Lotte's journal lists any occupation for Heinz Kochmann or his wife, or mentions any other living relatives. The Simon tree shows the deaths of Heinz's parents in occupied Poland in 1942 and of his sister Marie in the Warsaw Ghetto in 1943. One older sister, Hanna, escaped to Paraguay. The three deaths are duly inscribed in the Yad Vashem Database, each documented by a careful page of testimony from Hanna's daughter, neatly printed in Spanish on a bilingual Spanish–Hebrew form, and sent to Jerusalem from Asunción in July 1986.

During the two weeks in Israel, Franz also gave four formal lectures. Two days after arrival, he spoke at the institute on "Nuclear Power." On 22 March, at the Technion in Haifa, his topic was "Absolute Zero," and this topic was repeated three days later at the institute in Rehovoth. On the 28th, as their departure date neared, he lectured on "Atomic Energy" in Jerusalem. These lectures were clearly subordinate to the interviews and meetings on policy and planning, which continued up to the evening before their departure.

While Franz focused on meetings and lectures, Lotte had more freedom to look around, sometimes on her own—a walk to the village of Rehovoth, a bus to Tel Aviv, a lift by car with Berenblum to Jerusalem—and also more distant visits with Franz to Haifa, Acre and the north, Galilee, the Jordan Valley, the Negev, and Jerusalem again. She kept her own separate journal on this trip to record impressions of two lands never previously seen. It is her journal that captures more fully than Franz's diary the human drama of a new country, born in hope and turmoil and war just five years before their visit. To some things she responds very positively: walking and taking photographs in beautiful weather in nearby orange groves; "the very sweet heavy scent of orange blossom" along the road to Tel Aviv; visiting an older, well-established kibbutz; admiring "lots of newly planted trees" along the road to Jerusalem; seeing fertile fields reclaimed from swamps "by Russian settlers 80 years ago" in the valley of Jesreel below Nazareth; visiting Roman and medieval ruins at Caesarea and Ashkelon; and viewing architecture and monuments of any period wherever she went.

Of some things she was more critical, including her first views of nearby towns and villages: "Tel Aviv very ugly, very crowded with cars & people . . . . All colours from very fair to nearly black skins . . . . Jaffa old Arab town, narrow streets, miserable hovels partly in ruins but all inhabited some by Arabs, some by African Jews." At Rehovoth village she saw "miserable looking shops" and a "picturesque little market," but even in Tel Aviv she found the shops "not very exciting" apart from some "very nice Yemenite silver things." A darker side was frequent evidence of recent fighting, especially the Arab villages through the countryside, some deserted, some destroyed, some occupied by newly arrived immigrants. En route to Jerusalem on 25 March, she notes "many burnt out cars by roadside as memorials to the siege of Jerusalem."

Once arrived in West Jerusalem, she describes the city, its architecture, and its inhabitants at some length. She surveys the city wall, roofs, and domes of the Old City across the armistice line under Jordanian rule, but also notes: "Frustrating not to be able to go there." In the Mea Shearim district she details features of the ultra-Orthodox houses and dress, but finds the comportment of the young men and boys irksome: "Their swaying gait gives impression of indescribable arrogance." In a private journal, she felt free to speak with total frankness, but she also remained open to revising her first impressions. On the drive from central Israel south to Beersheba, she noted: "Small 'wild West' type of town." Later, after visiting the primitive new kibbutz, Sde Boker, further south in the Negev desert where Ben Gurion was living, Beersheba "seemed far less desolate but rather a centre of civilisation by comparison."

As their departure approached, formalities reappeared. On Sunday 28 March, they had a day in Jerusalem—a second visit for Lotte—which featured lunch with the director of the Prime Minister's office and his wife. Lotte notes: "Very pleasant modern flat in Rehavia, intelligent youngish man & wife." Their host was Theodor (or Teddy) Kollek, who ran the office for 12 years under Ben Gurion, and who became mayor of West Jerusalem in 1965 and of united Jerusalem in 1967. Governmental science policy was coordinated at this date largely through the Prime Minister's office.

After lunch, they were taken for a brief visit to the house of the philosopher Martin Buber and his wife. Lotte notes: "Wonderful old man, long conversation about Ebeth & H[enschel] with him and his wife." At 5 o'clock Franz gave the last of his public lectures in the Hebrew University's physics laboratory, and afterwards met for half an hour with a member of the Atomic Energy Commission, Ernst David Bergmann. Bergmann had studied and worked at the University of Berlin as an organic chemist until forced into exile in 1933. He went to

Palestine, did research at Rehovoth, becoming a government scientist after independence and first chairman of the Israel Atomic Energy Commission in 1952. The Simons had a private dinner with friends that evening, stayed overnight in the King David Hotel, and did more sightseeing the next morning before their return to Rehovoth.

At mid-day on Monday, they were picked up in Vera Weizmann's car and went with her to a diplomatic luncheon with a few British and American officials at the home of the British ambassador, Sir Francis Evans. Franz notes that the discussion included both physics and "pleasant talk." At 3 p.m. they returned with Mrs Weizmann to Rehovoth. The ceremonial side was over, but one more working meeting was still scheduled. The Simons had dinner with three senior scientists and their wives, but the men left early. Lotte notes "Men had long meeting"; Franz writes "8^{30} Mtg. on Physics, rather lively . . . - 10^{45}." Up early on Tuesday morning, they left Rehovoth at 9.30, for a take-off at 10.45 and flight of two and a half hours to Athens.

The two weeks in Israel had been strenuous for Franz. They were also rewarding, for the visit came at a critical stage in the scientific development of a new country. For Lotte, the weeks in Israel offered more time for exploring a new society and for retracing the careers of Ebeth and her husband in Palestine. In Greece, for reasons probably unforeseen at the planning stage, the work component would almost disappear, and the Simons would become serious tourists for almost all of their stay.

Their itinerary allowed for 16 days in Greece. For both, it was a first visit to a country familiar since their school days as one of the cradles of Western civilization. On arrival, they went to the British Council office and developed plans: a two-day trip to Delphi; a drive to Cape Sounion south of Athens; a two-day excursion through the isthmus of Corinth into the Peloponnesus to Nauplion; a one-day trip by boat from Piraeus to Aegina to visit the temple of Athena; a three-day visit, with travel by air, to Crete; and time in Athens for the Acropolis, current excavations, museums, and other sites in the city. Franz also agreed to give one public lecture at the university; this was arranged for 7 April, midway through their stay.

The plans for a public lecture developed complications traceable to international tensions between Britain and Greece. The global wave of decolonization had awakened in Cyprus a movement for enosis, or union of Cyprus with Greece. From a British standpoint, Cyprus remained strategically important as a Crown colony. A series of unstable Conservative governments in Greece, which depended on British and American economic and military assistance in defending a long northern border against Balkan neighbours, was reluctant to challenge Britain over Cyprus. At the time of the Simon visit, the British and Greek governments were both exercising caution to avoid any action that might inflame a potentially dangerous situation. In contrast with their visit to Israel, no British ambassador or embassy official appears during these two weeks. They did meet extensively, however, with British Council staff, who made every effort to make their visit memorable in cultural terms.

In both diaries, the description of the public lecture at the university is brief, and slightly odd. It was scheduled for late afternoon on 7 April. In mid-afternoon they were picked up and driven to the university to have tea with the professors. Both diaries then note: "Police for protection!" Both note that the lecture came next, but Franz adds a detail: "finally abbrev[iated] lecture," which suggests a demonstration or delay. Untypically, he makes no comment on the lecture topic or its reception. Both diaries move directly to their unrelated dinner arrangement in town that evening. The only other contact with faculty members occurred off campus a week later, at a private party in their honour at a professor's house on the eve of their departure. Franz sums it up: "Party for us. Wonderful flat and most excellent cold Buffet. About 40 people, professors. Talk to many." Their visit had come at an inauspicious time, and this may have been the best that could be expected.

If university contacts were limited, there were compensations elsewhere. On two of their excursions, to Epidaurus and Nauplion in the Peloponnesus and to Phaestos in Crete, they shared a car and driver with the Oxford bookseller Basil Blackwell and his wife, who were visiting Greece at the same time. These shared experiences led to a lasting friendship. Among the Simons' contacts at the excavations of the Athenian Agora were the archaeologists Homer Thompson and his wife Dorothy, friends of Lotte from her wartime years in Toronto. On 14 April her journal records a visit to the excavation site and tea with the Thompsons afterwards.

In these two weeks in Greece, Lotte found herself at the heart of subjects for which she had first been educated. She was embracing at first hand the art, architecture, and archaeology of the first flowering of Western culture. She found examples in museums, at excavation sites, and on their excursions outside Athens. In her youth she had absorbed the story of Heinrich Schliemann, the businessman turned archaeologist who took Homer seriously and went in search of the sites that Homer described, starting with his excavations of Troy in the 1870s. In her old age, Lotte spoke admiringly of Schliemann's career. If we look more closely at the Simons' itinerary, the links with Lotte's education in Germany become clearer. Their bus tour on 1 April to Delphi, with its rich oracular tradition, was an obvious starting point. The boat trip to Aegina on 8 April was mainly to view the famed Temple of Athena (earlier of the goddess Aphaia), the pediments of which Lotte may have seen in a museum in Munich. Their visit on 5 April to the graves at Mycenae, one of the oldest cities in Greece, where Schliemann came to dig in 1876 after his first triumph in Troy, was directly in his footsteps. Sensationally successful in both places, Schliemann had boldly proclaimed his golden discoveries as "Priam's Treasure" and the "Mask of Agamemnon," respectively.

Other sites competed for Lotte's attention: the marble of the Acropolis, viewed in varying light and weather; archaeological work in the Agora—"enjoyed it even more than first visit"; the theatre at Epidaurus, with its phenomenal acoustics; the palace at Knossos; the many museums, with high points from each carefully noted. Her focus was not narrowly on antiquities. She was also open to later historical and cultural developments: the Byzantine Museum; "a lovely Byzantine church, brick pattern, dark inside, very small," in the old part of Athens; a remote Byzantine monastery, the search for which was called off owing to an impassible road. She notes the work of the Venetians in building defences against the Turks, including impressive walls and fortifications at Corinth and a Venetian "fortress with lions" at the harbour in Heraklion. Perhaps most of all, her journal is constantly sensitive to the natural beauty of the settings and seasonal beauty of abundant spring flowers in many places that they visited. At Phaestos in Crete, with its view of Mount Ida with snow, she summarizes in the diary: "Excavations not particularly interesting. Country beautiful."

Climacteric year: 1956

The Ancients had much to say about the stages and turning points of human life. Their beliefs on the subject were fed by traditions of astrology, numerology, and medicine. In numerology, numbers commonly deemed significant for human lives were 7, 9, or multiples of 7 or

9, or the product of 7 by 9, that is, 63. Though these notions had a long life in Western traditions and were echoed in Shakespeare's *Seven Ages of Man*, Franz was far removed from such beliefs. Had anyone ventured to suggest any special significance—good or bad—for his sixty-third year, he would undoubtedly have dismissed it, as any trained scientist would, by a simple reference to a mortality table or similar statistics. At its beginning, 1956 looked likely to be a year similar to any other, and just as busy.

Franz's outlook and plan for the months ahead are briefly summarized in a letter he sent to Max von Laue dated 2 February:

> Now to your question whether I would come along and give a lecture in Berlin. In principle I would be very much interested, particularly as I was born in Berlin and lived there more or less for the first 40 years of my life, and I have not been back since 1933. The difficulty is to find the time. At the moment I am busy preparing the Guthrie Lecture which takes place in the middle of March and then I want to go to Sicily and have a holiday. Perhaps we could arrange something for the summer or early autumn, before I go out to Pasadena where I shall be lecturing for a month.
>
> If sometime during the summer would suit you I shall try to find a suitable date. Incidentally it is some time since you visited us and I wonder when we could count on seeing you here again.[9]

Four points in these plans deserve notice. He is more open to the renewed invitation to visit Berlin, and security concerns are no longer an obstacle. The Guthrie Lecture, a major enterprise, has already been noted (see Chapter 9). It was delivered in London some six weeks later, on 13 March. The following morning, Franz and Lotte would fly from London to Rome to begin their planned Sicilian holiday. The continuing lure of California had become by 1955 a firm plan to lecture at Cal Tech for a month. This was then postponed to 1956, but not abandoned until after Simon's election to succeed Cherwell.

The visit to Sicily, though clearly designated as a holiday, was different from their usual summer holidays in several ways. First, it was shorter, some 22 days away from Oxford, and only 16 in Sicily itself. Second, they shared this holiday with Mimi and Ludwig Frank, Franz's sister and brother-in-law, the only time in the postwar period that the Franks and Simons holidayed together. Their daughter Kay, by this time a lecturer in art history, also joined the group. In a Mediterranean climate at this time of year—mid-March to early April—the spring weather was unpredictable. For Franz and Lotte, the Sicilian visit had more in common with their visit to Greece in 1954 than with any of their longer summer holidays, and the itinerary for Sicily may well have evolved from their Greek experience.

The tour of Sicily was an approximate circuit of the island, anticlockwise, starting from Palermo. Franz's diary spells out the daily details, and these include brief reactions to places visited as well as his usual record of weather, roads, hotels, and meals. On their departure from Oxford, Lotte began one of her typical travel journals, but this account breaks off abruptly and without explanation after entries on their first two days in Palermo and vicinity, leaving pages for the circuit of the island blank.

From the first day in Palermo, the visitors had a clear plan for what they wanted to see. After surveying the city's famed setting—Monte Pellegrino and the Mondello beach area—they headed for the eleventh-century church of St John of the Hermits, with its combination of Norman and Arab architecture. From there they went on to the cathedral or Duomo, founded and built in the Norman period, but enlarged and restyled several times in later centuries. On the second day, they visited other villages close to Palermo, the highlight being the cathedral at Monreale, with its Benedictine cloister. On the third day they visited Palermo's archaeological museum, and the Royal Palace with its rich collections of Byzantine, Arab, and Norman art.

On 21 March, the travellers headed west from Palermo towards the northwest corner of the island. They stopped for two hours at Segesta to see and photograph the magnificent Greek temple, a large Doric temple of the fifth century BC, remarkably well preserved but never completely finished. They continued to Erice on the coast (Eryx in classical times) and stopped overnight at Trapani, founded by Greek colonists as Drepanon, later under Carthaginian and then Roman rule. In the morning they continued southward around the coast heading for Agrigento. Along the way, just after lunch, they reached Selinunte, site of one of the largest Greek temples in the world. In the seventh century BC, Greek colonists had built a rich, walled city that flourished for four centuries before its destruction around 250 BC in the First Punic War. Franz's diary notes: "Sun finally out, beautiful position, stay about 2 hours ... very bad roads, but beautiful country." Reaching Agrigento around 7, they went to the Jolly Hotel. After dinner, the diary notes: "Meet Br[itish] couple from Salerno, exchange experiences." They were like explorers, meeting and comparing discoveries, in an unknown land.

Selinunte and Agrigento each had ruins of several temples close by. The difference was that at Agrigento the temples were close to a sizable town. On the morning of 23 March, the travellers drove out in mid-morning to the Valley of the Temples, where a unique cluster of Greek temples and ruins of temples stands outside the modern town. By noon the sun had disappeared. After a picnic lunch in the car, they headed eastward, in high winds and rain, towards central Sicily, aiming for Enna and Piazza Armerina. The diary best conveys the scene:

> Rain, slippery road, lorry forces me into ditch. 3 cars stop, help us out! fantastic lightning ... appalling storm rises. Turn back about 4 km. before Enna ... road to Piazza Armerina. Storm worsens. Sheets of water coming down. Finally around 7 P. Arm[erina]. All hotels full! At last Jolly hotel arr[anges] things.

Eventually they were fitted into the full hotel by displacing a group of young geologists for one night.

The next morning they drove a few kilometres to see the large Roman "Villa del Casale," built in the fourth century AD, and renowned for its extensive mosaics, which Franz notes as "very beautiful." The weather then turned bad and the travellers visited Caltagirone, a town in the mountains known for its ceramics. By noon, they had a picnic in the car and headed eastward towards the coast. Around 4.30 p.m. they reached Siracusa, or Syracuse, founded by Greek colonists from Corinth in the eight century BC. In time it became the most powerful Greek city in the Mediterranean world, visited more than once by Plato in the fourth century, and home to Archimedes in the third.

The travellers spent a week on the east coast: two days in Syracuse, 4 in Taormina, and a day driving up the coast between them, with a stop and "short stroll" in Catania. Their sightseeing in and around Syracuse included a visit to the Eurialo, a massive fortification started by Dionysius the Elder in the fourth century BC to defend the city against the Carthaginians. They also drove to Noto, 35 kilometres to the southwest, a town rebuilt completely in Baroque style on a new site after its predecessor, Noto Antico, had been totally destroyed in a violent earthquake in 1693. In Syracuse and Taormina they saw impressive Greek theatres. The weather for most of this week was cloudy or "indifferent." Their much-anticipated viewing of Mount Etna from different angles was reduced to occasional glimpses, or "very faint views" through cloudy skies.

By 1 April, it was time to begin their return. At 10 a.m., after a last visit to the Greek theatre, they left Taormina for Messina. From there they headed up the mountain, down to the north coast, and west towards Tindari, founded by Dionysius the Elder in 396 BC, the last Greek city to be founded in Sicily. After a stop there, they drove "along beautiful coast, stop from time to time," until they arrived in Cefalú, where they stayed overnight. The next morning they visited the town's twelfth-century Norman cathedral before continuing towards Palermo, where they arrived just after noon. Since the Franks were staying on longer there, a priority was to see them into their hotel. The afternoon was still free for one more look at favourite places seen before: the Eremite church, Monreale, the Duomo. By 6.30 p.m. the car was loaded on the boat for Naples. The diary records "dinner at 8 . . . sleep quite well," and waking up "at Naples harbour" on 3 April.

Disembarked by 6.30 a.m., they had most of the day available for sightseeing. After breakfast in town, Franz drove Kay to Pompeii. The diary continues: "Return to Herculaneum, up Vesuvio, cloudy. Walk up to crater, quite impressive, from above. Down, on to Pompeii around 1.45. Short look at excavations. [Rejoin] Kay." The crater that Franz found "quite impressive" had been formed by a major eruption of Vesuvius in 1944. At its formation, it was about 300 metres deep by 500 metres in diameter, and massive lava flows from the eruption swallowed up two villages on the slopes below.

At 3.15 they returned from Pompeii to Naples, and soon after left Naples for the journey north to Rome. They stopped overnight at a hotel on the coast at Formia, almost half way to Rome, and had a walk on the beach before dinner. In the morning they completed the drive to Rome's Ciampino airport, with time for a "very good" lunch and a stroll in nearby Frascati. After take-off at 3.15, they had a "quiet flight [and] fine view of Alps," reaching Heathrow at 6 p.m. local time. This spring holiday, short as it was, maintained the Simons' tradition of finding out-of-the-way destinations not yet developed for mass tourism. They had encountered challenges—bad roads, indifferent weather—but had come through in good shape, physically intact, intellectually and culturally stimulated.

The day after their return from Sicily, 5 April, was Cherwell's birthday. Franz's calendar has a lone entry: "Cherwell 70." The diary, terse and factual as usual, has just two lines. Back at the laboratory after lunch at home, he notes: "Cherwell. 70th birthday, keeps it quiet." On a separate line above this, he notes: "Cherwell tells me that I shall probably be his successor." On the next afternoon, he mentions a longer-than-usual discussion with Cherwell from 4 to 6 p.m. on "Sicily, politics etc." Since Franz had been absent and remote from news for over three weeks, their talk probably ranged widely.

Cherwell's concern for his own retirement, and for the continuing welfare of the Clarendon Laboratory, had begun in the autumn of 1955. His appointment to the chair in 1919 had been without limit of time, and was not affected by a university statute of 1925 that set a fixed retirement age for faculty. At the same time, he could not forget the conspicuous damage that physics in Oxford had suffered as a result of two predecessors who had lingered too long: the 50-year tenure of Robert Clifton at the Clarendon from 1865 to 1915 and the 41-year tenure of John Townsend at the Electrical Laboratory from 1900 to 1941.[10]

Townsend, who had a promising early career, became increasingly marginalized after 1910, when he rejected quantum physics and related developments.

A precipitating factor in Cherwell's decision was the loss of two leading Clarendon physicists to other institutions. Harrod's memoir of Cherwell describes a discussion of these losses and the need for their replacement. Cherwell "pointed out to me that it was impossible. No first-rate physicists would be attracted to Oxford without some security of tenure of the position offered."[11] Such an assurance could only come from Cherwell's successor. Though Harrod does not name the two who left, one can deduce that they were Maurice Pryce, who was invited in 1954 to succeed Nevill Mott at Bristol, and Hans Halban, who moved to Paris in 1955 at the invitation of the French Prime Minister to build France's nuclear research centre at Saclay.

One major obstacle delayed arrangements for Cherwell's retirement. Though his chair was formally attached to Wadham College, he had lived since 1922 in a comfortable set of rooms in Christ Church overlooking Christ Church meadow. At 70 he could not bear the thought of leaving them and rearranging his living conditions so fundamentally. As his biographer notes, "he embarked on a long correspondence with the college and university authorities, and, in the end, by a unique decision, the Prof was allowed to keep his rooms after retirement, paying, at his own request, rent for them."[12] Such an honour, Harrod notes, "had never been done before within living memory."[13] The special arrangements included continuing membership in the college's senior common room.

By the time that Franz returned from Sicily, the university registrar had initiated steps to choose Cherwell's successor. A notice dated 14 March appeared in the *Oxford University Gazette*, announcing the expected vacancy and inviting applications. In late March the date for the electors to meet was set for 29 May. On 9 April one of the electors, Sir Lawrence Bragg, sent regrets that he could not attend this meeting, but at the registrar's suggestion he agreed to send his views in writing through another elector, Sir George Thomson, a colleague from Imperial College who had become master of a Cambridge college (Corpus Christi).

As the date for the meeting approached, the tempo speeded up. On 21 May, the Vice-Chancellor sent the electors an agenda, attaching to it the minutes of two meetings in February 1919 that had elected Lindemann to the chair. On the 22nd, the registrar informed the electors that the Warden of Wadham, Maurice Bowra, "has stated that he would be grateful if Sir Francis Simon could be considered as a candidate."[14] On the 24th, Bragg wrote to Thomson to announce his support for Simon: "He has now become such an international figure, and done so much to build up the Oxford school."[15]

The board met on 29 May. Its sparse minutes reveal that Simon was elected without recorded discussion or formal vote, apparently by consensus. In choosing Simon, the electors downplayed Cherwell's concern for long-run stability, for Simon himself would reach the university's statutory retirement age in four years. The board paid more attention to other matters. It recommended that the money for Simon's former Chair in Thermodynamics be retained in the department and "be available for a post of similar status." The members also agreed to be consulted again "about the field in other branches of physics before the other principal posts in the laboratory were filled."

On 30 May, the day after his election, Simon prepared two letters.[16] The first was to the university registrar, to acknowledge "your letter informing me of my election to the Dr. Lee's Professorship, which I have great pleasure in accepting." He moves immediately to the financial outlook for the next quinquennium: provided that present budget demands are met, "I will have no major requests to make." He does, however, have an important proposal to introduce:

> I intend to continue to look after the Low Temperature Department in the Clarendon Laboratory and thus there would be no point in appointing a new Professor of Thermodynamics. What we need most at the moment is someone to take care of the nuclear physics department and I very much hope that a position similar to the one which I have held till now can be created for a nuclear physicist.

The second letter, briefer, and marked "STRICTLY PERSONAL," is a note to Cockcroft:

> The electors to the Dr. Lee's Chair have elected me yesterday to succeed Cherwell on October 1st. We have now to do some thinking about the future of nuclear physics and I would very much like to discuss this whole complex question with you.

He proposes dates for a possible early meeting. The brevity of the second note suggests that Cockcroft needs no further explanation of the issues for discussion. Indeed, the speed of Simon's actions suggests that the plan for a Chair in Nuclear Physics has been worked out in advance, approved by Cherwell and discussed with others, and that it needed only a decision of the electors for Dr Lee's Chair before it could be set in motion. Further, Simon had in view a strong candidate for such a chair, Denys Wilkinson, currently Reader in Nuclear Physics at Cambridge. On 11 July, Simon writes another note to Cockcroft:

> I saw Wilkinson in Cambridge last weekend and he seems to be very interested indeed in the Oxford position. He has a few questions concerning the van de Graaff which I would like to discuss with you rather soon.

However, Simon explains that a personal problem has just arisen:

> Unfortunately, I had a haemorrage in the right eye a short time ago and have been ordered by the doctor to keep very quiet for the next few weeks, thus I am going to stay at home and do my work from a deck chair in the garden.[17]

He asks Cockcroft to look in if he will be in Oxford soon. The diary shows a meeting with Cockcroft on 14 July.

Forty-six years later, on 21 July 2002, a ceremony was organized in Oxford to rename the university's Nuclear Physics Laboratory as the "Denys Wilkinson Building." On that occasion, Wilkinson, who had been knighted in 1974, reviewed in some detail the history of the laboratory, beginning from his years in Cambridge, his recruitment to Oxford, and the protracted struggle over the next 14 years for sufficient resources to build one of the leading centres for nuclear physics research in Europe. In particular, he describes Simon's first letter of June 1956. It offered a professorship, but stressed that "the future of nuclear physics here is intimately bound up with very close collaboration with Harwell." Simon's well-timed offer coincided with Wilkinson's growing realization that his application for an American-built accelerator for Cambridge, though scientifically approved by DSIR, had little or no possibility of adequate funding. "I told Simon that I was interested in Oxford's proposal . . . . Towards the end of July 1956 I decided that I would indeed move to Oxford. I also decided that I would not press for an accelerator there but rather rely on the Harwell and Aldermaston machines."[18] Both of these centres had promised that their large accelerators would be made available to university researchers.

A further step remained. The university's Hebdomadal Council requested the electoral board for Dr Lee's chair to reconvene and to give its opinion on filling the new chair to be created from Simon's former Chair in Thermodynamics. The move to recruit Wilkinson, planned and launched before Simon's formal accession to Dr Lee's chair on 1 October, opened the way to major future developments in nuclear physics in Oxford that would be comparable to Lindemann's adroit recruitment of the low-temperature group from Breslau in 1933.

Simon's letter to Cockcroft of 11 July, which explains the eye problem and the treatment for it, is dated on the first day of the new, "very quiet" regime marked as such in his calendar. From this date until the beginning of term in October, the calendar designates each day as "at home," or later "in bed," or later still, "in chair." Some eye trouble had been noted late in April. On 30 April a specialist diagnosed a thrombosis in the right eye. Franz then received injections during May and June, and was treated—successfully, in his own view—for glaucoma. Then a second eye incident, in Cambridge early in July, led the doctors to impose a quiet period for several weeks. In the 1950s, a treatment of quiet rest was deemed appropriate for the eye problem, and also for his chronic heart condition.

What Franz understood by "very quiet" was a reduction and curtailment of physical activity. He applied no similar restriction to his mental and intellectual activity. He continued to think, plan, discuss, and write about his usual activities and new responsibilities. The daily timetable shows periods for preparing letters with his secretary, for handling other business with his current assistant, John Wilks, and for visits to the house from colleagues, students, and others. At weekends, academic visitors were replaced by relatives and family friends. Diary entries for these months also show more time than usual spent on casual reading or radio listening.

The top issue through the summer and autumn was the recruitment of Wilkinson. Arrangements were still incomplete when Franz returned to the laboratory, for the first time in more than three months, on 15 October. He met with Wilkinson on the 15th, and again on the 16th, for a wide-ranging discussion of the move to Oxford, personnel needs, machines, a college connection, housing, and schools. On the 18th a meeting followed with the registrar to discuss space and funding implications. The so-called Keble triangle, bounded by Banbury Road, Parks Road, and Keble Road, had been designated by this date for future science needs, and the new nuclear physics laboratory, to be built in the early 1960s, would occupy the southwest corner of the triangle. Though Wilkinson and his family would not move to Oxford until the summer of 1957, the foundations for the move had been explored in detail before the electoral board met again to approve the appointment formally on 27 October.

Another task of the summer was Franz's part in preparations for the next international conference on low-temperature physics, scheduled for the summer of 1957 in Madison, Wisconsin. By this date the low-temperature section of the International Union of Pure and Applied Physics (IUPAP) had developed a pattern of sponsoring and supporting a biennial conference on low-temperature research. Franz had been active at previous conferences, notably at Cambridge, Massachusetts, in 1949 and Oxford in 1951. For 1957 Simon was the representative of IUPAP's low-temperature commission to the Wisconsin organizing group, and this involved not only dealing with IUPAP but also approaching other bodies, such as the IIF, for additional support. On 12 October, he could report to the Wisconsin group a tangible success: a financial grant from IUPAP towards costs of the 1957 meeting.

As a principal organizer of earlier meetings, he had every intention of attending this conference in person. Faced with his new obligations, he had reluctantly abandoned his earlier plan for a month-long visit to Cal Tech in Pasadena. The Wisconsin conference became a new anchorage point for building another visit to North America.

Other activities in this summer of eye problems and working at home demonstrate the continuing diversity of Franz's interests. Beyond routine administration at the Clarendon, he gave prompt and caring attention to all correspondence, whether from friends, colleague, or strangers, no matter how trivial or complex the issues. On 5 June, in the midst of his eye injections, a London rental agency wrote to ask for an opinion, in confidence, on Michael Perrin, who was proposing to lease a house: would he be "a respectable and responsible tenant and fully able to pay" a rental of £350 per annum? In 1953 Perrin had left the public sector and ICI to become chairman of the Wellcome Foundation, which under his guidance over 17 years became one of the major British charities. By return of post, Franz assured the agency on all three counts.[19]

On occasion, Franz initiates an exchange. In August, he noticed a publisher's announcement of a forthcoming volume on popular science projected into the near future, a view of the world as it might be in 1982. The author and his wife, Egon and Ursula Larsen, were both known to Franz already. Egon Larsen (1904–1990) was a German science journalist, who, after being forbidden to write by the Nazis, moved to Prague in 1935 and to London in 1938. His numerous books on science were translated into many languages, and Franz thought it important to avoid the dissemination of errors. On 30 August Simon wrote Larsen a two-page letter, gently chiding him for the utopian social assumptions of the announcement, but challenging more seriously two unrealistic scientific predictions. The first projected "coastal power stations [to] tap the energy of the oceans." The second questions the author's intention to conclude the volume with a report on the "first voyage to the moon." It would be a pity, Franz advises, to spoil an otherwise valuable book by "such over-statements."[20]

Larsen replied, with gratitude and at length, on 13 September. The societal utopianism, he explained, was the publisher's idea, not his own. His imagined coastal power stations were neither tidal nor wave-based, but "of the thermal type," a type of which Franz is equally dismissive in a further note of 1 October. They have "no long range future."

Their respective positions on a moon landing are more interesting. Franz's original letter argued against its likelihood partly on technical grounds, but also—and more strongly—as a misplaced priority:

> Even if it were technically possible by that time – which I very much doubt – the effort would be so enormous that nobody could justify such an enterprise when so much remains to be done to improve the lot of people living on earth.

Larsen had a more detached view of human nature that time would vindicate:

> Well, that voyage to the moon . . . . I am afraid it will come, mankind being what it is . . . . Men, alas, love the spectacular stunt.

Early in October, Franz received a request for an interview from a young physicist with a PhD. from the University of London, currently lecturing at Battersea Polytech. The writer, Wolfgang Rothenstein, was aspiring to work in physics in Israel, and was aware of Simon's "considerable interest in the work of the Weizmann Institute and other scientific projects in

Israel." Could Simon meet him to discuss the prospects there? Franz replied on 10 October. He was still working from home, but expecting to be back in the laboratory the following week: "Perhaps we could provisionally fix Wednesday 31st October," he suggested, with a definite time to be set later.[21]

It was an appointment that could not be kept. On the 29th, Simon's secretary wrote to Rothenstein to explain: "Yesterday afternoon he was admitted to the Radcliffe Infirmary here for observation and of course we do not know just how long he will be kept there." The appointment would have to be rescheduled. But this too would not happen, for Simon would die, unexpectedly, two days later.

The diverse summer correspondence, some willingly accepted and some initiated by Simon himself, contains an obvious message. He is determined to continue as before his broader roles as public educator, mentor to individuals, and reliable friend. The message is reinforced by a short draft article found among Franz's papers, probably the last piece that he wrote. Untitled, undated, and apparently never published, it revisits the energy outlook for Britain after the Suez crisis. In essence, Britain will face after the crisis a dangerous energy dependency on Middle East oil, and supplies will be vulnerable to interruptions. The gap will remain acute until about the end of the century, when expansion of nuclear power can reduce dependency on foreign oil. Until then the gap will be dangerous, but much of the dependency could be reduced or even eliminated by a serious national fuel policy to improve the efficiency of coal usage in British homes, offices, and factories. The message is unchanged, but Suez has given it a new urgency.

On his return to the Clarendon in the week of 15 October, Franz looked forward with optimism as he replied to recent correspondence. To his former student and colleague Keith MacDonald in Ottawa, he announced on the 17th that he had been "back in the lab for the last two days, but only in the morning." He is regretful that "I have to miss the Calder Hall opening today,"[22] the inauguration of Britain's first nuclear power plant to be linked with the national electricity grid.

On the 19th, he wrote in similar vein to two long-term friends, to von Laue in Berlin and to Max Born, now settled in a new house in Bad Pyrmont near Göttingen. To Born, a close confidant on numerous past occasions, he writes of his condition more openly than to others:

> I should have written to you a long time ago, but as you know if one does not feel too well, it always takes an effort to do even the simplest things. My eye is now getting much better, but of course I am still rather weak from having been immobilised for nearly three months. However, I have started coming to the laboratory this week in the morning only, and so please forgive me if I am very brief and reply only to your question about Industrial Distributors . . . . I shall write again as soon as I have got through the bulk of the work here.
>
> Dorothy and her baby left 10 days ago and are now back in Canada where the next baby is soon to arrive. Kay has a job in London so that fortunately we see more of her now.
>
> Lotte joins me in sending our kindest regards to both of you.
>
> Yours ever,[23]

The optimism and promise of the first week back in the laboratory did not hold up. On Monday 22 October, there was a lunch in Christ Church for some Russian visitors, attended by Cherwell, Cockcroft, and Simon. After lunch, Simon took the Russians back to the Clarendon for a tour, but arranged for his colleague Keeley to give them the tour while he went home, in discomfort, to rest. The next day he felt better, but worked from home, following an order from Ludwig Frank to "stay very quiet." The week continued in this way, with work and meetings shifted to Belbroughton Road. The diary entries show that four of the nights in this week were less comfortable than the days, requiring pain relief from nitroglycerin pills. Saturday 27 October—the day of Wilkinson's formal election—is the last entry in Franz's diary. The calendar lists entries for Sunday 28 October—"To hospital, Doctors' palaver."—and for the 29th—"Radcliffe." After this point there is no written record.

On 1 and 2 November, newspapers in Britain and abroad carried the stunning news that Simon had died during the night of 31 October. The early reports of his death emphasized its extreme untimeliness and shocking suddenness, barely a month after formally assuming his new post. In time, longer obituaries would add details, and some perspective. As Nicholas Kurti reports: "In the summer of 1956 he was taken ill with coronary heart disease, a condition which had first shown itself 10 years earlier. The seriousness of his illness was known neither to his friends and colleagues, nor, fortunately, to him."[24] Over 10 years, since his first heart episode in December 1946, he had learned how to manage his condition and keep trouble at bay. When symptoms reappeared in 1956, he fully expected that the same remedies would work again. The diary of these months is a chronicle of ups and downs, good days and bad days, general optimism and occasional depression, but never a hint of surrender to his disease.

It fell to the 80-year-old Clemens Schaefer, in an obituary for a German science journal, to capture the full depth of the tragedy. He did so by reviewing the four references to Simon that appeared in the pages of *Nature* in 1956: an opening address in March to a low-temperature crystallography conference held at the Clarendon; Cherwell's retirement and Simon's appointment to Cherwell's chair; his team's spectacular breakthrough in nuclear cooling to approach absolute zero; and the notice of his death. Three spectacular triumphs, then a "shattering *memento mori*,"[25] all in a single climacteric year.

APPENDIX: SIMON'S REPORT TO PERRIN ABOUT HEISENBERG, 11 MARCH 1948

The following pages show a full text of Simon's letter to Michael Perrin following Heisenberg's visit to Oxford on 8 and 9 March 1948. For the context of this letter, see chapter 10. (Source: Simon Papers).

CONFIDENTIAL

11th March, 1948

Dear Perrin,

Heisenberg, who as you know was staying in Cambridge during last term, has just been to Oxford on a visit for two days. I had a number of talks with him and some points may be of interest to you. He wanted to talk to me about Goudsmits' book, which obviously infuriated him very much; he asked me to be quite frank about the matter – and I was!

Heisenberg claims that German scientists had no other wish then to prevent Hitler from getting the bomb. They knew about everything, including the fast neutron reaction and the possibility of using plutonium, but all their actions were determined by their aim to mislead Hitler and the "high ups" about the possibilities of a bomb. He said that if he had gone to Hitler at the beginning of the war and told him what he knew, then he was quite sure that Germany could have developed the atomic bomb just like the Allies!

I replied that judging from the reports which came into our hands, this did not seem very plausible and we had more or less the same impression about the German effort that was expressed in Goudsmits' book. In particular, no serious mention of the plutonium possibilities was made in the German reports, nor about the possibility of a fast neutron bomb. The German scientists thought of a bomb as a whole heavy water pile which one had to drop – this incidentally was confirmed by Bonhoeffer when I saw him in Oxford last year. Heisenberg tried to ridicule this. In particular he emphasised that he always knew that the bomb would be of the size of a pineapple. He actually had expressed this opinion to many people including one of the ministers (not quite consistent with his former remarks!) and that though nothing written could be produced about it, his secretary could confirm it!

Naturally I could not tell him about some of our sources of information from which we know for certain that his story does not correspond with the facts. I therefore, took the line that his interpretation did not seem to me quite consistent with the general attitude of the great majority of German scientists before the war. At the end of our lengthy discussion of this point he admitted that this was not without justification; the German scientists had not behaved very well except for a few, for instance Hahn and Laue, and more or less himself.

I am quite sure that Heisenberg, like many other Germans, is a strictly honest person in his private life, but as soon as the greater glory of the "fatherland" is involved – and perhaps also his glory as a scientist – it is quite a different matter. Whether he now deliberately tells these falsehoods I cannot say. It is quite possible that he has constructed post festum a picture of the state of affairs as he would have liked to see it and that he has so persuaded himself that this picture is correct that he now seriously believes in it. One should also compare what Heisenberg says in his "Naturwissenschaften" article. There he also emphasised that the German scientists did not bother much about the bomb, but he gave quite a different reason, namely, that they knew that the industrial effort was too much for Germany. He had not the moral courage to assume responsibility before his countrymen for what now might be called – an "unpatriotic" act. Moral courage is not the strong side of the Germans – though I am quite sure that in a battle Heisenberg would have shown great personal courage.

Two things are clear:

(a) That what he says does not correspond with the facts, and
(b) that he tells these stories with an air of complete sincerity and conviction.

It must be extremely difficult for a man who is not conversant with the facts and who is not familiar with the workings of this corner of the German mind – or should one say soul – to get a true picture of the situation. I can see that it must be well nigh impossible for a man like Fraser to grope through this maze. (Incidentally, Heisenberg has, of course, the highest praise for Fraser – small wonder as he has accepted Heisenberg's point of view).

Heisenberg already exerts a considerable influence and as far as scientific matters are concerned this is all right; but he also exerts a great influence on matters of policy, and it seems very probable to me that his influence is going to increase in the near future. Perhaps this is inevitable as there are few people about with his personality and reputation and who are young enough to continue to play an important part for some time. But I think it is essential that the people who have to deal with him in Germany should know the general picture.

I am sending a copy of this [to] Cockcroft, who I know will be interested.

Yours sincerely,

NOTES

Chapter 1. Growing Up into a World at War

1. "During the journey . . . . Jewish population." Herold, 1927, 31.
2. Losses, Battle of the Somme. Herold, 1927, 58.
3. "I was with . . . . Charlotte Münchhausen." Eisner, 1984, 54.
4. Escape from Gurs. Eisner, 1984, 202.
5. "Louise Escoffier." Eisner, 1984, 206–208.

Chapter 3. Breslau 1931–1933.

1. "a bearable handicap." Peierls, 1985, 6.
2. AAAS meetings. Reported in *Science*, 75:1936, February 5, 1932.
3. "And they discussed . . . . work." BBC transcript, 6.
4. Lindemann interviews in Germany, 1933. Cherwell Papers, D95/4.
5. Kuhn interview. Kuhn's account in *Balliol College Record*, 1987, 48–49.
6. ICI project outline. Cherwell Papers, D95/7–9, and approvals list, D95/1–2.
7. "the dominant response . . . . German universities." Remy, 2002, 18.

Chapter 4. Oxford 1933–1939.

1. "clergymen . . . on less." BBC transcript, 13.
2. "We must . . . really like." BBC transcript, 14.
3. "I still remember . . . . scrambled egg." BBC transcript, 18.
4. "If one has children . . . sense of security." BBC transcript, 19.
5. Autobiographical memoir. Cassirer, 1981, 206–216.
6. "I arrived . . . . And I did not." Sanders, 2000, 305.
7. "Offer it to me . . . the position." Moore, 1989, 269.
8. "Cher ami . . . about your things." Cherwell Papers, D224/8.
9. Mond opening. *Nature*, 131:210–211, 1933.
10. Liquefier in Oxford. *Nature*, 131:191–192, and Arms, 1966, 66.
11. "capricious . . . systematic oppression." Kater, 1989, 199, and cf. 195–200.
12. Kater also notes. Kater, 1989, 184–185.
13 "cancer is curable . . . early." Proctor, 1999, 28.
14 Lindemann research proposal. Cherwell Papers, D78/2–12.
15 "now appears . . . left behind." Quoted in Arms, 1966, 89.
16 "So we started . . . undesirable foreigners." Ruhemann, undated manuscript. "Half a Life."
17 "repentant confession." Text in Moore, 1989, 337.
18. Had seen neither. They were found in private correspondence, not at the Royal Society.
19. "Then suddenly . . . as Simon." Moore, 1989, 345ff.
20. "the effects . . . our grandchildren." Cherwell Papers, D170/11.
21. "takes time." Cherwell Papers, D170/5 and D170/8.
22. Nernsts in Oxford. Mendelssohn, 1973, 174; telephone conversation, Charles Cahn, 12 March 2006.
23. "I have been intending . . . present government." Arms, 1966, 88.
24. "the membership . . . withdrawal from the Society." Translation in Hentschel, 1996, 181–182.
25. "We did not succeed . . . tell in a letter." Simon to A.C. Menzies, 29 March 1939. Simon Papers.
26. "up to 1000 times." "Congrès international du magnétisme." *Les dernières nouvelles de Strasbourg*, 26 May 1939.

Chapter 5. Any Capable Physicist 1939–1941.

1. Churchill version. Churchill, 1948–54, 1:301. Cf. Cherwell Papers, D230/2–3.
2. "In all probability . . . than others." Simon to Demuth, quoted in Arms, 1966, 100–101.
3. "foreign nations." Gowing, 1964, 36.
4. "I think one can say . . . calculations." Gowing, 1964, 38–39.
5. "I gather . . . in our beds." Gowing, 1964, 39.
6. "One might think . . . for the bomb." Gowing, 1964, 390.
7. When Germany invaded Denmark, Lise Meitner was visiting Bohr's institute in Copenhagen, and at his request she sent a telegram from Stockholm to England to her nephew Frisch to say that Bohr and his family were safe. Duly censored, the telegram ended: "Tell Cockcroft and Maud Ray Kent." Some scientists construed this as an anagram for "radium taken" or something similar. After the war, it was revealed that the Bohr family's governess had been Maud Ray, who lived in Kent, at an address deleted by the censor.
8. "It will interest you . . . Birmingham people." Simon to Born, June 4, 1940. Simon Papers.
9. "Dear Bridgman . . . F.E. Simon." Arms, 1966, 93–94.
10. "There is one . . . origin." Simon to Bridgman, Stern, and Van Vleck, 27 June 1940. Simon Papers.
11. "On account . . . right thing." Simon to Born, 8 July 1940. Simon Papers.
12. "I think . . . I would do it." Born to Simon, 10 July 1940. Simon Papers.
13. Nazi Black List. Simon's identity is confirmed in a letter from L.A. Jackets (Air Historical Branch, Air Ministry ref. A.H.B.2/440) to Nicholas Kurti, 22 September 1960 (copy in Simon family files).
14. "Let me also . . . my own career." Jones, 1965, 120, and cf. Jones, 1978, 26–28.
15. Expulsion of Hans Thost. Barnes and Barnes, 2005, ch. 5.
16. "Hitler's oldest enemies." Arms, 1966, 97, 104.
17. "The public danger . . . now receding." Churchill, 1948–54, 2:627.
18. "The remark . . . in this direction." Simon to Lindemann, 7 May 1940. Cherwell Papers, D230/1. In the summer of 1940 Simon began to keep a diary on nuclear matters at the laboratory. It begins with an undated page headed "Early Development." The first entry begins: "Discussions with Lindemann about isotope separation since beginning of 1939." Clarendon diary, 1.
19. "At the beginning . . . known at present." Simon to Lindemann, 14 June 1940. Cherwell Papers, D229/14.
20. "to put it . . . made to work." Chadwick to Lindemann, 28 June 1940. Cherwell Papers, D230/8.
21. "It seems . . . June 4th." Simon to Born, 4 September 1940. Simon Papers.
22. "There is an urgent need . . . get it quickly." Simon to Lindemann, 19 November 1940. Cherwell Papers, D230/4–5.
23. "95 per cent certain." Clark, 1961, 119–120.
24. Postwar sources. Clark, 1961, 130.
25. Clark gives details. Clark, 1961, 131–132.
26. Clark on 19 May meeting. Clark, 1961, 189.
27. "A uranium bomb . . . experimental work." Gowing, 1964, 164, 398.
28. "The lines . . . any capable physicist." Gowing, 1964, 395.
29. Neptunium and plutonium. Clark, 1961, 124.
30. SAC panel report. Summarized in Gowing, 1964, 105.
31. "Although personally . . . think. Churchill, 1948–54, 3:730.
32. "It would be unforgivable . . . had been defeated." Gowing, 1964, 96. For more on Lindemann's minute and its context, see Fort, 2003, 308–309.
33. "This organisation . . . mismanaged." Gowing, 1964, 111.
34. "no longer free . . . our research." Simon to Lord McGowan, 9 December 1941. Copy in Cherwell Papers, D104/2.
35. Kowarski: heavy water. Clark, 1961, 169.
36. "Several members . . . by the Nazis." Thomson to Lindemann, 4 July 1941. Cherwell Papers, D230/6–7. "Veblen von Neumann," possibly a code name or combination, seems unidentifiable further.
37. "it would be best . . . jointly in Canada." Hewlett and Anderson, 1962, 46.
38. "in order that . . . jointly conducted." Hewlett and Anderson, 1962, 259.
39. Third report. Hewlett and Anderson, 1962, 47–48.
40. "end of the beginning." Hewlett and Anderson, 1962, 52.
41. Missed opportunity. Gowing, 1964, 106. Cf. also Goldschmidt, 1987, 164–168.
42. Debye at Cornell. Simon was soon aware of these changes. He received a postcard dated 19 March 1940, from his colleague Peter Paul Ewald in Belfast: "I only want you to know that Peter the Great has gone to

Cornell for this term only, accompanied by his son, but not by his wife, who preferred to stay at home." Simon Papers. In the summer, when Lotte was planning a visit from Toronto to friends in the United States, Franz warned her in a letter dated 16 August: "If you should see Debye, be careful, I don't trust him; but this you know yourself."

43. Heisenberg's two reports. Walker, 1989, 21; G-39 and G-40 of the German restricted series.
44. Harteck's separation tube. Walker, 1989, 29–32.
45. "Discuss methods . . . no objections." Clarendon Diary, 17 September 1940.
46. Laurence's story. Holloway, 1994, 59–60.
47. "nothing happened for two months." Laurence, 1961, 43.
48. "without any great urgency." Holloway, 1994, 68.
49. "I think now . . . in the XIXth." Holloway, 1994, 61.
50. "the dormant period." Kramish, 1959, 48–62.
51. Japan: Army moves. Rhodes, 1988, 327, 346.
52. "If in some way . . . a super explosion." Rhodes, 1988, 375, with an amended translation from *Bulletin of the Atomic Scientists*, 56/4, July–August 2000, Letters section. Rhodes's source was a slightly flawed handwritten summary of the lecture prepared for an army laboratory, which confuses two almost identical kanji characters of different meanings. The printed text suggests a uranium explosion using hydrogen as a moderator, not a hydrogen bomb as Rhodes's version suggests.

Chapter 6. Industrial Plants . . . Heretofore Deemed Impossible 1942–1945

1. "full of spies . . . transients." Peierls, 1985, 169.
2. "Jointly or separately . . . going on." Peierls, 1985, 170.
3. "The time available . . . a month from now." Gowing, 1964, 139.
4. "It is still . . . Anglo-American effort." Full text Gowing, 1964, Appendix 3, 437–438.
5. "there were now . . . full speed." Hewlett and Anderson, 1962, 69.
6. "I fear we are in the soup." Groves, 1983, 20.
7. "literally hundreds . . . barriers." Hewlett and Anderson, 1962, 126.
8. "Both types . . . end of the war." Hewlett and Anderson, 1962, 101.
9. Columbia research described in detail. Hewlett and Anderson, 1962, 99–101, 122–125.
10. A statement of needs. Hewlett and Anderson, 1962, 83.
11. Mansion on Simpson Street. Goldschmidt, 1987, 210–211.
12. "I would guess . . . something similar." Gowing, 1964, 148.
13. "Already . . . electromagnetic separation." Gowing, 1964, 151.
14. "I do not know . . . about them." Simon to Cherwell, 10 December 1942. Cherwell Papers, D231/3.
15. "from the Americans." Gowing, 1964, 222.
16. "basic principle . . . or Canadians." Full text in Gowing, 1964, 156.
17. "a convenient . . . restrictive policy." Eggleston, 1965, 69.
18. "from the top . . . to the top." Hewlett and Anderson, 1962, 266–268.
19. "I hear . . . with empty pockets." Goldschmidt, 1987, 220–221.
20. "not discuss barrier manufacture." Hewlett and Anderson, 1962, 269.
21. Heavy water veto. Hewlett and Anderson, 1962, 268, 270.
22. "you made a firm . . . go through with it." Hewlett and Anderson, 1962, 274.
23. Quebec Agreement. Full text in Gowing, 1964, Appendix 4, 439–440.
24. "stormy conference." Hewlett and Anderson, 1962, 268–269.
25. "a set . . . entire war effort." Hewlett and Anderson, 1962, 272.
26. "The President agreed . . . to be informed." Churchill, 1948–54, 4:723.
27. "needed all the help it could get." Hewlett and Anderson, 1962, 278 (citing Conant to Bush, 30 July, 3 and 6 August, 1943).
28. "If you want it . . . you can have it." Wilson, undated, 5.
29. "We were a large party . . . to Los Alamos." Peierls,1985, 182–183.
30. Joint meetings, December 1943 to January 1944. Hewlett and Anderson, 1962, 135.
31. The official histories diverge. Hewlett and Anderson, 1962, 137–138; Gowing, 1964, 252–254.
32. "British design and technology." Hewlett and Anderson, 1962, 134.
33. "Their report . . . worthwhile." Hewlett and Anderson, 1962, 282.
34. "in Washington in May 1944." Gowing, 1964, 324.

35. "On 19 May . . . future U.K. plans . . . ." Quoted in Eggleston, 1965, 129. The date here seems erroneous. The Simon diary lists Cockcroft in Washington on the evening of Friday 12 May, then 13, 14, 15, and 16 May, but not on 17 May.
36. "compromise solution." Gowing, 1964, 328.
37. Adjustments to Clarendon team. Arms, 1966, 120–121.
38. "still greater effort . . . bour social community." *New York Times*, 1 January 1945, 1, 4. Full text in English on page 4.
39. Liberation of Paris. Rhodes, 1988, 606.
40. Initial operative results. Hewlett and Anderson, 1962, 295–296.
41. "My first impression . . . being prosecuted." Groves, 1983, 23.
42. Army–Navy tensions. Hewlett and Anderson, 1962, 171–172.
43. "would be available . . . June, 1945." Hewlett and Anderson, 1962, 252–253.
44. "nothing went very smoothly." Hewlett and Anderson, 1962, 249.
45. "the second half of 1945." Hewlett and Anderson, 1962, 321.
46. "certainly possible, but not close at hand." Walker, 1989, 49.
47. Rust's dismissals. Beyerchen, 1977, 44–45.
48. Heisenberg's lecture. Summarized in Walker, 1989, 56–58.
49. "She said . . . back in order." Beyerchen, 1977, 159.
50. "Because . . . scientists involved." Original and translation in Goudsmit, 1947, 117–119.
51. "specifically and typically Jewish affair." Beyerchen, 1977, 166.
52. "Heisenberg certainly . . . received them." Walker, 1989, 74.
53. Letter to Himmler. Heisenberg to Himmler, 4 February 1943. With thanks to Mark Walker for copies of this letter and SS reply (Brandt) of 15 February. See also Cassidy, 1991, 463–464.
54. "I assume . . . Heil Hitler!" R. Brandt to Heisenberg, 15 February 1943.
55. "America . . . without Einstein." Quoted in Rose, 1998, 270.
56. Meeting of 4 June 1942. Summarized in Powers, 1993, 146–151.
57. Krupp inquiry. Walker, 1989, 86–87.
58. "might turn out poorly." Walker, 1989, 104.
59. "Dear Party Brother . . . in store for us." Original and translation in Goudsmit, 1947, 4–5.
60. "strongly opposed" Irving, 1967, 231.
61. Postwar British inquiry. Beyerchen, 1977, 190, citing intelligence report by Major E.W.B. Gill.
62. "I direct you to cease . . . scientific research." Irving, 1967, 231.
63. Volkssturm draft of scientists. Irving, 1967, 233.
64. Harteck's centrifuges. Walker, 1989, 146–149.
65. "a wise government." Powers, 1993, 409.
66. "exceptional significance . . . great expectations" Walker, 1989, 136.
67. Gerlach's optimism. Powers, 1993, 409.
68. Perrin opens the pit. Powers, 1993, 419–420.
69. "successful to date . . . good frame of mind." *Operation Epsilon*, 1993, 30.
70. "I wonder . . . old fashioned in that respect." *Operation Epsilon*, 1993, 33.
71. "Sam . . . Fred." Full text in Irving, 1967, 251–252.
72. "from a political . . . certainly take it." Complete text in Churchill, 1948–54, 6:406–407.
73. Perrin in Germany. Irving, 1967, 253–260. Powers, 1993, 419–423.
74. "You will be glad . . . M.W.P." Perrin file, social correspondence (in family files).
75. "is a hundred times more than we had." *Operation Epsilon*, 1993, 75.

Chapter 7. Why Manhattan?

1. "either unfilled . . . by 1938." Beyerchen, 1977, 173.
2. "harassed . . . continually." Walker, 1989, 63.
3. "nuclear fission." Frisch, 1979, 117.
4. "Of course . . . quite differently." *Operation Epsilon*, 1993, 81.
5. "But such experiments . . . best thing." *Operation Epsilon*, 1993, 81–82.
6. "They have decided . . . didn't succeed." *Operation Epsilon*, 1993, 144.
7. "Hitler's gift." The title of Medawar and Pyke's (2000) study of refugee scientists from Nazi Germany. Alert reviewers were quick to note the primary meaning of *gift* in German as "poison" or "venom."

8. "neither Army . . . great pressure." Walker, 1989, 44.
9. "particularly urgent . . . defense." Neufeld, 1995, 119.
10. "lukewarm at best." Neufeld, 1995, 130.
11. "But what . . . (*vernichtende Wirkung*)." Neufeld, 1995, 192, and cf. Dornberger, 1981, 118.
12. "The Army rocket . . . atomic bomb effort." Neufeld, 1995, 170.
13. About 3,200 V2s . . . 5,000 civilians. Neufeld, 1995, 264.
14. Elting's data. See "Costs, Casualties and Other Data" (http://gi.grolier.com/wwii/wwii_16.html). Accessed 14/04/2010.
15. "At least twenty . . . a kilometer." Neufeld, 1995, 225.
16. A bargaining chip. For a postwar overview, see Neufeld, 1995, 267–272.
17. "I had been . . . course of the war." Speer, 1970, 226.
18. "Thanks to . . . for the war." Cited in Sereny, 1995, 318. This passage is not included in the published *Spandau: The Secret Diaries* (Speer 1976).
19. In five revealing pages. Speer, 1970, 225–229.
20. "the switch . . . five years." Speer, 1976, 382.
21. "Korsching: That shows . . . we didn't succeed." *Operation Epsilon*, 1993, 75–77.
22. "a somewhat disturbed night." *Operation Epsilon*, 1993, 91.
23. "History will record . . . weapon of war." *Operation Epsilon*, 1993, 92.
24. "It seems paradoxical . . . new weapon." Jungk, 1970, 102.
25. "I think it is absurd . . . all of us." *Operation Epsilon*, 1993, 90.
26. "Someone said . . . it *must* be done." *Operation Epsilon*, 1993, 85.
27. "When I think . . . against Heisenberg's wishes." *Operation Epsilon*, 1993, 90.
28. "Did they use . . . the whole explosion." *Operation Epsilon*, 1993, 71–72.
29. "I still don't believe . . . for the moment." *Operation Epsilon*, 1993, 72–73.
30. "There are so many . . . isotopes before." *Operation Epsilon*, 1993, 73–74.
31. "There is a great difference . . . difficult business." *Operation Epsilon*, 1993, 74.
32. "Well, how have they . . . all other methods." *Operation Epsilon*, 1993, 118–120.
33. "It is still not clear . . . element 94 with machines." *Operation Epsilon*, 1993, 121.
34. "the guests . . . with great avidity." *Operation Epsilon*, 1993, 91.
35. "and said they hoped . . . such statements." *Operation Epsilon*, 1993, 93.
36. "considerable discussion on the wording." Summarized in *Operation Epsilon*, 1993, 93–94. Goudsmit (1947: 201) identifies Schwab as the SS General who directed the SS's own technical research laboratories.
37. 1941 bomb patent. Rose, 1998, 146–154.
38. Cyclotron at Miersdorf. On Post Office funding, see Irving, 1967, 70–72. On the move to the Soviet Union, see Oleynikov, 2000, 6–7, 11–12.
39. Memorandum text. *Operation Epsilon*, 1993, 105–106; German original, 102–104.
40. "Sometimes . . . does the job." Speer, 1970, 210.
41. "The failure . . . in which it lived." Goudsmit, 1947, xi.
42. Encounters with the Gestapo. See *Operation Epsilon*, 1993, 75, 81 (Hahn, Gerlach); Goudsmit, 1947, 105 (Laue); Powers, 1993, 406 (Heisenberg); Walker, 1989, 127–128 (Harteck).
43. von Braun . . . arrested and jailed. Neufeld, 1995, 213–220.
44. "mediocre physicist." Goudsmit, 1947, 142–144.
45. "We must take off . . . heads cut off." *Operation Epsilon*, 1993, 85–86.
46. "living in the same . . . monosyllables." Goudsmit, 1947, 122.
47. "He also said . . . *since their detention*." *Operation Epsilon*, 1993, 28 (italics added).
48. "complete enigma." *Operation Epsilon*, 1993, 101.
49. Endorsed . . . by Wirtz and by Bagge. *Operation Epsilon*, 1993, 86, 90.
50. ". . . we also . . . he complained." Goudsmit, 1947, 104.
51. "Gerlach is not . . . in Germany." *Operation Epsilon*, 1993, 90.
52. "Our national policies . . . a few years!" Hartshorne, 1937, 112; Beyerchen, 1977, 43.
53. "outmoded large-scale uranium plate." Walker, 1989, 150.
54. "who – if anybody . . . heavy water?" Walker, 1989, 55.
55. "When I first . . . could have built it." *Operation Epsilon*, 1993, 114.
56. Heavy water issue in 1944. Walker, 1989, 137–146.
57. Quantitative comparison. Dahl, 1999, 184, 286.
58. Graphite issue. Irving, 1967, 78; Walker, 1989, 26–27; Dahl, 1999, 180–181.

59. "was shipped in . . . as well." Walker, 1989, 144.
60. Harteck in Norway. Walker, 1989, 138, 140.
61. "booty camp." Walker, 1989, 147.
62. "had to go . . . never did." Walker, 1989, 149.
63. Casablanca directive. Combined Chiefs of Staff, Memorandum C.C.S. 166/1/D, 21 January 1943.
64. Coal shipments. Mierzejewski, 1988, 191, 193.
65. "at my request . . . comfortable quarters." Documented in Powers, 1993, 338–339 and note 17.
66. "Early in 1943 . . . throughout the world." Groves, 1983, 180.
67. Oranienburg air raid. Groves, 1983, 230–231; Walker, 1989, 156.
68. "The backyard . . . high-grade achievement." Goudsmit, 1947, 125–127.
69. "physicist's symbol . . . Nazism." Goudsmit, 1947, 127.
70. Flerov's campaign. Holloway, 1994, 76–79.
71. "I was left . . . impression on me." Holloway, 1994, 88.
72. Letters to Beria, Stalin. Holloway, 1994, 102–103, 114–115.
73. Stalin at Potsdam. Holloway, 1994, 116–118.
74. "Ask for . . . won't be refused." Holloway, 1994, 132.
75. Kigoshi interview. Wilcox, 1985, 105–108.
76. "atomic bomb . . . and Germany." Wilcox, 1985, 96.
77. Appeals to Germany. Wilcox, 1985, 102–104.
78. "under one percent." Wilcox, 1985, 139.
79. Separators at Osaka. Wilcox, 1985, 115–117.
80. "Industries . . . to Korea." Wilcox, 1985, 147, citing OSS report of 9 February 1945.
81. "an enormous . . . Army." Wilcox, 1985, citing OSS report of 2 March 1945.
82. "Research . . . has been made." *The Day Man Lost*, 1981, 26.
83. "because of . . . 'Jewish scientists'." Wilcox, 1985, 76.
84. "The best . . . as correct." *The Day Man Lost*, 1981, 35.
85. "had replied . . . the next war." *The Day Man Lost*, 1981, 183.
86. "the most important . . . 1945." Wilcox, 1985, 145.
87. Search for uranium. Wilcox, 1985, 150–154.
88. Centrifuge rumors. Wilcox, 1985, 164–165.
89. Suzuki's separators. Wilcox, 1985, 147.
90. "people's volunteer corps." *The Day Man Lost*, 1981, 111.
91. "in the research . . . decisive weapon." *The Day Man Lost*, 1981, 183.
92. "was concerned . . . they dispersed . . ." *The Day Man Lost*, 1981, 202.
93. ". . . we believe . . . background." Wilcox, 1985, 179.
94. Interview with Sagane. Wilcox, 1985, 180.
95. Interview with Nishina. Wilcox, 1985, 181–182.
96. "The following . . . produce a bomb." Wilcox, 1985, 182.
97. Interview with Arakatsu and Yukawa. Wilcox, 1985, 182–183.
98. "three tons . . . experimentation." Wilcox, 1985, 186.
99. "at the level of 1942 . . . anything of importance." Wilcox, 1985, 187–188.
100. Interviews, Army and Navy representatives. Wilcox, 1985, 189–190.
101. "a full report . . . all documents." Wilcox, 1985, 190.
102. "We did not make . . . get wind of it . . . ." Groves, 1983, 187, and cf. 141.
103. Destruction of the cyclotrons. Groves, 1983, 367–372.
104. "The Japanese . . . to make war." President Truman, Statement, 6 August 1945.
105. "such *ad hoc* . . . earliest moment." Gowing, 1964, 440.
106. "work on the planning . . . background." Gowing, 1964, 336.
107. "Certainly . . . correct." Hewlett and Anderson, 1962, 282.
108. "The gaseous diffusion . . . completion of the plant." Groves, 1983, 117–118.
109. "any serious differences among its members." Groves, 1983, 136.
110. Combined Development Trust. Groves, 1983, ch. 12, 170–184.
111. British contributions to Manhattan. Groves, 1983, 406–407.
112. "sent meticulously . . . Lend-Lease." Gowing, 1964, 256.
113. "On the whole . . . in his debt." Groves, 1983, 408.
114. A long memory. His 700-page autobiography begins with a tribute to his great grandfather, who at 16 had just missed the Battle of Bunker Hill on 17 June 1775, but enlisted in Washington's army a few days later on reaching his 17th birthday. Conant, 1970, 3–4.

115. Two historians. Sherwin, 1977, ch. 3 (quotations at 80, 81); Hershberg, 1993, ch. 10 (quotations at 180, 182).
116. "Just why . . . I do not know." Groves, 1983, 120.
117. "We were not . . . for peace." *Operation Epsilon*, 1993, 176.
118. "I feel . . . what will be done." *Operation Epsilon*, 1993, 222.
119. "Their appreciation . . . lie with them" *Operation Epsilon*, 1993, 264.
120. "To do modern physics . . . Russians after all." *Operation Epsilon*, 1993, 202–203.
121. "wanted to stay . . . his children." Summarized in Walker, 1989, 184–185.
122. Discussions with Blackett. *Operation Epsilon*, 1993, 189–191.
123. "During the war . . . he was murdered." *Operation Epsilon*, 1993, 55.
124. "Throughout his life . . . sought to help." Cassidy, 1991, 483–484.
125. "Since only . . . from the beautiful . . . ." Cassidy, 1991, 308.
126. " . . . I must be satisfied . . . work is beautiful." Cassidy, 1991, 330.

Chapter 8. Something Reasonable Again.

1. "Dear Joel . . . reasonable again." Simon to Hildebrand, 17 February 1945. Simon Papers.
1a. As one of the earliest refugees to be knighted, he anglicized his first name to become Sir Francis. By family members he was already referred to often as Franciscus.
2. "Gentlemen . . . unfortunately secret." Cherwell to Nobel Committee, 7 January 1946. Cherwell Papers, B48/2.
3. Strasbourg paper. Simon, 1939, 17–23.
4. "Urey told me . . . little exaggeration." Arms, 1966, 135.
5. "had become . . . in the world." Kurti, 1958, 232.
6. Simon's experimental work. Kurti, 1958, 233–242; Bridgman, 1960.
7. 1927 restatement of the third law. Kurti, 1958, 235.
8. "a period of utter confusion." Simon, 1956, 1.
9. "very chequered history." Simon, 1956, 1.
10. "I hope . . . as we know it to-day." Simon, 1956, 21.
11. ". . . in May 1944 . . . provide estimates." Gowing, 1964, 336.
12. "compromise solution." Gowing, 1964, 328.
13. "put on ice." Gowing, 1964, 336.
14. "A decision . . . is imperative." Cathcart, 1994, 8.
15. Gen 75. Gowing, 1974, 1:21–22.
16. HER. Gowing, 1974, 1:210.
17. "in order . . . rather than scores." Gowing, 1974, 1:169.
18. Area and population comparisons. Data in Barnhart, 1954, 1:654, 3:3955.
19. "The discriminative test . . . atomic bomb." Gowing, 1974, 2:500.
20. *Test of Greatness*. Cathcart, 1994.
21. Portal's veto. Gowing, 1974, 1:191–193.
22. "Calder Hall . . . factory." Pocock, 1977, 30.
23. "atomic energy committees." Cherwell Papers, D232/5.
24. IPDC. Gowing, 1974, 2:428.
25. Chemical separation plants. Gowing, 1974, 2:402–423.
26. "The impact . . . Research Programme." Mark Newton to Kay Baxandall, 16 March 2010. Letter, in family files.
27. "Simon liked . . . the Clarendon." Arms, 1966, 138.
28. Articles written later. For a select list, see Arms, 1966, 166.
29. "We should at least . . . at once." Simon, 1951, 16.
30. "at the expense . . . weaker countries." Simon, 1951, 37.
31. "The shortage of scientific manpower," Reprint from *Financial Times Annual Review of British Industry*, 1955.
32. "Technologists and technicians." Reprint from *Sunday Times*, 4 March 1956.
33. "Why waste coal?" Simon, 1951, 47–52. Quotations, 49, 50.
34. "Economical heat production." Simon, 1951, 53–62. Quotations, 54, 56.
35. "Wanted: a national fuel policy." Simon, 1951, 75–87. Quotations, 75, 78, 86–87.
36. "Prospects of atomic power." Simon, 1951, 63–68.
37. "Power sources of the future." Simon, 1951, 69–74. Quotations, 73, 74.
38. "Reactors . . . English Channel. Simon, 1955, 6.
39. White Paper forecast for 1975. Simon, 1955, 7.
40. "last decade of the century." Simon, 1954b, 257.
41. Electricity sources. In 1975: Pocock, 1977, 261; in 2004: multiple Internet sources.

42. "... the first half ... but come it will." Reprint from "Nuclear power: how soon?" *Financial Times*, 25 November 1953.
43. "The compromises ... wartime projects." Simon, 1951, 89–90.
44. USSR bomb timetable. Simon, 1951, 90–92.
45. "the large reactors ... plutonium. ..." Simon, 1951, 93.
46. "If we weigh ... many other things." Simon, 1951, 95.
47. "The atomic rivals." Reprinted in *Atomic Scientists' Journal*, 4(3), 155–160. Original in *Financial Times*, 6 August 1954.
48. "A Ministry of Science?" Reprint from *The Sunday Times*, 7 August 1955.
49. "The soviet bid for technological leadership." *The Listener*, 19 January 1956, and reprinted in *State Service*, April 1956.
50. Soviet delegation to Geneva. Holloway, 1994, 352–353.
51. Holloway's account. Holloway, 1994, 355–363. Quotations, 355, 363. **The physicists' autonomy was precarious. In March 1949, 5 months before the first bomb test, Stalin cancelled at the last minute a major congress organized to examine Soviet physics for its conformity with dialectical materialism, as had been done for biology in 1948. To avoid possible delay for those in the bomb project, Stalin reportedly instructed Beria: "Leave them in peace. We can always shoot them later" (Holloway, 1994, 211).**
52. "Modern drugs." Reprint from *Financial Times*, undated (September 1949?).
53. "Vitamins of the soil." Reprint from *Financial Times*, 17 March 1949.
54. "Some reflections on accuracy." Simon, 1951, 134–138.
55. "Can the world's population be fed?" Simon, 1951, 128–133.

Chapter 9. Security Lapses.

1. "There is no reason ... time and effort." Perrin to Simon, 10 May 1951. **Simon Papers.**
2. "Another reason ... their own work." Simon, 1951, 92n.
3. "... we found him ... in Edinburgh." Born, 1978, 284.
4. "X is quite ... simply 0.00." Born to Simon, 24 February 1941. **Simon Papers. Born** was conscientious in supporting his students and finding them appropriate employment. **At this point Fuchs had lost out to X [name withheld] in a competition for a mathematics lectureship at a London women's college temporarily evacuated to Oxford.** "I am awfully disappointed," Born writes, " ... if you could find out how it happened that Fuchs was not even asked for an interview I should be grateful." Three months later, Fuchs began working for Peierls in Birmingham.
5. "as efficient ... and even-tempered." Peierls, 1985, 163.
6. Fuchs's security reviews. Gowing, 1974, 2:145–149.
7. "I suppose ... I *am* Harwell." Gowing, 1974, 2: 149.
8. "Cockcroft ... pleasant lunch." Born, 1978, 287.
9. "a non-American ... in the United States." Cathcart, 1994, 101.
10. A radioactive cloud. Cathcart, 1994, 107–108.
11. General Nichols' visit. Gowing, 1974, 1:293–294.
12. Interrogations by Skardon. Williams, 1987, 121–123.
13. Fuchs's statement to Perrin. Williams, 1987, 188–194.
14. Gold's statements to the FBI. Williams, 1987, 195–220.
15. "controlled schizophrenia." Williams, 1987, 184.
16. "I would like to add ... brushed this aside." Williams, 1987, 218–219. **Gold's biographer traces this exchange to experiments in the sugar company's laboratories in the 1930s to separate carbon dioxide from flue gases by thermal diffusion in order to make dry ice. Any link with the Navy Yard or Abelson seems ruled out. In 1937 and 1940, Gold's Soviet handlers asked him to move to a job at the Navy Yard for better opportunities for espionage, but Gold twice refused to leave Pennsylvania Sugar. Hornblum, 2010, 68–69, 89–90, 115–116.**
17. "Finally ... passed to them." Williams, 193.
18. "the girl from Banbury." Williams, 1987, ch. 5.
19. Flight from England. Werner, 1991, 288–290.
20. "When I met Klaus ... believe they were used." Werner, 1991, 251–252. **The references to Fuchs do not appear in the original 1977 German edition, as Fuchs was still alive.**
21. "took place ... New York." Williams, 1987, 204.
22. "Kurchatov sat ... those materials." Holloway, 1994, 90–91.

23. "Wonderful . . . what we are lacking." Holloway, 1994, 95.
24. Memoranda of 7 March and 22 March 1943. Holloway, 1994, 91–95.
25. Memorandum of July 1943. Holloway, 1994, 103–104.
26. "Kurchatov's memoranda . . . as a spy." Holloway, 1994, 108.
27. Packets at Santa Fe meetings. Williams, 1987, 214, 217.
28. Report sent to Beria. Holloway, 1994, 138.
29. "killed stone dead." Cathcart, 1994, 114.
30. "We are awfully sorry . . . poor security risk." Urey to Simon, 10 May 1950. Simon Papers.
31. "Yes, the Fuchs case . . . foresworn it." Simon to Urey, 15 May 1950. Simon Papers.
32. "dissolute and insulting behaviour." *Oxford Dictionary of National Biography*, 2004, s.v. Burgess, Guy Francis de Moncy.
33. Secret meeting in London. Holloway, 1994, 82–83. See also Rhodes, 1995, 52–53.
34. 5,832 documents. *Oxford Dictionary of National Biography*, 2004, s.v. Cairncross, John.
35. Kuczynskis. Williams, 1987, 49–50 and passim; Jurgen's obituary, *The Independent*, 13 August 1997 (online edition).

Chapter 10. Germany in the Balance.

1. "My dear Michels . . . glad to do this." Simon to Michels, 12 June 1945. Simon Papers.
2. "from the soul." von Simson to Simon, undated [December], 1946. Simon Papers.
2a. Kaete Gerstel had been a secretary in the Berlin laboratory. She accompanied the Simons in their move to Oxford and lived with the family in the first years there.
3. "to have . . . news from you." Simon to von Simson, 23 December 1946. Simon Papers.
4. Schaefer survival signal. Schaefer to Simon, 5 November 1946. Simon Papers.
5. Simon's reply. Simon to Schaefer, 2 January 1947. Simon Papers.
6. "many unseemly . . . things." Schaefer to Simon, February 10, 1947. Simon Papers.
7. "behaved . . . respect." Simon to Schaefer, 7 March 1947. Simon Papers.
8. "beyond reproach." Schaefer to Simon, 10 April 1947. Simon Papers.
9. "We want . . . in Germany." Simon to Schaefer, 12 April 1948. Simon Papers.
10. "and I am afraid . . . can do it." Gutsche to Simon, 31 December 1946. Simon Papers.
11. "I can vouch . . . of the Nazis." Testimonial for Gutsche, 7 January 1947. Simon Papers.
12. "I am writing you . . . of Antisemitism." Meissner to Simon, 15 December 1947. Simon Papers.
13. "I know personally . . . intervention. . . ." Simon to Meissner, 10 March 1947. Simon Papers.
14. "could let me know more concrete data." Simon to Laue, 11 March 1947. Simon Papers.
15. Letter from Harteck. Harteck to Simon, 29 November 1946. Simon Papers.
16. "I will try . . . for him." Simon to Harteck, 11 December 1946. Simon Papers.
17. ". . . I think . . . help in this matter." Simon to Stern, 20 December 1946. Simon Papers.
18. "an extended stay . . . and family." Harteck to Simon, 23 January 1947. Simon Papers.
19. "Bonhoeffer's visit . . . pound per day!!" Simon to Jost, 24 June 1947. Simon Papers.
20. "By the way . . . Quakers in Germany." Simon to Jost, 24 June 1947. Simon Papers.
21. "I was a fool . . . than I am now." Quoted from Jost in Simon to Pakenham, 28 January 1948. Franz notes that this case is in the American zone.
22. "full of hate . . . something positive." Ann Jost to Simon, 9 February 1949. Simon Papers.
23. "every non-nationalist . . . this matter?" Simon to Education Branch, OMGUS, 22 February, 1949. Simon Papers.
24. "to take . . . this office." Montgomery to Simon, 25 March 1949. Simon Papers.
25. "only deaf ears . . . Occupation Forces." Simon to Pakenham, 4 August 1947. Simon Papers.
26. "He tells me . . . do something about this." Simon to Blount, 7 May 1946. Simon Papers.
27. ". . . is still . . . most welcome." Murray to Simon, 9 December 1946. Simon Papers.
28. "our feelings . . . Jewish race." Günthermann to Simon, 14 June 1933. Simon Papers.
29. "first sign . . . much worse off." Günthermann to Simon, 7 June 1949. Simon Papers.
30. "I have had . . . always declined." Simon to Günthermann, 11 June 1949. Simon Papers.
31. "your long report . . . use of it." Simon to Jost, 24 June 1947. Simon Papers.
32. "From letters . . . Nazi government was." Simon to Pakenham, 4 August 1947. Simon Papers. Most of the Pakenham file was received at the Royal Society in a new accession in 1993.
33. "biggest problem . . . people can be found." Pakenham to Simon, 8 September 1947. Simon Papers.
34. "They emphasized . . . realised at the time." Simon to Pakenham, 30 September 1947. Simon Papers.

35. "People who are still . . . party insignia as 'change'." The Martius article, "Videant consules," is reprinted in full in Hoffmann and Walker, 2007, 636–640. Quotation, 636.
36. "national honour . . . cleansing almost impossible." Extract from Martius to Simon, 30 November 1947, enclosed in Simon to Pakenham, 28 January 1948. Simon Papers.
37. "absolutely exceptional case." Hoffmann and Walker, 2007, 391.
38. "research and politics . . . (*Bereiche*)." Hoffmann and Walker, 2007, 416.
39. "This generation . . . member of parliament." Martius, in Hoffmann and Walker, 2007, 637–638.
40. "We were not . . . situation." Martius interview, 4 June 2008.
41. "went so far . . . these things." Extract from Martius, to Simon, enclosed in Simon to Pakenham, 28 January 1948. Simon Papers.
42. "watershed . . . absolutely clear." Martius interview, 4 June 2008.
43. "hitherto insufficiently explored." Hoffmann and Walker, 2007, 419–420.
44. "That it will . . . and do nothing." Extract from von Simson to Simon, enclosed in Simon to Pakenham, 28 January 1948. Simon Papers.
45. "I deplore . . . in Germany." Pakenham to Simon, undated [February 1948]. Simon Papers.
46. "I was very interested . . . before the Nazis." Simon to Pakenham, 11 March 1948. Simon Papers.
47. "halo for eccentricity." Longford, 1986, 231.
48. "Now Lord Pakenham . . . in the clouds." Longford, 1986, 265–266.
49. "had lost . . . authority." Longford, 1986, 265.
50. "Dear Pakenham . . . mischievousness." Simon to Pakenham, 7 March 1949. Simon Papers.
51. *Oliver Twist* riot. See *Time*, 28 February 1949, and 4 October 1948; *Der Spiegel*, 26 February 1949; *International Herald Tribune*, 24 August 2005.
52. "Now as much . . . adversaries of Nationalism." Werfl to Herzberg, 28 February 1948. German text and English translation with names deleted. Translation enclosed in Simon to Pakenham, 4 November 1949. Simon Papers.
53. Enquiry to Herzberg. Stoicheff, 2002,194.
54. "usually." Ebsworth, 1960, 19, and cf. 210.
55. "Dear Pakenham . . . first hand experience." Simon to Pakenham, 18 October 1952.
56. Landmark study. See Garner, 1995 (online version).
57. "was rendered . . . Nazi persecution." Garner notes the vast disparity in numbers deemed eligible for benefits. Article 131 claimants numbered over 430,000; the restitution law applied to fewer than 1,000. Garner, 1995, 49–50.
58. Compliance levels. Garner, 1995, 51.
59. "Dear Simon . . . willing to give." Fraser to Simon, 12 May 1947. Simon Papers.
60. "bring along . . . material." Simon to Fraser, 30 May 1947. Simon Papers.
61. "Dear Simon . . . important and sound." Mott to Simon, 15 September 1947. Simon Papers.
62. Mott at Göttingen. Hoffmann and Walker, 2007, 376–377, 382–383.
63. "Heisenberg's British minder." Frayn, 2000, 95 and 107.
64. "Many thanks . . . another question. . . ." Simon to Mott, 16 September 1947. Simon Papers.
65. "some Christian associations." Rosbaud to Goudsmit, 25 April 1948. Cited in Rose, 1998, 320.
66. "Dear Simon . . . Yours ever." Mott to Simon, 6 October 1947. Simon Papers.
67. "Dear Mott . . . Yours ever." Simon to Mott, 8 October 1947. Simon Papers.
68. "but I have . . . coming alone." Mott to Simon, 12 November 1947.
69. "as pleasant . . . noticeably 'Nazified'." *Born–Einstein Letters*, 2005, 163.
70. "It seems to me . . . case of Justi," Simon to Dale (draft), 15 October 1947. Simon Papers.
71. "I remember . . . Nazi government was." Simon to Pakenham, 28 January 1948. Simon Papers.
72. "a word direct . . . desirable." Darwin to Simon, 8 November 1947. Simon Papers.
73. "Actually . . . Mr. Worsfold." Simon to Darwin, 10 November 1947. Simon Papers.
74. "We are very sensible . . . afraid." Worsfold to Darwin, 7 November 1947. Simon Papers.
75. "Clusius . . . no objection." Simon to Nancy Parkinson (British Council), 3 September 1949. Cf. Simon to Clusius, 12 September 1949. Simon Papers.
76. "Heisenberg claims . . . like the Allies!" Simon to Perrin, 11 March 1948. Simon Papers.
77. Article of 1946. *Die Naturwissenschaften*, 33: 325–329.
78. "the foremost German . . . of interest to you." Simon to Pakenham, 11 March 1948. Simon Papers.
79. "I had a long . . . reading." Simon to Jost, 16 March 1948. Simon Papers.
80. "Heisenberg thought . . . had suffered." Walker, 1989, 191 and note 38.

81. "an undertone . . . old times." Heisenberg to Sommerfeld, 5 January 1948. With thanks to Mark Walker for a copy.
82. Missing letter. Heisenberg to Sommerfeld, 31 March 1948 (no copy found). ". . . most of the estate material of Heisenberg has gone to the Max-Planck Archive in Berlin." Helmut Rechenberg to author, 4 May 2011.
83. Inability to forget. On Holocaust losses in the Peierls family, see Peierls, 1985, 141.
84. "Heisenberg has been . . . soon be forgotten." Rosbaud to Goudsmit, 25 April 1948. Quoted in Rose, 1998, 320.
85. "I went . . . quite reasonable." Simon to Born, 11 August 1951. Simon Papers.
86. Other reports. See Rose, 1998, 314.
87. Heisenberg on revolution. Rose, 1998, 311, quoting Goudsmit to Rosbaud, 20 October 1950.
88. "How fine . . . won this war." Goudsmit, 1947, 114; Cassidy, 1992, 492–493; cf. James Glanz in *The New York Times*, 21 March 2000, online ed.
89. Control over science. Cassidy, 1992, 531–537. Quotations at 534–537. Heisenberg's view of democracy was at best unconventional.
90. "Fraser seems . . . superiour Blount." Simon to Born, 18 October 1948. Simon Papers.
91. "I do not think . . . to Schroedinger." Born to Simon, 20 October 1948. Simon Papers.
92. "Your version . . . mental stature. . . ." Fraser to Irving, 27 August 1966. Quoted in Rose, 1998, 323.
93. "I used . . . prison to the west." Fraser to Irving, 27 August 1966. Quoted in Rose, 1998, 323.
94. "It is a very . . . disloyal to their country." Rosbaud to Goudsmit, 22 June 1947. Quoted in Rose, 1998, 321.
95. "You asked me . . . the right spirit." Simon to Jost, 5 August 1947. Simon Papers.
96. "in view of . . . in this country." B.D. MacLean to Simon, 9 August 1949. Simon Papers.
97. "I have now had . . . senior colleague." Simon to B.D. MacLean, 20 August 1949. Simon Papers.
98. "In the meantime . . . the invitation. Simon to B.D. MacLean, 10 February 1950. Simon Papers.
99. "rather than . . . your programme." B.D. MacLean to Simon, 18 February 1950. Simon Papers.
100. "accumulation of work . . . kind invitation." Simon to Rector, Berlin Technical University, 25 February 1950.
101. "The question . . . beyond repair." Simon to Bonhoeffer, 18 March 1950. Simon Papers.
102. "In my opinion . . . please correct me." Simon to Bonhoeffer, 22 March 1951. Simon Papers. See also Szöllösi-Janze, 2001, 271.
103. "It would be nice . . . to know why." Simon to Jost, 7 April 1952. Simon Papers.
104. "Yes, I shall go . . . for you alone." Simon to Laue, 5 July 1952. Simon Papers
105. "I was particularly . . . some publicity." Simon to Bonhoeffer, 6 October 1952. Simon Papers.
106. "We did not go back . . . hundreds of people." BBC transcript, 33–34.
107. Pollitzer case. Simon Papers: Simon to Paul Schuftan, 1 December 1938; Schuftan to Simon, 8 December 1938 and 5 March 1947. See also Central Database, Yad Vashem, Jerusalem, Israel.
108. Impressions of visit. Simon to Bonhoeffer, 6 October 1952; Simon to Jost, 27 October 1952. Simon Papers. Lotte's interview, BBC transcript, 34.
109. " . . . with regard . . . of any use." Simon to Bonhoeffer, 21 April 1954. Simon Papers.
110. "I believe . . . have not our experience." Simon to Brickwedde, 12 October 1954. Simon Papers.

Chapter 11. A Rounded Life.

1. "mainly . . . three years." Simon to Bridgman, 2 March 1953. Simon Papers.
2. "an opulent dinner." Simon, 1953. Bill of Fare: reprinted in Shapley, 1953, 61–62.
3. "the academicians . . . wassailing." Shapley, 1953, 256.
4. Note in *Nature*. Simon, 1953.
5. "Simon was . . . all his writings." Arms, 1966, 139.
6. Coffee making. Brown, 1979, 290–293.
7. "We wish . . . subjects, etc." Gillis to Simon, 15 May 1953. Simon Papers.
8. "We all appreciate . . . visit to us." B.M. Bloch to Simon, 14 December 1953. Simon Papers.
9. "Now to your question . . . here again." Simon to Laue, February 2, 1956. Simon Papers.
10. Two predecessors. Fox and Gooday, 2005, 77–79, 114–118 (Clifton); 226–232 (Townsend).
11. "pointed out . . . position offered." Harrod, 1959, 267.
12. "he embarked . . . rent for them." Birkenhead, 1961, 326.
13. "had never . . . living memory." Harrod, 1959, 268.
14. "has stated . . . as a candidate." Veale to W. L. Bragg, 22 May 1956. William Lawrence Bragg Papers, 38A/72. For the full set of papers sent to Bragg as an elector, see William Lawrence Bragg Papers, 38A/63–81.

15. "He has now . . . Oxford school." W.L. Bragg to G.P. Thomson, 24 May 1956. William Lawrence Bragg Papers, 38A/73.
16. Two letters. Simon to Registrar, 30 May 1956; Simon to Cockcroft, 30 May 1956. Simon Papers.
17. "I saw Wilkinson . . . in the garden." Simon to Cockcroft, 11 July 1956. Simon Papers.
18. Wilkinson's recollections. Wilkinson, 2002.
19. Perrin reference. Potters to Simon, 5 June 1956; Simon to Messrs Potters, 6 June 1956. Simon Papers.
20. Larsen project. Simon to Larsen, 30 August 1956; Larsen to Simon, 13 September 1956; Simon to Larsen, 1 October 1956. Simon Papers.
21. Rothenstein project. Rothenstein to Simon, 8 October 1956; Simon to Rothenstein, 10 October 1956; Secretary to Rothenstein, 29 October 1956. Simon Papers. Rothenstein realized his goal, becoming professor of nuclear engineering at the Technion in Haifa.
22. "back in the lab . . . opening today." Simon to MacDonald, 17 October 1956. Simon Papers.
23. "I should have . . . Yours ever." Simon to Born, 19 October 1956. Simon Papers.
24. "In the summer . . . to him." Kurti, 1958, 233.
25. "Shattering *memento mori*." Schaefer, 1958, 406. Cf. *Nature*, 1956, 177:1067–1069; 177:1155–1156; 178:450–453; 178:1434–1435. Schaefer could also have noted the Guthrie Lecture, delivered in March 1956 but announced in *Nature* in 1955 (176:241).

BIBLIOGRAPHY

Archival collections

Cherwell Papers, Nuffield College, Oxford.
Simon Papers, Royal Society Library, London.
William Lawrence Bragg Papers, The Royal Institution, London.

Diaries and calendar

Simon, F. E. 1940–1942. Diary. National Archives, Kew. In the series Clarendon Laboratory Atomic Energy Research, AB3/113. Cited as Clarendon Diary.
Simon, F. E. 1942–1956. Diary and calendar. Loose-leaf pages, in LEFAX format. Cited as Diary or Calendar.

Other documents

BBC transcript. Interview of Lady Simon for the Imperial War Museum, c. 1979. 38 pages. Cited as BBC transcript.
Henschel-Simon, Elisabeth. Untitled reports circulated to friends and family. Typewritten.
   1. Berlin to Haifa, undated
   2. First weeks in Palestine, 22 Feb. 1934.
Ruhemann, Martin, undated. Half a Life. Unpublished partial autobiography.
Simon, Charlotte. 1951. Untitled (travel journal, mainly on Italy). 39 pages.
Simon, Charlotte. 1954. Untitled (travel journal, Israel and Greece). 47 pages.

Interviews

Cahn, Charles. Telephone interview, 12 March 2006.
Franklin, Ursula Martius. Interview, Toronto, 4 June 2008.

Books and articles

Archives du Conseil de Flandre (*Raad van Vlaanderen*). 1928. Brussels: Dewarichet.
Arms, Nancy. 1966. *A Prophet in Two Countries: The Life of F.E. Simon*. Oxford: Pergamon.
Barnes, James J. and Barnes, Patience P. 2005. *Nazis in Pre-War London 1930–1939*. Brighton: Sussex University Press.
Barnhart, Clarence L., ed. 1954. *The New Century Cyclopedia of Names*. 3 vols. New York: Appleton-Century-Crofts.
Bernstein, Jeremy, ed. 2001. *Hitler's Uranium Club: The Secret Recordings at Farm Hall*. 2nd ed. New York: Copernicus.
Bertin, Leonard. 1955. *Atom Harvest*. London: Secker and Warburg.
Beyerchen, Alan D. 1977. *Scientists under Hitler*. New Haven: Yale University Press.
Birkenhead, Frederick, Earl of. 1961. *The Prof in Two Worlds*. London: Collins.
Born, Max. 1978. *My Life: Recollections of a Nobel Laureate*. New York: Charles Scribner's Sons.
Born–Einstein Letters. Revised edition, 2005. Basingstoke: Macmillan.
Bridgman, P.W. 1960. "Sir Francis Simon," *Science*, 131:1647–1654.
Brown, Sanborn C. 1979. *Benjamin Thompson, Count Rumford*. Cambridge, MA: MIT Press.
Cassidy, David C. 1992. *Uncertainty: The Life and Science of Werner Heisenberg*. New York: W.H. Freeman.
Cassirer, Toni. 1981. *Mein Leben mit Ernst Cassirer*. Hildesheim: Gerstenberg.
Cathcart, Brian, 1994. *Test of Greatness: Britain's Struggle for the Atom Bomb*. London: John Murray.
Churchill, Winston S. 1948–54. *The Second World War*. 6 vols. London: Cassell.
Clark, Ronald W. 1961. *The Birth of the Bomb*. London: Phoenix.
Compton, Arthur H. 1956. *Atomic Quest*. London: Oxford University Press.
Conant, James B. 1970. *My Several Lives*. New York: Harper and Row.

Dahl, Per F. 1999. *Heavy Water and the Wartime Race for Nuclear Energy*. Bristol: Institute of Physics Publishing.
The Day Man Lost: Hiroshima, 6 August 1945. 1981. Kodansha International: Tokyo, New York, London. Prepared by the Pacific War Research Society.
Dornberger, Walter. 1981. *Peenemünde: Die Geschichte der V-Waffen*. 3rd ed. Esslingen: Bechtle.
Ebsworth, Raymond. 1960. *Restoring Democracy in Germany*. London: Stevens; New York: Praeger.
Eggleston, Wilfrid. 1965. *Canada's Nuclear Story*. Toronto: Clarke, Irwin.
Eisner, Lotte H. 1984. *Ich hatte einst ein schönes Vaterland: Memoiren*. Heidelberg: Wunderhorn.
Elston, D.R. 1963. *Israel: The Making of a Nation*, London: Oxford University Press.
Fort, Adrian. 2003. *Prof: The Life of Frederick Lindemann*. London: Jonathan Cape.
Fox, Robert, and Gooday, Graeme, eds. 2005. *Physics in Oxford, 1839–1939*. Oxford: Oxford University Press.
Frayn, Michael. 2000. *Copenhagen*. Revised edition. London: Methuen.
Frisch, Otto. 1979. *What Little I Remember*. Cambridge: Cambridge University Press.
Führer durch die Sowjetunion. 1929. Berlin: Neuer Deutscher Verlag.
Garner, Curt. 1995. Federal public services personnel in West Germany in the 1950s, *Journal of Social History*, 29:1, 25–80. Online edition.
Goldschmidt, Bertrand. 1987. *Pionniers de l'atome*. Paris: Stock.
Goudsmit, Samuel A. 1947. *Alsos: The Failure in German Science*. London: Sigma Books.
Gowing, Margaret. 1964. *Britain and Atomic Energy, 1939–1945*. London: Macmillan; New York: St. Martin's Press.
Gowing, Margaret. 1974. *Independence and Deterrence: Britain and Atomic Energy, 1945–1952*. 2 vols. London: Macmillan.
Greenspan, Nancy T. 2005. *The End of the Certain World: The Life and Science of Max Born*. New York: Basic Books.
Groves, Leslie R. 1993. *Now It Can be Told*. New York: Da Capo Press.
Harrod, R.F. 1959. *The Prof: A personal memoir of Lord Cherwell*. London: Macmillan.
Hartshorne, Edward Y. 1937. *The German Universities and National Socialism*. Cambridge: Harvard University Press.
Hentschel, Klaus, ed. 1996. *Physics and National Socialism: An Anthology of Primary Sources*. Basel: Birkhäuser.
Herold, Eduard von. 1927. *Das K.B. Reserve-Feldartillerie-Regiment Nr. 8*. Munich: Verlag Max Schick. Erinnerungsblätter deutscher Regimenter, Bayerische Armee, 49.
Heisenberg, Werner. 1946. "Über die Arbeiten zur technischen Ausnutzung der Atomkernenergie in Deutschland." *Naturwissenschaften*, 33:325–329. English translation by *Die* Ronald Fraser in *Nature*, 1947 4059:211–215.
Hershberg, James G. 1993. *James B. Conant: Harvard to Hiroshima and the making of the nuclear age*. New York: Knopf.
Hewlett, R.G., and Anderson, O.E. 1962. *The New World, 1939–1946*. University Park, PA: Pennsylvania State University Press.
Hoffmann, Dieter, and Walker, Mark, eds. 2007. *Physiker Zwischen Autonomie und Anpassung*. Weinheim: Wiley-VCH.
Holloway, David. *Stalin and the Bomb*. 1994. New Haven: Yale University Press.
Hornblum, Allen M. 2010. *The Invisible Harry Gold*. New Haven: Yale University Press.
Irving, David. 1967. *The Virus House*. London: William Kimber.
Jersch-Wenzel, Stefi, and John, Barbara, eds. 1990. *Von Zuwanderern zu Einheimischen: Hugenotten, Juden, Böhmen, Polen in Berlin*. Berlin: Nicolai.
Jones, R.V. 1965. Impotence and achievement in physics and technology, *Science*, 207:120–125.
Jones, R.V. 1978. *Most Secret War: British Scientific Intelligence, 1939–1945*. London: Hamish Hamilton.
Jungk, Robert. 1970 (1st German ed. 1956). *Brighter than a Thousand Suns*. Harmondsworth: Penguin Books.
Kater, Michael. 1989. *Doctors Under Hitler*. Chapel Hill: University of North Carolina Press.
Kramish, Arnold. 1959. *Atomic Energy in the Soviet Union*. Stanford: Stanford University Press; London: Oxford University Press.
Kuhn, Heinrich. 1987. Interview with Heini Kuhn, *Balliol College Annual Record*, 42–53.
Kurti, Nicholas. 1958. Franz Eugen Simon. *Biographical Memoirs of Fellows of the Royal Society*, 4:224–256.
Laurence, William L. 1961. *Men & Atoms*. London: Hodder and Stoughton.
List of Displaced German Scholars 1936. With *Supplementary List of Displaced German Scholars*. 1937. London: Notgemeinschaft Deutscher Wissenschaftler im Ausland.
London, Louise. 2000. *Whitehall and the Jews, 1933–1948*. Cambridge: Cambridge University Press.
Longford, Elizabeth, Countess of. 1986. *The Pebbled Shore*. London: Weidenfeld and Nicholson.
Macrakis, Kristie. 1993. *Surviving the Swastika: Scientific Research in Nazi Germany*. New York: Oxford University Press.

McRae, K.D. 2004. Balliol in wartime: letters of F.E. Simon, *Balliol College Annual Record*, 24–28.
Medawar, Jean, and Pyke, David. 2000. *Hitler's Gift: Scientists Who Fled Nazi Germany*. London: Richard Cohen.
Mendelssohn, Kurt. 1973. *The World of Walther Nernst: The Rise and Fall of German Science, 1864–1941*. London: Macmillan.
Mierzejewski, Alfred C. 1988. *The Collapse of the German War Economy, 1944–1945*. Chapel Hill: University of North Carolina Press.
Moore, Walter. 1989. *Schrödinger: Life and Thought*. Cambridge: Cambridge University Press.
Mott, Nevill, and Peierls, Rudolf. 1977. Werner Heisenberg. *Biographical Memoirs of Fellows of the Royal Society*, 23:213–251.
Neufeld, Michael. 1995. *The Rocket and the Reich*. New York: The Free Press.
Oleynikov, Pavel V. 2000. German scientists in the Soviet atomic Project, *Nonproliferation Review*, 7 (no. 2):1–30.
Operation Epsilon: The Farm Hall Transcripts. 1993. Bristol: Institute of Physics Publishing.
Oxford. Special Number, February 1937. Oxford: University Press for the Oxford Society.
Oxford Dictionary of National Biography. Ed. H. C. G. Matthew and B. Harrison. 2004. Oxford: Oxford University Press.
Peierls, Rudolf. 1985. *Bird of Passage*. Princeton: Princeton University Press.
Pocock, R.F. 1977. *Nuclear Power: Its Development in the United Kingdom*. Old Woking: Unwin.
Powers, Thomas. 1993. *Heisenberg's War*. New York: Knopf.
Proctor, Robert. 1999. *The Nazi War on Cancer*. Princeton: Princeton University Press.
Remy, Steven P. 2002. *The Heidelberg Myth: The Nazification and Denazification of a German University*. Cambridge, MA: Harvard University Press.
Rhodes, Richard. 1988. *The Making of the Atomic Bomb*. New York: Simon and Schuster.
Rhodes, Richard. 1995. *Dark Sun: The Making of the Hydrogen Bomb*. New York: Simon and Schuster.
Rose, Paul Lawrence, 1998. *Heisenberg and the Nazi Atomic Bomb Project: A Study in German Culture*. Berkeley: University of California Press.
Sanders, J.H. 2000. "Nicholas Kurti, C.B.E." *Biographical Memoirs of Fellows of the Royal Society*, 46:299–325.
Schaefer, Clemens. 1958. Erinnerungen an Franz Simon, Reprint from *Naturwissenschaftliche Rundschau*, 1958:405–406.
Sereny, Gitta. 1995. *Albert Speer: His Battle with Truth*. New York: Knopf.
Shapley, Deborah. 1978. Nuclear weapons history: Japan's wartime bomb projects revealed, *Science* 199:152–157.
Shapley, Harlow, ed. 1953. The Rumford bicentennial, *Proceedings of the American Academy of Arts and Sciences*, 82 (no. 7):253–364.
Sherwin, Martin J. 1977. *A World Destroyed: The Atomic Bomb and the Grand Alliance*. New York: Vintage Books.
Simon, F.E. 1939. Possibilités et limites de la méthode de refroidissement. Réunion d'études sur le magnétisme, Strasbourg, 21–25 mai 1939. Mimeograph.
Simon, F.E. 1951. *The Neglect of Science*. Oxford: Blackwell.
Simon, F.E. 1953. Rumford Bicentennial Symposia and Awards, *Nature* 171:915.
Simon, F.E. 1954a. *Waste: The Threat to Our Natural Resources*. 34th Earl Grey Memorial Lecture. King's College, Newcastle.
Simon, F.E. 1954b. Power from Atomic Energy, *Atomic Scientists' Journal*, 3(no. 5):247–263.
Simon, F.E. 1955. *Nuclear Energy and the Future*. Reprint from *Lloyds Bank Review*, April.
Simon, F.E. 1956. *The Third Law of Thermodynamics: An Historical Survey*. Reprinted from *Year Book of the Physical Society*, 1956.
Smyth, Henry D. 1946. *Atomic Energy for Military Purposes* fifth ed. Princeton: Princeton University Press.
Speer, Albert. 1970. *Inside the Third Reich*. New York: Macmillan.
Speer, Albert. 1976. *Spandau: The Secret Diaries*. New York: Macmillan.
Speer, Albert. 1981. *Infiltration*. New York: Macmillan. Original title *Der Sclavenstaat*, English translation by Joachim Neugroschel.
Stoicheff, Boris. 2002. *Gerhard Herzberg: An Illustrious Life in Science*. Montreal: McGill-Queen's University Press; Ottawa: NRC Press.
Szöllösi-Janze, Margit. 2001. *Science in the Third Reich*. Oxford: Berg.
Walker, Mark. 1989. *German National Socialism and the Quest for Nuclear Power, 1939–1949*. Cambridge: Cambridge University Press.
Walker, Mark. 1995. *Nazi Science: Myth, Truth, and the German Atomic Bomb*. New York: Plenum Press.
Warmbrunn, Werner. 1993. *The German Occupation of Belgium, 1940–1944*. New York: Peter Lang.

Werner, Ruth (pseudonym). 1991. *Sonya's Report*. London: Chatto and Windus. Translated by Renate Simpson. First published in German as *Sonjas Rapport*, 1977.

Wilcox, Robert K. 1985. *Japan's Secret War*. New York: William Morrow.

Wilkinson, Denys. 2002. Creation of the nuclear physics Laboratory. (www.physics.ox.ac.uk/pp/dwb//Wilkinson-talk.htm). Accessed 22/07/2009.

Williams, Robert C. 1987. *Klaus Fuchs, Atom Spy*. Cambridge, MA: Harvard University Press.

Wilson, Richard. Undated. *A Brief History of the Harvard University Cyclotrons*. (www.physics.harvard.edu/~wilson/cyclotron/history.html). Accessed 03/01/2010.

INDEX

Page numbers in *italic* refer to illustrations and captions

A
abdication crisis 52–3
Abelson, Philip 117–19, 169
absolute zero temperature 176–8
Adler, Edward 93
Akers, Wallace 77, 133, 182
   Simon's reports to 92, 94
   in US during war 79, 90, 91, 95, 105, 106, 109, 110, 113, 114, 115
   and US nuclear security policy 96, 99, 101
Alsos Mission 116–17, 130–3, 156–8
   Alsos-type mission to Japan 163–5
'Amama' *see* Simon, Anna
American Association for the Advancement of Science (AAAS) 20, 239
Amsterdam, research visits 39–40
Anderson, Sir John 77, 90–1, 94, 179
   and US nuclear security policy 99, 100, 102, 103
anti-intellectualism 31
antisemitism 19
   anti-Jewish legislation 29, 45
   attacks on Jewish influences in science 52, 122–4, 137–8, 150
   burnings of books by Jews 31–2, 138
   dismissals of Jewish academics 27, 31, 121–2, 137
   early Nazi action against Jews 27, 28–9, 31–2, 45–6
   Kristallnacht 55
   measures against Jewish doctors 45–6
   Nazi agreement on Final Solution 81
   schoolboy attacks on Stephan Frank 45
   in Toronto 62
Arakatsu, Bunsaku 162, 163, 164
Arandora Star 64
Ardenne, Manfred von 148, 159
Arms, Nancy 54, 96–7, 239–40
Arms, Shull 65, 68, 115, 133

Army Ordnance, Germany 84–5, 86–7, 120–1, 122, 139, 140, 153
Ash, Claire 55–6
Atomic Energy Research Establishment *see* Harwell
Attlee, Clement 168, 179
Auger, Pierre 95–6, 99, 114
Auschwitz 231, 232
Austria, annexation of 49, 53

B
Bagge, Erich
   detained at end of war 130–3, 144, 145, 150, 151
   Heisenberg's view of 172
   isotope sluice research 85, 129, 156
Bailey, Cyril 23
Bainbridge, Kenneth 73
Balliol College, Oxford 37, 51, 69, 112
Banting Institute, Toronto 66–7
Banting, Sir Frederick 67
barrier research 92–3, 97, 99, 102–3, 107–8, 109–12
Battle of Britain 67
Baxandall, Kathrin *see* Simon, Kathrin
Beams, Jesse 81, 90, 104
Belgian uranium stocks 58, 59–60, 152
Bell, Kenneth 61
Bell, Ronald 108, 113, 134
Bell Telephone Laboratories, New Jersey 94, 107, 110, 111
Bellevue laboratory, Paris 39–40, 53
Berenblum, Isaac 241
Bergmann, Ernst David 243–4
Berlin University 11–12, 16
biomedical research 46
Birley, Robert 210–11
Birmingham University
   Frisch–Peierls paper 59, 60

Birmingham (*continued*)
  Fuchs affair 192, 194
  nuclear fission research 59, 60, 65, 68, 78, 102, 115
Blackett, Patrick 171, 172
Blitzkrieg 61
Bohr, Aage 108–9
Bohr, Niels 81
  Heisenberg's visits to 220, 228
  meetings with Simon 108, 110, 113, 226, 240
  visits to Oxford 108–9, 193
bombing raids, Allied 152, 154–6, 205–6
Bonhoeffer, Karl-Friedrich 205–6, 230, 231, 232, 233
book burnings 31–2, 138
Booth, Eugene 111–12
Bormann, Martin 128, 129
Born, Max 173
  and Fuchs 192
  Heisenberg's visits 222, 224, 226
  postwar correspondence with Simon 177, 226, 228, 253
  prewar correspondence with Simon 50, 54
  wartime correspondence with Simon 60–1, 62, 68, 74
Bothe, Walther 84, 122, 126, 130, 145, 153
Braun, Wernher von 141, 150, 190
Breslau *see* Technische Hochschule, Breslau
Brickwedde, Ferdinand G. 94, 233–4, 238
Bridgman, Percy 20, 61–2, 74, 105, 177, 239
Briggs Committee 81
Briggs, Lyman 81, 94
British Central Scientific Office (BCSO) 94
British nuclear project
  Birmingham research 59, 60, 65, 68, 78, 102, 115
  Cambridge research 68, 71, 75–6, 78
  Clarendon research 60, 65, 73, 97, 115
  cooperation with US 92–6, 98, 105–8, 109–12, 115, 166–9
  Fuchs affair 191–7, 198, 199
  joint British-US separation plant proposal 76, 82–3, 90–1, 94
  low-separation plant proposal 115, 178
  Maud Committee 59, 60, 68–9, 70–3, 82, 86, 166
  Maud Report 75–7, 82, 201
  Montreal Laboratory 95–6, 98, 99, 105–6, 114, 179
  plans for British separation plant 72, 73, 75, 76, 100
  postwar 176, 178–82
  Quebec Agreement 99, 100–1, 105, 166–7, 168
  Rhydwmwyn Valley pilot plant 96–7, 98, 102, 115, 167
  rift with US 98–103, 170, 240
  Tube Alloys created 77
  Tube Alloys Washington meeting 114–15, 178
Buber, Martin 243
Burgess, Guy 200
Burton, Eli 67, 95
Bush, Vannevar 78, 81, 82–3, 91–2, 94, 118
  nuclear security policy 96, 99, 101, 169

C
Cairncross, John 201
Calder Hall 180–1, 253
California Institute of Technology, Pasadena 21, 246, 252
Cambridge Five spies 200–1
Cambridge University
  nuclear fission research 68, 71, 75–6, 78
  Royal Society Mond Laboratory 39
Canada
  evacuation of children to 61–2, 70
  heavy water 99, 100, 105–6
  Montreal Laboratory 95–6, 98, 99, 105–6, 114, 179
  as site for separation plant 76
  statement on Hiroshima 135
  see also Toronto, Canada
cancer, German campaign against 45
Cassirer, Ernst 36–7
Cassirer, Toni 36–7
Censor, the 66
central heating 34, 36
centrifuge research
  Germany 85, 129, 133, 154, 156
  Japan 162, 163
  United States 90, 91, 104
Chadwick, James 59, 65, 76, 78, 99, 178
  relationship with Groves 115, 168
  in US during war 105, 106, 109, 110, 113, 114, 115, 168
Chamberlain, Neville 54
Cherwell, Lord *see* Lindemann, Frederick, Lord Cherwell

Cherwell Papers, Oxford 30, 57
Chicago group 90, 95, 98, 99, 103, 105, 153–4
child evacuation programmes 61–2, 70
Christ Church College, Oxford 112, 175, 210, 215, 249
Churchill, Winston 52, 54, 57, 134, 168–9, 180
  Anderson's minute to 90–1
  decision based on Maud Report 76–7
  internment of Germans 63, 64
  letter on uranium 58
  Lindemann's minutes to 58, 76–7
  Quebec Agreement 100–1, 105
  and Roosevelt 83, 100, 102
  statement on Hiroshima 134–5
  and Truman 120
City of Benares 70
Civil Service Restoration Act (1933) 29
Clapham, Michael 92, 93
Clarendon Laboratory, Oxford xxi, 29, 37–9, 53
  Bohr's visit 108–9
  Lindemann's retirement 248–51
  low-temperature research 29, 37, 39, 176–8, 202, 237
  nuclear medicine 46
  recruitment of displaced Jewish scientists 29–30, 38
  Simon moves to 29–30, 32–3
  Simon's wartime research 60, 65, 73, 97, 115
Clifton, Robert 38
Clusius–Dickel thermal diffusion 85, 86, 160–1
Clusius, Klaus 85, 122, 127, 146–7, 156, 223
coal, Simon's articles on 185–6
Cockcroft, John 113, 114, 133, 179, 192, 193, 250
Cohn, Ernst 27
Columbia University 90, 93, 103, 110
Combined Development Trust 167, 168
Combined Policy Committee (CPC) 100–1, 105, 106, 167, 168, 200
Committee on the Scientific Survey of Air Defence 58–9
compartmentalization policy *see* nuclear security policy, US
Compton, Arthur 37, 94
Conant, James 78, 82, 94, 105

  letter to Navy 118
  nuclear security policy 96, 98–9, 101, 103, 169, 240
concentration camps viii–ix, xv, 28, 202, 231, 232
Conti, Leonardo 45
Cotton, Aimé 39, 53
Crowther, James 48
Currie, Lauchlin 110
cyclotrons
  Germany 85, 125, 126, 130, 148
  Japan 88, 165
  Los Alamos 105
  Soviet Union 87, 158, 199
Czechoslovakia 54, 157

D
D-Day landings 116
Dale, Sir Henry 222, 223
Darwin, Sir Charles 139, 221, 222, 223
de Valera, Eamon 49
Debye, Peter 39, 84, 122, 239
Department of Scientific and Industrial Research (DSIR) 59, 77, 106, 182
Depression, Germany 12–14, 19, 20, 25, 26
Deutsche Bunsengesellschaft 16
Deutsche Physik faction 84, 122, 123, 124, 137
  see also Stark, Johannes
diamond industry 183
Dickel, Gerhard 85
Diebner, Kurt
  detained at end of war 130–3, 144, 148, 151
  evacuation of wartime research 127, 129, 156
  head of Army Ordnance nuclear group 84, 85, 122, 151
  Heisenberg's view of 172
  wartime nuclear research 126
diffusion *see* gaseous diffusion; thermal diffusion
Dirac, Paul 38, 80, 108
Döpel, Robert 125–6
Dornberger, Walter 140, 141
Dr Lee's Professor of Physical Chemistry 248, 249–50
Dublin, Schrödinger's move to 49, 50
Dunkirk 61
Dunning, John R. 95, 110

E

economy, Germany 12–14, 19, 20, 25, 26
education, Simon's articles on 184–5
Egerton, Alfred 37
Einstein, Albert 29, 81, 124, 177, 222, 241
  Nazi attacks on 21, 124, 138, 150
  refuses to return to Germany 22, 138
  Simon visits *xix*, 21–2, 238, 240
Eisner, Lotte 8–10
Elbe, Günther von 24
electromagnetic isotope separation 85, 91, 103–4
emigration from Germany
  friends 55–6
  Jewish academics 29–30, 37–8, 48–50, 139
  other family members 40–8, 80–1, 242
  Simon and family 29–30, 32–3, 34–7
Enabling Act, Germany (1933) 28
entropy 177–8
Esau, Abraham 84, 121, 122, 125, 126, 127, 150
espionage, Soviet 191–201
Eucken, Arnold 18

F

Farkas, Ladislaus 135
Farm Hall detainees 130–3, 143–9, 170–3
  create myth of wartime capability 144, 145, 224–5, 255–6
  Heisenberg–Gerlach memorandum 147–9
  reaction to Hiroshima bomb 134, 135, 138–9, 143–7, 150, 151
FBI 193, 195, 197
Fermi, Enrico 99–100, 103, 153–4, 239
financial crisis, Germany 12–14, 19, 20, 25, 26
Financial Times 183–4, 187, 190
First World War *xvii*, 2–8
Flerov, Georgii 88, 148, 158
Franck, James 22, 30, 86, 203, 238
Frank, Ludwig 44, 45–7, 69, 108, 246, 254
Frank, Mimi (née Simon) *xvi*, 15, 40, 44, 45–6, 69, 246
Frank, Stephan 40, 44, 45, 46, 69
Frank, Thomas *xxi*, 40, 44, 47
Franlin, Ursula *see* Martius, Ursula
Fraser, Ronald 219–20, 221, 222–3, 225, 227, 228
Friedländer, Lili 35, 44

Frisch, Otto 59, 60, 109, 138
Fuchs, Emil 194, 195
Fuchs, Klaus 191–7, 198, 199
Fues, Erwin 231–2
Furman, Robert 163–5

G

gaseous diffusion
  barrier research 92–3, 97, 99, 102–3, 107–8, 109–12
  British project 92–3, 96–7, 98, 99, 102–3
  British–US cooperation 92–3, 98, 106–8, 109–12, 167–8
  Hertz's research 85, 86, 139
  Simon's research 60, 65, 73, 97, 115
  US project 91, 98, 103, 107–8, 109–12, 117, 119–20
Geiger, Hans 122, 145
Gentner, Wolfgang 130
Gerlach, Walther
  detained at end of war 130–3, 138–9, 143–4, 148, 151
  nuclear plenipotentiary 128, 129, 150, 153, 156, 227
German nuclear research 83–7, 120–2, 124–9
  effect of loss of Jewish scientists 138–40
  effect of missiles programme 142–3
  effects of enemy action 154–8
  isotope separation 85–6, 129, 133, 139, 145, 146–7, 154, 156
  lack of cooperation with Japan 170
  organizational factors 149–52
  raw materials and priorities 152–4
  reasons for failure 173–4
German Physical Society 55
German scientists
  conscription of 128, 144–5
  detained at end of war 117, 130–3
  killed in action 97, 145
  postwar invitations to Britain 219–23
  postwar preference for ex-Nazis 208, 209–17, 219–23, 227–9
  Simon's postwar assistance to former colleagues 202–8
  see also Farm Hall detainees; Jewish academics
Gerstel, Kaete 202, 203, 206
Giauque, William 22, 39
Gill, E.W.B. 133
Goebbels, Joseph 31, 97

Gold, Harry 195–6, 197, 199
Goldschmidt, Bertrand 95–6, 99
Goldschmidt, V.M. 47
Göring, Hermann 125, 126, 128
Göttingen University 1, 29, 30, 127, 130, 156, 173
Goudsmit, Samuel 148, 150, 151, 170
   Alsos Mission 130, 132, 156, 157–8, 214, 224, 225, 255
   correspondence with Rosbaud 226, 227, 229
graphite
   as moderator 73, 85, 153–4, 174
   uranium–graphite reactors 91, 104, 106, 158, 179, 199
Greece, holiday 244–5
Griffiths, W.T. 99
Groth, Wilhelm 84, 85, 133
Groves, Leslie 155–6
   Alsos Mission 116–17, 130–2, 156–7
   British-US cooperation 106, 115, 167–9
   Japan 163–5
   Manhattan Project 91–2, 103–5, 110, 111, 117–19, 120, 166
   nuclear security policy 96, 99, 101, 106, 169
Gunn, Ross 81
Günthermann, Leonhard 209
Guthrie Lecture 177–8
Gutsche, Hugo 204

H

Hagiwara, Tokutaro 88
Hahn, Otto 16, 84, 122, 150, 204
   detained at end of war 130–3, 138–9, 144, 145, 146, 170, 171
   evacuation of wartime research 127, 155–6
   Hahn–Strassmann uranium experiments 58, 81, 138
   protactinium 147
Halban, Hans von 16, 77, 249
   evacuation from France 65
   heavy water supplies 152
   proposal for low-separation plant 115, 178
   research in Cambridge 71, 72, 75–6, 78, 100, 198
   research in Montreal 95–6, 99, 105–6, 114
   and US nuclear security policy 99–100
   wartime visits to US 79, 82, 90
Hankey, Maurice, Lord 59, 76, 201
Hanle, Wilhelm 83

Harteck, Paul 84, 85, 87, 122, 150
   detained at end of war 130–3, 138–9, 144, 147, 148
   evacuation of wartime research 127, 156
   heavy water research 153, 154
   isotope separation research 84, 85–6, 129, 154, 156, 170
   postwar correspondence with Simon 205–6
Hartmann, Helmut 208
Harwell 179, 182, 192, 193, 194, 196, 250
heavy water 59, 71, 73, 75–6, 78, 85
   Canada 99, 100, 105–6
   evacuation from France 65, 152
   Germany 126, 148, 152–3, 154, 174
   Norway 60, 79, 152, 153, 154
   United States 76, 91, 98, 174
heavy water reactors 86, 106, 114, 128, 158, 174
Heidelberg University 31, 84, 127, 130, 156
Heisenberg, Werner
   character 172–3
   detained at end of war 130–3, 144, 146, 147–8, 151, 170, 171–3
   evacuation of wartime research 127, 128, 155–6
   Nazi attacks on 52, 122–4, 138, 150
   offer from Soviet Union 172
   postwar visits and attitude 220, 222, 224–8
   Simon's report to Perrin on 224–5, 255–6
   wartime nuclear research and leadership 84–5, 86, 87, 122, 125–8, 142, 145–6, 152
Heitler, Walter 222
helium liquefaction 16, 29, 39
Henschel, Albert 40–3, 44
Henschel-Simon, Elisabeth (Ebeth) xvi, 11, 40–3, 44, 48, 135–6, 203
Hertz, Gustav 85, 86, 138–9, 148, 159
Herzberg, Gerhard 216–17
Hess, Rudolf 80, 123, 143
Hildebrand, Joel 19, 113, 175
Hilpert, R.S. 207–8, 210
Himmler, Heinrich 122–4, 128, 150
Hindenburg exemptions 29, 45
Hindenburg, Paul von 26, 28, 29
Hiroshima bomb 134–5
   reaction of German detainees 134, 135, 138–9, 143–7, 150, 151
   Stalin's reaction to 159–60
Hitler, Adolf 28–9, 116, 128, 133, 140–1, 142–3, 149, 151–2

Holbrook, Richard 23
holidays 47, 53, 233, 235–7, 244–5, 246–8
Holocaust victims
  family members vii, viii–ix, *xv*, 218, 233–4, 242
  Franz Pollitzer 204, 205, 231, 232
  Gerstel, Kaete's mother 202
Hopkins, Harry 100
hydrogen bomb 180
hydrogen liquefaction 29
  see also liquid hydrogen aircraft fuel

I

ICI (Imperial Chemical Industries)
  grants for displaced Jewish scientists 30, 34, 37, 38, 39, 48, 77
  Maud Report 75, 76
  Metals Division 92, 102
  work on nuclear project 70–1, 72, 73, 77, 101, 115, 178, 179
inflation, Germany 12–14, 20
international rivalry in science, Simon's articles on 187–9
International Union of Pure and Applied Physics (IUPAP) 251
internment, of Germans in Britain 63–4
Iron Cross 6, 175
Isotope Plant Design Committee (IPDC) 182
isotope separation plants
  British low-separation plant 115, 178
  British pilot, Rhydwmwyn Valley 96–7, 98, 102, 115, 167
  British plans 72, 73, 75, 76, 100
  British postwar 182
  proposal for joint British-US 76, 82–3, 90–1, 94
  United States 91, 103, 117, 119–20
isotope separation research
  centrifuges *see* centrifuge research
  Clusius–Dickel thermal diffusion 85, 86, 160–1
  discussed by Maud Committee 86
  electromagnetic methods 85, 91, 103–4
  gaseous diffusion *see* gaseous diffusion
  Germany 85–6, 129, 133, 139, 145, 146–7, 154, 156
  isotope sluice 85, 129, 156
  Japan 160–1, 162, 163

liquid thermal diffusion 117–19, 120, 129, 169
question of German capability 64, 79
isotope sluice 85, 129, 156
Israel
  postwar visit 241–4
  see also Palestine
Italy
  holidays 236–7, 248
  Lotte's 1921 tour 8–10

J

Jaffé, Fritz 66
Japanese nuclear research 88, 160–6, 170
Jewish academics
  emigration from Germany 29–30, 37–8, 48–50, 139
  Nazi dismissals of 27, 31, 121–2, 137
Jews
  American 21
  see also antisemitism; Holocaust victims
Joffé, Abram 22, 88
Joliot-Curie, Jean Frédéric 58, 117, 157
Jones, G.O. 115
Jones, R.V. 37, 63, 131
Joos, Georg 83
Jost, Wilhelm 206–7, 210, 225, 229, 231, 233
Jungk, Robert 145
Justi, Eduard 211, 213, 220, 221

K

Kaiser Wilhelm Institute of Chemistry 127
Kaiser Wilhelm Institute of Physics 79, 84, 121, 127, 148, 157
  Heisenberg as director 122, 123–4, 171, 227
Kallmann, Hartmut 210, 212, 213
Kapitza, Peter 39
Kearton, Frank 113, 114
Keith, Percival 94, 103, 107, 110, 238
Kellex Company 99, 103, 107, 110–11, 114
Kellogg Corporation 93–4, 107
Kennedy, John F. 59
Khariton, Julii 199
kidnap risk, postwar Berlin 230–1
Kigoshi, Kunihiko 160
Kindertransports 62
Kochmann, Heinz 242
Korea 88, 161, 162, 164
Korsching, Horst 129, 130–3, 143–4, 145, 151, 172

Kowarski, Lew 65, 76, 198
Kramers, Hendrik 22
Kristallnacht 55
Kuczynzki, Ursula (Sonya) 196–7, 201
Kuhn, Heinrich 30, 37
Kurchatov, Igor 158, 159, 189, 198–9
Kurti, Nicholas *xxviii*, 78, 203, 238, 254
  collaboration with Simon 27, 39, 65, 68, 71, 177
  move to Oxford 30, 37–8
  on Nazi Black List 63
Kyoto University 88, 162, 163

L

Langevin, Paul 70
Larsen, Egon 252
Laue, Max von 150, 204, 212, 221, 228
  detained at end of war 130–3, 170, 171
  postwar correspondence with Simon 205, 231, 246, 253
  prewar correspondence with Simon 51, 52
Laurence, William 87
Lauritsen, Charles 75, 82
Law for the Restoration of the Professional Civil Service (1933) 29
Lawrence, Ernest 23, 82, 93, 94, 96
Lenard, Philipp 84
Linde Medal 231–2
Lindemann, Charles *xxiii*, 67, 113
Lindemann, Frederick, Lord Cherwell *xx*, 223, 224
  development of Clarendon 38–9
  and Einstein 21
  first meeting with Simon 29
  interventions for Ludwig Frank 46
  minutes to Churchill 58, 76–7
  nominates Simon for Nobel Prize 176
  recruitment of displaced Jewish scientists 29–30, 38
  retirement from Clarendon 248–9
  and Schrödinger 38, 49, 50
  and US nuclear security policy 99, 100
  wartime correspondence with Simon 64–5, 70–1, 97
  wartime meetings with Simon 67, 79, 97, 101, 102, 112
Lindsay, A.D. 47
Lipmann, Carl 15, 80–1
Lipmann, Curt (Peter) 40, 69, 80

Lipmann, Ellie 15, 80–1
liquid hydrogen aircraft fuel 60, 65, 67, 68
liquid thermal diffusion 117–19, 120, 129, 169
London, Fritz 37, 38, 48, 203
London, Heinz 37, 48, 115, 203
Los Alamos 104–5
  British scientists move to 106, 109, 111, 167
  fissile material supplies 117, 119–20
  Fuchs affair 194–5, 199
  weapons technology research 117, 120
low-temperature research
  at Bellevue, Paris 39–40, 53, 56
  at Berkeley 22–3
  at Berlin 16
  at Breslau 18, 27
  at Clarendon Laboratory 29, 37, 39, 176–8, 202, 237
  doctoral thesis 11
  lectures in Leningrad 17
  postwar conferences 182, 238, 251–2
  Strasbourg conference 56

M

McCarthy, Joseph 238
Mackenzie, C.J. 95, 99, 106
Maclean, Donald 200
Magdalen College, Oxford 38
magnetic cooling 39, 56, 176
Manhattan Project 89, 91–6, 103–5, 117–20, 153–4, 166–70
  Alsos-type mission to Japan 163–5
  Alsos Mission 116–17, 130–3, 156–8
  cooperation with Britain 92–6, 98, 105–8, 109–12, 115, 166–9
  early US nuclear research 81–3
  reasons for success 174
  rift with Britain 98–103, 170, 240
  security policy 96, 98–103, 169, 240
Marshall, James 91
Martius, Ursula (Ursula Franklin) 211–13, 238
Maud Committee 59, 60, 68–9, 70–3, 82, 86, 166
Maud Report 75–7, 82, 201
medical profession, anti-Jewish measures 45–6
Meissner, Walther 204–5, 238
Meitner, Lise 138, 147, 150
membrane research *see* barrier research
Mendel, Bruno *xxiv*, 66–7
Mendelssohn, Elisabeth 48

Mendelssohn, Heinrich 242
Mendelssohn, Kurt 27, 29, 37, 39, 48, 203, 224
Mentzel, Rudolf 121, 122, 124, 125, 127, 150
Merker, Aarlo 3
MI5 192–3, 194
Michaelis, Leonor 1, 24
Michels, Antonius M.J.F. 39, 202
Military Policy Committee 92, 94, 100, 103, 106
Millikan, Robert 21
Ministry of Aircraft Production 61, 68, 69, 76
missiles programme, Germany 140–3
moderators *see* graphite; heavy water
Molotov, Vyacheslav 158, 198
Mond, Henry, Lord Melchett 71, 72, 73, 77
Mond Nickel Company 97, 102, 108
Montreal Laboratory 95–6, 98, 99, 105–6, 114, 179
moon landings 252
Morgenroth, Julius 7, 66
Morrison, Philip 164
Mott, Nevill 192, 220, 221–2
Münchhausen, Margarete (née Lipmann), 'Omama' *xix*, 40, 44, 47, 69
Münchhausen, Pauline vii, viii, 47
Münchhausen, Rosa vii, viii–ix, 47
Münchhausen, Sigmund, 'Opapa' vii–viii, 40, 44, 47, 69
Munich crisis 49, 54–5

N

National Defense Research Committee 81, 82
National Physico-Technical Institute (*Reichsanstalt*) 52
Navy, Japanese 161–2, 165
Navy, US 117–19
Nazi Black List 63
Nazi Party 53
  agreement on Final Solution 81
  attacks on Jewish influences in science 52, 122–4, 137–8, 150
  burnings of books by Jews 31–2, 138
  dismissals of Jewish academics 27, 31, 121–2, 137
  early action against Jews 27, 28–9, 31–2, 45–6
  election gains 26, 28
  Kristallnacht 55
  legislation 28, 29, 45

Night of the Long Knives 44
Nazis, postwar
  preference for ex-Nazi scientists 208, 209–17, 219–23, 227–9
  re-nazification of public service 217, 218–19
Neglect of Science, The 183–4, 188, 190, 191–2
Nernst, Walther 11, 16, 29, 39, 79
  honorary doctorate from Oxford *xxi*, 51–2
  third law of thermodynamics 177, 178
Neumann, Bernhard 26, 31
neutron moderators *see* graphite; heavy water
newspaper articles 183–90
Nichols, Kenneth 193
nickel barriers 85, 93, 94, 97, 99, 102, 107, 110
  Norris–Adler 93, 103, 107, 110, 111
Night of the Long Knives 44
Nininger, Robert 164
Nishina, Yoshio 88, 160–1, 164
Nobel Prizes 122
  Hertz and Franck 86
  Schrödinger and Dirac 38
  Simon nominated 176
Normandy landings 116
Norris–Adler barrier 93, 103, 107, 110, 111
Norris, Edward 93
Norway
  Crown Prince Olav 69–70
  heavy water 60, 79, 152, 153, 154
nuclear bomb
  British postwar development 179–80
  British wartime prospects 73, 75, 76
  correspondence on effectiveness of uranium 57–8, 59, 60
  Fuchs passes secrets to USSR 194–5, 199
  German wartime prospects 125
  Hiroshima *see* Hiroshima bomb
  Japanese wartime prospects 161–2
  Soviet development 180, 187–8, 191–2, 193, 199–200
  triggering technology 117
  US development 82, 83, 117, 119, 120
  see also nuclear research
nuclear intelligence *see* Alsos Mission
nuclear medicine 46
nuclear power
  Britain 72, 73, 180–1, 187
  Germany 86
  Maud Report recommendations 75–6
  Simon's articles on 186–7

nuclear research
  Britain *see* British nuclear project
  Germany *see* German nuclear research
  Japan 88, 160–6, 170
  question of German capability 57–8, 64, 79
  Soviet Union 87–8, 158–60
  United States *see* Manhattan Project
  see also isotope separation research; nuclear bomb
nuclear security policy, US 96, 98–103, 169, 240
Nuremberg racial laws (1935) 45

O
Oak Ridge, Tennessee 103, 104, 117, 119–20
Office of Scientific Research and Development (OSRD), USA 82, 91
  see also S-1 Committee, USA
Olav, Crown Prince of Norway 69–70
Olden, Rudolf 70
Oliphant, Mark 59, 65, 77, 82, 106, 113, 114
Oliver Twist (film) 215–16
'Omama' *see* Münchhausen, Margarete
'Opapa' *see* Münchhausen, Sigmund
Operation Epsilon *see* Farm Hall detainees
Oppenheimer, Robert 105, 117, 240
Oxford
  10 Belbroughton Road v–vii, ix–x, *xx*, 34
  Simon and family move to 29–30, 32–3, 34–7
  see also Clarendon Laboratory, Oxford

P
Pakenham, Frank, 7th Earl of Longford 210–19, 222–3, 225, 238
Palestine 40–3
Papen, Franz von 26
Paris, research visits 39–40, 53
Pash, Boris 117, 130
Paton, Herbert James 53
Pearl Harbor 80, 87, 89, 166
Pegram, George 78, 81
Peierls, Rudolf 19, 65, 77, 78, 108, 182
  Frisch–Peierls paper 59, 60
  and Fuchs 192, 194, 200
  and Heisenberg 226
  moves to Kellex and Los Alamos 111, 113, 114
  wartime visits to US 79, 90, 99, 105–7, 109–11
Perrin, Michael 72, 134, 252

with Alsos Mission 130, 132–3
and Fuchs 191, 192–3, 194, 196
Simon's report on Heisenberg 224–5, 255–6
Tube Alloys 77, 90, 113, 114
Philby, Kim 200
philosophy, in Germany 25, 53
photography, Simon's interest in *xvi*, 1, 15
Planck, Max 16, 84, 151–2
plutonium 75, 86, 99, 115, 198–9
  British postwar production 179, 180, 182
  US production 104, 117
plutonium bomb
  British development 180
  Fuchs passes secrets to USSR 194–5, 199
  Soviet development 199–200
  triggering technology 117
  US development 117, 119, 120
Poland 54, 57, 58, 87, 242
Polanyi, Michael 16, 74, 205
police surveillance, Germany 150
politics, correspondence on 52–3
Pollitzer, Franz 204, 205, 231, 232
Pontecorvo, Bruno 200
population, Simon's article on 190
Potsdam Conference 134, 159
protactinium 147
Pryce, Maurice 193, 249
Pye, David 65

Q
quantum electrodynamics 80
Quebec Agreement 99, 100–1, 105, 166–7, 168

R
radiation effects research 76
radioactive isotopes, biomedical research 46
recognition and awards 175–6
  C.B.E. 175
  Chair in Thermodynamics 175
  elected to Dr Lee's Professorship 248, 249–50
  elected to Royal Society 74
  Iron Cross 6, 175
  Linde Medal 231–2
  nominated for Nobel Prize 176
  Reader in Thermodynamics 37, 52
  Rumford medal 239
Reich Research Council 84, 121, 122, 124, 125, 128

Reichenbach, Hans 25
Reichsanstalt (National Physico-Technical Institute) 52
Riken research institute, Tokyo 88, 160–1
Rittner, T.H. 131, 147, 151, 170
rocket development, Germany 140–3
Roosevelt, Franklin D. 81, 82–3, 100, 102, 105, 118, 132, 166
Rosbaud, Paul 16, 129, 211, 213, 226, 227, 229
Roskill, Stephen 113
Rothenstein, Wolfgang 252–3
Royal Society ix, 80, 97, 175, 182
   Heisenberg elected to 228
   Rumford medal 239
   Simon elected to 74
   Simon's committee proposal 213, 214
   Walter Heitler nominated 222
Royal Society Mond Laboratory, Cambridge 39
Ruhemann, Barbara 25, 48
Ruhemann, Martin 25, 48, 203
Rumford, Count *see* Thompson, Benjamin, Count Rumford
Rumford medal 239
Rust, Bernhard 27, 28, 32, 121, 122, 123, 125

S
S-1 Committee, USA 82, 91, 92, 94, 98, 118
Sagane, Ryokichi 163–4
satellites 189–90
Schaefer, Clemens 203–4, 207, 254
Scherrer, Paul 221
Schlick, Moritz 25
Schmidt, Wilhelm 3
Schrödinger, Erwin *xx*, 38, 49–50
Schumann, Erich 86, 121, 122, 148, 150, 212
Schütze, Werner 139
Schwarze Korps, Das 52, 53, 122, 123
Scientific Advisory Committee (SAC) 76
security policy *see* nuclear security policy, US
Sellafield 180, 182
Sicily, holiday 246–8
Simon, Anna (née Mendelssohn), 'Amama' *xvi, xvii*, 15, 35, 40, 44, 69, 203
Simon, Dorothea, 'Dor' *xviii, xix, xxi*, 203, 239, 240, 253
   birth and childhood 15–16
distracts Fuchs 193
effect of move to Oxford 35
evacuation to Toronto 61–2, 112
relocation to Switzerland 19–20
returns from Toronto 116
Simon, Ernst *xv, xvi, xvii*, 1, 12, 14
Simon, Franz (Sir Francis) *xvi–xix, xxi–xxiii, xxv–xxvii*
   awards *see* recognition and awards
   death 181, 253, 254
   early married and family life 12–16
   education 1, 11
   health 108, 237, 250, 251, 253–4
   holidays 47, 53, 233, 235–7, 244–5, 246–8
   interest in photography *xvi*, 1, 15
   kidnap risk in Berlin 230–1
   meets Lotte 9
   military service *xvii*, 2–7
   on Nazi Black List 63
   Nazi student action against 31
   newspaper articles 183–90
   postwar assistance to friends in Germany 202–9
   postwar committees 183
   postwar visits to Germany 229–34
   wounded in action *xvii*, 3–4, 6–7
Simon, Kathrin, 'Kay' (later Kay Baxandall) *xviii, xix, xxi, xxiv*, 203, 246, 248, 253
   birth and childhood 14–16
   effect of move to Oxford 35
   evacuation to Toronto 61–2
   relocation to Switzerland 19–20
   returns from Toronto 112–13, 116
Simon, Lotte (née Münchhausen) *xviii, xix, xxii, xxiii, xxviii*
   death v
   early married and family life 12–16
   education 1–2, 245
   evacuation to Toronto 62, 65–7
   First World War experiences 7–8
   holidays 47, 53, 233, 235–7, 244–5
   Italian tour in 1921 8–10
   meets Simon 9
   musical and theatrical interests 8, 13, 14
   relocation to Switzerland 19–20
   returns from Toronto 112–13, 116
   visits Berlin in 1934 44
Simson, Clara von 202–3, 213

Skardon, William 194
Skinner, Herbert 182, 200
Slade, R.E. 72, 75, 77
Smith, S.S. 92, 93
Smyth Report 135, 187
Somme, Battle of the 4–5
Sommerfeld, Arnold 123, 225–6
Soviet Union
   Alsos-type mission 159
   development of nuclear bomb 180, 187–8, 191–2, 193
   espionage activities 191–201
   Fuchs affair 191–7, 198, 199
   nuclear research 87–8, 158–60
   Ruhemann's move to 25, 48
   satellites 189
   Simon's 1930 visit 16–17
   Simon's articles on 187–9, 191–2
   war on Russian front 80, 88, 97–8
Spanish Civil War 53
Speer, Albert 124–5, 126, 128, 129, 140, 142–3, 149, 150
Stalin, Joseph 158, 159–60, 192
Stalin purges 48
Stark, Johannes 52, 84, 86, 122, 123, 138
Stern, Otto 22, 61–2, 85, 138, 203, 205
Stimson, Henry 83, 100
Stransky, Iwan 212
Strasbourg conference (1939) 56
Strassmann, Fritz 58, 81, 138
Sudetenland 54, 55
Suhrmann, Rudolf 25, 50–1
Suzuki, Tatsusaburo 88, 160–1
Switzerland
   family's relocation 19–20
   holidays 47, 53, 233, 236
Szilard, Leo 24, 81

T
Taylor, Geoffrey I. 113, 114
Taylor, Hugh 110, 205
Technical Committees v
   Maud 72
   postwar 181, 182, 193, 196
   Tube Alloys 77, 90, 100, 102, 108, 110, 114–15, 132, 178
   Washington meeting 114–15, 178
Technische Hochschule, Breslau 16, 18–19, 26–7, 28, 31, 32

technological education, Simon's articles on 184–5
Teller, Eduard 30, 81
Theresienstadt Camp ix, *xv*, 202
thermal diffusion
   Clusius–Dickel 85, 86, 160–1
   liquid 117–19, 120, 129, 169
Thermodynamics Chair 175
Thermodynamics, Reader in 37, 52
thermodynamics, third law of 177–8
Thompson, Benjamin, Count Rumford 238–40
Thompson, J.W.R. 217
Thomson, George 58, 59, 60, 79, 249
   invites Simon to join British nuclear project 67–8, 69
   Maud Report 75, 78, 82
thorium 157
Thost, Hans 63
Tizard, Sir Henry 58–9, 76, 109
Tolman, Richard 106, 107, 108
Toronto, Canada
   antisemitism 62
   family's evacuation to 61–2, 65–7
   family's return from 112–13, 116
transuranic elements 75, 86, 198–9
   see also plutonium
Truman, Harry S. 134, 159, 166, 168
Tube Alloys *see* British nuclear project
Tuve, Merle 81
Tyndall, Arthur 59

U
UNESCO 217
Union Carbide Corporation 103, 110
United States
   nuclear programme *see* Manhattan Project
   Simon's 1932 visit 19, 20–6
   Simon's postwar visits 237–41
   Simon's wartime visits *xxii*, 78–9, 89–90, 92, 93–6, 105–7, 109–11, 113–15
University of California, Berkeley 19, 22–3, 90, 104
University of Jerusalem 43
University of Virginia 24, 90, 104
uranium
   Belgian stocks 58, 59–60, 152
   Canadian supplies 105–6

uranium (*continued*)
  correspondence on effectiveness for bomb 57–8, 59, 60
  German supplies 152
  Japanese supplies 160, 162
  Soviet supplies 87, 158–9
  US removal of German stocks 130–1
  US supplies 92
  see also isotope separation research
Uranium Club *see* German nuclear research
uranium–graphite reactors 91, 104, 106, 158, 179, 199
uranium hexafluoride 70–1, 85, 92, 115, 118, 160
uranium oxide 58, 59–60, 71
Uranium Technical Subcommittee *see* Maud Committee
Urey, Harold 70, 81, 99, 177
  correspondence on Fuchs 200
  diffusion project 93, 95, 100, 103, 107, 110
  postwar assistance to German scientists 205, 206
  wartime visits to Britain 77–8, 111–12
USSR *see* Soviet Union

V

Van Vleck, John 61–2
Vernadskii, Vladimir 87, 88
Vögler, Albert 121, 122, 148

W

Waetzmann, Erich 27
Wallace, Henry A. 82, 83
Warburg, Otto 66, 212
Wardenburg, Fred 132
Weizmann, Chaim 241, 242
Weizsäcker, Carl Friedrich von 87, 126, 220
  detained at end of war 130–3, 139, 144, 145, 146–7, 151
Werfl, Walter 216–17
Weyl, Hermann 240–1
Wigner, Eugene 81
Wilkinson, Denys 250–1
Windscale 180, 181, 182, 196
Wirtz, Karl 126, 129, 130–3, 146, 148, 151, 173

Y

Yad Vashem Database viii, 242
Yukawa, Hideki 163, 164